RELATED KAPLAN BOOKS

College Admissions and Financial Aid
Guide to the Best Colleges in the U.S.
Kaplan/Newsweek College Catalog
Parent's Guide to College Admissions
Scholarships
What to Study: 101 Fields in a Flash
Yale Daily News Guide to Internships
Yale Daily News Guide to Succeeding in College
You Can Afford College

Test Preparation
ACT
Fast Track ACT
SAT & PSAT
Fast Track SAT & PSAT
SAT Math Mania
SAT Math Workbook
SAT or ACT? Test Your Best
SAT II: Chemistry
SAT II: Mathematics
SAT II: Writing
SAT Verbal Velocity
SAT Verbal Workbook

SAT* II
BIOLOGY
E/M
Fourth Edition

By
Claire Aldridge, Ph.D.
Glenn E. Croston, Ph.D.
and the
Staff of Kaplan, Inc.

Simon & Schuster

NEW YORK · LONDON · SINGAPORE · SYDNEY · TORONTO

Kaplan Publishing
Published by Simon & Schuster
1230 Avenue of the Americas
New York, New York 10020

For bulk sales to schools, colleges, and universities, please contact Order Department, Simon & Schuster, 100 Front Street, Riverside, NJ 08075. Phone: (800) 223-2336. Fax: (800) 943-9831.

The material in this book is up-to-date at the time of publication. The College Entrance Examination Board may have instituted changes in the test after this book was published. Please read all materials you receive regarding the SAT II: Biology E/M Subject Test carefully.

Contributing Editors: Marc Bernstein and Marcy Bullmaster
Project Editor: Larissa Shmailo
Cover Design: Cheung Tai
Interior Page Design: Jobim Rose
Production Editor: Maude Spekes
Editorial Coordinator: Dea Alessandro
Executive Editor: Del Franz

Special thanks to: Ruth Baygell, Gordon Drummond, Craig Dubois, Dr. James Major, Sara Pearl, and Rudy Robles

Manufactured in the United States of America.
Published simultaneously in Canada.

March 2001
10 9 8 7 6 5 4 3 2 1

ISBN 0-7432-0532-4
ISSN 1096-4800

CONTENTS

About the Authors

Claire Aldridge, Ph.D. received her bachelor of science degree in biomedical science from Texas A&M University; she graduated magna cum laude. She earned her Ph.D. from the Department of Immunology and the University Program in Genetics at Duke University in 1996. Claire was an invited speaker at the Ninth International Congress of Immunology in San Francisco, CA, in July 1995. She is an author of *Microbiology and Immunology*, a Kaplan publication, and has written numerous scientific articles. She currently works for the Office for Technology Development at the University of Texas Southwestern Medical Center of Dallas.

Glenn E. Croston, Ph.D. is head of biomolecular screening at ACADIA Pharmaceuticals in San Diego, CA. He has worked as a research scientist and published extensively since earning his Ph.D. in biology at the University of California, San Diego in 1992. Glenn has also written and edited a wide range of MCAT biology preparative materials.

How to Use This Book

For more than 60 years, Kaplan has prepared students to take SATs. Our team of researchers and editors knows more about SAT preparation than anyone else, and you'll find their accumulated experience and knowledge in this book. As you work your way through the chapters, we'll show you precisely what knowledge and skills you'll need in order to do your very best on the SAT II: Biology E/M (Ecological/Molecular) Subject Test. You'll discover the most effective way to tackle each type of question, and you'll reinforce your studies with lots of practice questions. At the beginning of the biology review section, you'll find a lengthy diagnostic test, and each chapter in this section ends with a short quiz. At the back of the book you'll find two full-length, formatted tests with answer keys, scoring instructions, and detailed explanations. In addition, the "Stress Management" chapter contains helpful tips on beating test stress while you're preparing for the test, and "The Final Countdown" can help you in those final days and in pulling off a victory on Test Day.

Get Ready to Prep

If possible, work your way through this book bit by bit over the course of a few weeks. Cramming the week before the test is not a good idea; you probably won't absorb much information, and it's sure to make you more anxious.

If you find that your anxiety about the test is interfering with your ability to study, start off by reading the Stress Management chapter in this book. It provides many practical tips to help you stay calm and centered. Use these tips before, during, and after the test.

Learn the Basics

The first thing you need to do is find out what's on the SAT II: Biology E/M Test. In the first section of this book, "The Basics," we'll provide you with background information about the SAT II: Subject Test and what it's used for. We'll also give you the lowdown on all the typical kinds of questions that are asked on the test and how best to tackle them.

The Best Prep

Kaplan's three practice tests give you a great prep experience for the SAT II: Biology E/M Subject Test.

Biology Review

Once you have the big picture, it's time to focus on the biology that's tested. The second section of this book, "Biology Review," gives you a succinct review of the biology you need to know to answer questions on Test Day. Each chapter in this section deals with a major subdivision of biology and focuses on concepts crucial to a full understanding of organisms and their interactions with the environment.

Note that throughout these biology review chapters, we use a select group of organisms to illustrate a variety of different concepts. We have chosen these organisms, the hydra, the earthworm, the grasshopper, and the human, because they exemplify different evolutionary stages particularly well. The hydra belongs to the Cnidarians, which is a phylum of simple organisms that developed relatively early in the evolution of organisms. Earthworms of the phylum Annelida are characteristic of an intermediate, more complex evolutionary stage, while grasshoppers (phylum Arthropoda) developed later than Annelida and are somewhat more complicated in structure. Finally, humans, the most highly evolved organisms of phylum Vertebrata, made their appearance fairly recently in evolutionary history. In general, humans exemplify the most complex adaptations that organisms have made to their environment.

The biology review section begins with a lengthy diagnostic test. If your time is limited, you can use the diagnostic test to bypass the material you already know well enough and to zero in on what you need to work on.

This book can help you to decide whether you should choose the E-option or the M-option on your SAT II: Biology exam. The first Biology E/M practice test in this book, Test One, is a diagnostic test. After you take this test, read through the instructions on how to score it that follow the answer key. These instructions identify which questions on the test are ecological and which are molecular in orientation. If you perform well on the ecological questions on the diagnostic test, it might be a good idea for you to select the Biology-E option. Likewise, a good performance on the molecular questions could indicate that molecular biology might be your area of strength, and that you should choose the M-option.

The diagnostic test will also help you to identify areas of weakness that you should address in your SAT II preparation. For example, if you find that you have a tendency to get questions related to the endocrine system wrong, you would probably do well to pay close attention to the chapter on organismal biology in this book.

Each chapter in the Biology Review section ends with a follow-up quiz, complete with answers and explanations. When you feel you've mastered the material in a chapter, take the follow-up quiz to be sure. Don't forget to use the glossary at the end of the book to brush up on vital definitions, and the index to look up lengthier explanations of troublesome concepts.

Kaplan Practice Tests

When you feel you're really ready, try your luck at Tests Two and Three. Test Two is a Biology-E (ecological) test, and Test Three is a Biology-M (molecular) test. The best way to use these tests is to take them under testlike conditions. Don't just drop in and do a random question here and there. Use these tests to gain experience with the complete testing experience, including pacing and endurance. You can do these tests at any time. You don't have to save them all until after you've read this whole book. Just be sure to save at least one test for your dress rehearsal some time in the last week before Test Day.

You can test your knowledge of ecology by taking a crack at Test Two (a Biology-E test), while Test Three will help you to prepare for the Biology-M test. However, don't take just one of these tests; both of them contain general biology questions that you will need to know how to answer regardless of which test option you select.

Finding Your Way

How you use this book depends on how much time you have. Let's take a look at three typical students who are planning to take the SAT II: Biology E/M Test. Note that our hypothetical students use this book in three different ways. Which student's study plan best matches your situation?

 "I'm taking the Biology E/M Test a month from today."

Angela has plenty of time to prep for the test. If you're like Angela, and you have at least two weeks to prepare, then we recommend that you do everything in this book that relates to the test you're taking.

 "I'm taking the Biology E/M Test in a week."

If you're like Bill, you'll need a shortcut. If you have fewer than two weeks but more than two days to prepare, then we recommend that you use the diagnostic test to determine which chapters you can safely skim, or even skip.

 "Help! It's two days before Test Day!"

Eric is in a panic. But you don't need to freak, even if you're in Eric's situation. Use our Panic Plan to get through this book. If you have only a day or two to prepare for the test, then you don't have time to prepare thor-

College Board Publications

The College Board has released some test questions that you might want to look at. The free pamphlet *Taking the SAT II: Subject Tests* has a few sample questions, and the College Board's book *Real SAT II: Subject Tests* includes a Biology E/M minitest.

oughly. But that doesn't mean you should just give up and not prepare at all. There's still a lot you can do to improve your potential score. First and foremost, you should become familiar with the test. Read the introductory section in this book. And if you do nothing else, you should at least sit down and work through one of full-length tests at the back of this book under reasonably testlike conditions. Choose Test Two if you are taking the Biology-E test, and Test Three if you are taking the Biology-M test.

When you finish the practice test, check your answers and look at the explanations to the questions you didn't get right. When you come across a topic that you only half remember, turn to the appropriate chapter in this book for a quick review. When you come across a topic you don't remember or understand at all, skip it. You don't have time to learn and assimilate completely new material. At least you'll know to skip any similar question you might encounter on the actual SAT II: Biology E/M Test.

The Icons

As you work your way through this book, you'll see the following helpful icons used repeatedly in the side bars in the margins of the text. Here's what they mean.

Basic Concepts. This icon highlights basic concepts that are explained in more detail in the accompanying text. Use these sidebars to quickly determine what is being discussed in the text, and to get a wider perspective on the topic at hand.

A Closer Look. Check out the sidebars that display this icon for additional facts and/or examples associated with a particular topic. Often, you'll be surprised by what you read here.

Study Tips. This icon appears next to information that can help you to grasp biological concepts more easily. Here you'll find "Don't Mix These Up on Test Day" sidebars, which point out important, easily confused concepts that are very likely to turn up on your SAT II: Biology E/M test, as well as mnemonics that can help you to memorize important facts.

Test Strategies. This icon highlights Kaplan test-taking strategies that can help you boost your score.

Quiz. You'll find this icon with sidebars that feature "quick quizzes"—questions that will get your brain working on tough concepts even as you read about them.

Take a Break before Test Day

A day or two before the test, be sure to review chapter 9: The Final Countdown. It includes essential "how-to" tips that will help you get the upper hand on test day. Then, relax! Read a book or watch a movie. And most important, get a good night's sleep. How you approach the days leading up to the test really does matter!

On the morning of the test, eat a light breakfast (nothing too heavy to make you sleepy!) and quickly review a few questions if you feel like it (just enough to get you focused). Walk into the test center with confidence— you're ready for the challenge!

A Special Note for
International Students

If you are an international student considering attending an American university, you are not alone. Approximately 500,000 international students pursued academic degrees at the undergraduate, graduate, or professional school level at U.S. universities during the 1998-1999 academic year, according to the Institute of International Education's Open Doors report. Almost 50 percent of these students were studying for a bachelor's or first university degree. This number of international students pursuing higher education in the United States is expected to continue to grow. Business, management, engineering, and the physical and life sciences are particularly popular majors for students coming to the United States from other countries.

If you are not a U.S. citizen and you are interested in attending college or university in the United States, here is what you'll need to get started.

- If English is not your first language, you'll probably need to take the TOEFL (Test of English as a Foreign Language) or provide some other evidence that you are proficient in English. Colleges and universities in the United States will differ on what they consider to be an acceptable TOEFL score. A minimum TOEFL score of 213 (550 on the paper-based TOEFL) or better is often required by more prestigious and competitive institutions. Because American undergraduate programs require all students to take a certain number of general education courses, all students-even math and computer science students-need to be able to communicate well in spoken and written English.

- You may also need to take the SAT or the ACT. Many undergraduate institutions in the United States require both the SAT and TOEFL of international students.

- There are over 2,700 accredited colleges and universities in the United States, so selecting the correct undergraduate school can be a confusing task for anyone. You will need to get help from a good advisor or at least a good college guide that gives you detailed information on the different schools available. Since admission to many undergraduate programs is quite competitive, you may want to select three or four colleges and complete applications for each school.

- You should begin the application process at least a year in advance. An increasing number of schools accept applications year round. In any case, find out the application deadlines and plan accordingly. Although September (the fall semester) is the traditional time to begin university study in the United States, you can begin your studies at many schools in January (the spring semester).

In addition, you will need to obtain an I-20 Certificate of Eligibility from the school you plan to attend if you intend to apply for an F-1 Student Visa to study in the United States.

Kaplan International Programs

If you need more help with the complex process of university admissions, assistance preparing for the SAT, ACT, or TOEFL, or help building your English language skills in general, you may be interested in Kaplan's programs for international students.

Kaplan International Programs were designed to help students and professionals from outside the United States meet their educational and career goals. At locations throughout the United States, international students take advantage of Kaplan's programs to help them improve their academic and conversational English skills, raise their scores on the TOEFL, SAT, ACT, and other standardized exams, and gain admission to the schools of their choice. Our staff and instructors give international students the individualized attention they need to succeed. Here is a brief description of some of Kaplan's programs for international students:

General Intensive English

Kaplan's General Intensive English classes are designed to help you improve your skills in all areas of English and to increase your fluency in spoken and written English. Classes are available for beginning to advanced students, and the average class size is 12 students.

English for TOEFL and University Preparation

This course provides you with the skills you need to improve your TOEFL score and succeed in an American university or graduate program. It includes advanced reading, writing, listening, grammar and conversational English, plus university admissions counseling. You will also receive training for the TOEFL using Kaplan's exclusive computer-based practice materials.

SAT Test Preparation Course

The SAT is an important admission criterion for American colleges and universities. A high score can help you stand out from other applicants. This course includes the skills you need to succeed on each section of the SAT, as well as access to Kaplan's exclusive practice materials.

English & SAT

This course includes a combination of English instruction and SAT test preparation. Our English & SAT course is for students who need to boost their English skills while preparing for the SAT and admission into an American university.

Other Kaplan Programs

Since 1938, more than 3 million students have come to Kaplan to advance their studies, prepare for entry to American universities, and further their careers. In addition to the above programs, Kaplan offers courses to prepare for the ACT, GMAT, GRE, MCAT, DAT, USMLE, NCLEX, and other standardized exams at locations throughout the United States.

Applying to Kaplan International Programs

To get more information, or to apply for admission to any of Kaplan's programs for international students and professionals, contact us at:

Kaplan International Programs
370 Seventh Avenue, New York, NY 10001 USA
Telephone: (212) 492-5990 Fax: (917) 339-7505
E-mail: world@kaplan.com Web: www.kaptest.com

- Kaplan is authorized under federal law to enroll nonimmigrant alien students.

- Kaplan is authorized to issue Form IAP-66 needed for a J-1 (Exchange Visitor) visa.

- Kaplan is accredited by ACCET (Accrediting Council for Continuing Education and Training).

- Test names are registered trademarks of their respective owners.

THE BASICS

CHAPTER ONE

ABOUT THE SAT II: SUBJECT TESTS

You're serious about going to the college of your choice. You wouldn't have opened this book otherwise. You've made a wise choice, because this book can help you to achieve your goal. It'll show you how to score your best on the SAT II: Biology Subject Test. But before turning to the biology review, let's look at the SAT II as a whole.

Frequently Asked Questions

The following background information about the SAT II is important to keep in mind as you get ready to prep for the SAT II: Biology E/M Test. Remember, though, that sometimes the test makers change the test policies after a book has gone to press. The information here is accurate at the time of publication, but it's a good idea to check the test on the College Board Website at www.collegeboard.org.

What Is the SAT II?

Known until 1994 as the College Board Achievement Tests, the SAT II is actually a set of more than 20 different Subject Tests. These tests are designed to measure what you have learned in such subjects as literature, U.S. history, world history, math, biology, and Spanish. Each test lasts one hour and consists entirely of multiple-choice questions, except for the Writing Test, which has a 20-minute essay section and a 40-minute multiple-choice section. On any one test date, you can take up to three Subject Tests.

How Does the SAT II Differ from the SAT I?

SAT I is largely a test of verbal and math skills. True, you need to know some vocabulary and some formulas for the SAT I; but it's designed to measure how well you read and think rather than how much you remember. The SAT II tests are very different. They're designed to measure what you know about specific disciplines. Sure, critical reading and thinking skills play a part on these tests, but their main purpose is to determine exactly what you know about writing, math, history, chemistry, and so on.

"What Does That Spell?"

Originally, *SAT* stood for *Scholastic Aptitude Test*. When the test changed a few years ago, the official name was changed to *Scholastic Assessment Test*. In 1997, the test makers announced that *SAT* no longer stands for anything, officially.

Dual Role

Colleges use your SAT II scores in both admissions and placement decisions.

How Do Colleges Use the SAT II?

Many people will tell you that the SATs (I and II alike) measure only your ability to perform on standardized exams—that they measure neither your reading and thinking skills nor your level of knowledge. Maybe they're right. But these people don't work for colleges. Those schools that require SATs feel that they are an important indicator of your ability to succeed in college. Specifically, they use your scores in one or both of two ways: to help them make admissions and/or placement decisions.

Like the SAT I, the SAT II tests provide schools with a standard measure of academic performance, which they use to compare you with applicants from different high schools and different educational backgrounds. This information helps them to decide whether you're ready to handle their curriculum.

SAT II scores may also be used to decide what course of study is appropriate for you once you've been admitted. A low score on the Writing Test, for example, might mean that you have to take a remedial English course. Conversely, a high score on an SAT II: Mathematics Test might mean that you'll be exempted from an introductory math course.

Call Your Colleges

Many colleges require you to take certain SAT II tests. Check with all of the schools you're interested in applying to before deciding which tests to take.

Which SAT II Tests Should I Take?

The simple answer is: those that you'll do well on. High scores, after all, can only help your chances for admission. Unfortunately, many colleges demand that you take particular tests, usually the Writing Test and/or one of the Mathematics Tests. Some schools will give you a degree of choice in the matter, especially if they want you to take a total of three tests. Before you register to take any tests, therefore, check with the colleges you're interested in to find out exactly which tests they require. Don't rely on high school guidance counselors or admissions handbooks for this information. They might not give you accurate or current information.

Count to Three

You can take up to three SAT II tests in one day. The Writing Test must be taken first.

When Are the SAT II Tests Administered?

Most of the SAT II Tests are administered six times a year: in October, November, December, January, May, and June. A few of the tests are offered less frequently. Due to admissions deadlines, many colleges insist that you take the SAT II no later than December or January of your senior year in high school. You may even have to take it sooner if you're interested in applying for "early admission" to a school. Those schools that use scores for placement decisions only may allow you to take the SAT II as late as May or June of your senior year. You should check with colleges to find out which test dates are most appropriate for you.

How Do I Register for the SAT II?

The College Board administers the SAT II tests, so you must sign up for the tests with them. The easiest way to register is to obtain copies of the *SAT Registration Bulletin* and *Taking the SAT II: Subject Tests*. These publications contain all of the necessary information, including current test dates and fees. They can be obtained at any high school guidance office or directly from the College Board.

You can also register online. Visit the College Board's Website at www.collegeboard.org for more information. If you have previously registered for an SAT I or SAT II test, you can reregister by telephone. If you choose this option, you should still read the College Board publications carefully before you make any decisions.

How Are the SAT II Tests Scored?

Like the SAT I, the SAT II tests are scored on a 200–800 scale.

What's a "Good" Score?

That's a tricky question. The obvious answer is: the score that the colleges of your choice demand. Keep in mind, though, that SAT II scores are just one piece of information that colleges will use to evaluate you. The decision to accept or reject you will be based on many criteria, including your high school transcript, your SAT I scores, your recommendations, your personal statement, your interview (where applicable), your extracurricular activities, and the like. So, failure to achieve the necessary score doesn't automatically mean that your chances of getting in have been damaged. For those who really want a numerical benchmark, a score of 600 is considered very solid.

A College Board service known as Score Choice offers you the chance to see your scores before anyone else. If you're unhappy with a score, you don't have to send it along to colleges. If you decide to take advantage of this service, you'll need to take your SAT II tests well in advance of college deadlines. At the very least, using Score Choice will slow down the reporting process. You may also want to retake one or more tests. Two more points to bear in mind:

- Once you've released a score, it can't be withheld in the future.

- If you use Score Choice, you lose the privilege of having some scores sent to schools for free.

For more information about Score Choice, contact the College Board.

Do the Legwork

Want to register or get more info? You can get copies of the *SAT Registration Bulletin* and *Taking the SAT II: Subject Tests* from the College Board. If you have a credit card, you can also register for the SAT II online at www.collegeboard.org. You can register by phone *only* if you have registered for an SAT I or SAT II test in the past.

College Board SAT Program
P.O. Box 6200
Princeton, NJ 08541-6200
(609) 771-7600

Pack Your Bag

Gather your test materials the day before the test. You'll need:

- Your admission ticket

- A proper form of I.D.

- Some sharpened No. 2 pencils

- A good eraser

Don't Get Lost

Learn SAT II directions as you prepare for the tests. You'll have more time to spend answering the questions on Test Day.

What Should I Bring to the SAT II?

It's a good idea to get your test materials together the day before the tests. You'll need an admission ticket; a form of identification (check the *Registration Bulletin* to find out what is permissible); a few sharpened No. 2 pencils; a good eraser; and a scientific calculator (for Math Level IC or IIC). If you'll be registering as a standby, collect the appropriate forms beforehand. Also, make sure that you know how to get to the test center.

SAT II Mastery

Now that you know a little about the SAT II tests, it's time to let you in on a few basic test taking skills and strategies that can improve your performance on them. You should practice these skills and strategies as you prepare for the SAT II.

Use the Structure of the Test to Your Advantage

The SAT II tests are different from the tests that you're used to taking. On your high school tests, you probably go through the questions in order. You probably spend more time on hard questions than on easy ones, since hard questions are generally worth more points. And you often show your work, since your teachers tell you that how you approach questions is as important as getting the right answers.

None of this applies to the SAT II tests. You can benefit from moving around within the tests, hard questions are worth the same as easy ones, and it doesn't matter how you answer the questions—only what your answers are.

The SAT II tests are highly predictable. Because the format and directions of the SAT II tests remain unchanged from test to test, you can learn the setup of each test in advance. On Test Day, the various question types on each test shouldn't be new to you.

One of the easiest things you can do to help your performance on the SAT II tests is to understand the directions before taking the test. Since the instructions are always the same, there's no reason to waste a lot of time on Test Day reading them. Learn them beforehand as you work through this book and the College Board publications.

Not all of the questions on the SAT II tests are equally difficult. The questions often get harder as you work through different parts of a test. This pattern can work to your benefit. Try to be aware of where you are in a test.

When working on more basic problems, you can generally trust your first impulse—the obvious answer is likely to be correct. As you get to the end of a test section, you need to be a bit more suspicious. Now the answers probably won't come as quickly and easily—if they do, look again because

KAPLAN

the obvious answers may be wrong. Watch out for answers that just "look right." They may be *distractors*—wrong answer choices deliberately meant to entice you.

There's no mandatory order to the questions on the SAT II. You're allowed to skip around on the SAT II tests. High scorers know this fact. They move through the tests efficiently. They don't dwell on any one question, even a hard one, until they've tried every question at least once.

When you run into questions that look tough, circle them in your test booklet and skip them for the time being. Go back and try again after you've answered the easier ones if you've got time. After a second look, troublesome questions can turn out to be remarkably simple.

If you've started to answer a question but get confused, quit and go on to the next question. Persistence might pay off in high school, but it usually hurts your SAT II scores. Don't spend so much time answering one hard question that you use up three or four questions' worth of time. That'll cost you points, especially if you don't even get the hard question right.

You can use the so-called guessing penalty to your advantage. You might have heard it said that the SAT II has a "guessing penalty." That's a misnomer. It's really a *wrong-answer penalty*. If you guess wrong, you get a small penalty. If you guess right, you get full credit.

The fact is, if you can eliminate one or more answer choices as definitely wrong, you'll turn the odds in your favor and actually come out ahead by guessing. The fractional points that you lose are meant to offset the points you might get "accidentally" by guessing the correct answer. With practice, however, you'll see that it's often easy to eliminate *several* answer choices on some of the questions.

The answer grid has no heart. It sounds simple, but it's extremely important: Don't make mistakes filling out your answer grid. When time is short, it's easy to get confused going back and forth between your test booklet and your grid. If you know the answers, but misgrid, you won't get the points. Here's how to avoid mistakes.

Always circle the questions you skip. Put a big circle in your test booklet around any question numbers that you skip. When you go back, these questions will be easy to locate. Also, if you accidentally skip a box on the grid, you'll be able to check your grid against your booklet to see where you went wrong.

Always circle the answers you choose. Circling your answers in the test booklet makes it easier to check your grid against your booklet.

Grid five or more answers at once. Don't transfer your answers to the grid after every question. Transfer them after every five questions. That way,

Leap Ahead

Do the questions in the order that's best for you. Skip hard questions until you've gone through every question once. Don't pass up the opportunity to score easy points by wasting time on hard questions. Come back to them later.

Guessing Rule

Don't guess, unless you can eliminate at least one answer choice. Don't leave a question blank unless you have absolutely no idea how to answer it.

Hit the Spot

A common mistake is filling in all of the questions with the right answers—in the wrong spots. Whenever you skip a question, circle it in your test booklet and make doubly sure that you skip it on the answer grid as well.

Think First

Try to think of the answer to a question before you shop among the answer choices. If you have some idea of what you're looking for, you'll be less likely to be fooled by "trap" choices.

Speed Limit

Work quickly on easier questions to leave more time for harder questions. But not so quickly that you make careless errors. And it's okay to leave a few questions blank if you have to—you can still get a high score.

you won't keep breaking your concentration to mark the grid. You'll save time and gain accuracy.

Approaching SAT II Questions

Apart from knowing the setup of the SAT II tests that you'll be taking, you've got to have a system for attacking the questions. You wouldn't travel around an unfamiliar city without a map, and you shouldn't approach the SAT II without a plan. What follows is the best method for approaching SAT II questions systematically.

Think about the questions before you look at the answers. The test makers love to put distractors among the answer choices. If you jump right into the answer choices without thinking first about what you're looking for, you're much more likely to fall for one of these traps.

Guess—when you can eliminate at least one answer choice. You already know that the "guessing penalty" can work in your favor. Don't simply skip questions that you can't answer. Spend some time with them in order to see whether you can eliminate any of the answer choices. If you can, it pays for you to guess.

Pace yourself. The SAT II tests give you a lot of questions in a short period of time. To get through the tests, you can't spend too much time on any single question. Keep moving through the tests at a good speed. If you run into a hard question, circle it in your test booklet, skip it, and come back to it later if you have time.

You don't have to spend the same amount of time on every question. Ideally, you should be able to work through the easier questions at a brisk, steady clip, and use a little more time on the harder questions. One caution: Don't rush through basic questions just to save time for the harder ones. The basic questions are points in your pocket, and you're better off not getting to some harder questions if it means losing easy points because of careless mistakes. Remember, you don't earn any extra credit for answering hard questions.

Locate quick points if you're running out of time. Some questions can be done more quickly than others because they require less work or because choices can be eliminated more easily. If you start to run out of time, look for these quicker questions.

When you take the SAT II: Subject Tests, you have one clear objective in mind: to score as many points as you can. It's that simple. The rest of this book is dedicated to helping you to do that on the SAT II: Biology E/M Subject Test.

CHAPTER TWO

GETTING READY FOR THE SAT II: BIOLOGY E/M TEST

Now that you know the basics about the SAT II: Subject Tests, it's time to focus on the Biology E/M test. What's on it? How is it scored? After reading this chapter, you'll know just what to expect on Test Day.

Test Content

Skills Tested

Three basic skills are tested on the SAT II: Biology E/M exam. First of all, the *ability to recall knowledge* will test your ability to remember specific facts, your mastery of terminology, and your comfort with straightforward knowledge.

The second skill tested is your ability to *apply your biology knowledge to unfamiliar situations*. These questions will test how well you understand concepts and your ability to reformulate information in a variety of ways. In other words, you will be required to express a given piece of information in different forms; you may be asked to compare graphical data to written data, for example. These questions will also test how well you can solve problems, particularly those dealing with mathematical relationships.

The third question type explores your proficiency at *synthesizing biological information*. These questions will require you to make inferences and deductions from qualitative and quantitative data, such as data you might accumulate doing an experiment in the laboratory, and to then integrate that data to form conclusions. The data may be in paragraph form, like those word problems you hated in fourth grade, or it may be in graph or chart form. These questions will also examine your ability to recognize unstated assumptions—you will need to be prepared to think about what is implied in the setup of the experiment or the question stem.

Format

The Biology E/M Test has a total of 80 multiple-choice questions that you are given one hour to answer: 60 "common core" questions and 20 ecolog-

Basic Skills

Three basic skills are tested on the SAT II: Biology E/M Test:

- Recalling information
- Applying knowledge
- Synthesizing information

More Info on the Biology E/M Test

For more information on the Biology E/M Test, check out *Taking the SAT II: Subject Tests*. Your high school guidance counselor should have a copy of this booklet. You can also have one sent to you by calling the College Board at (609) 771-7600.

ical or molecular questions. The purpose of this exam is to evaluate students' mastery of basic biological principles. The E/M Test places particular emphasis on the fields of ecology and molecular biology.

You will answer *either* the "E" (ecological) *or* the "M" (molecular) section of the exam on Test Day. When this big day rolls around, you will be given the opportunity to indicate whether you want to take the Biology-E or the Biology-M option; you may not take both on the same day.

Both of these options also assume a solid understanding of evolution and diversity. The ecological section may draw from the field of classical genetics, and you will probably find questions on molecular and bacterial genetics in the molecular section of the exam.

Here is a breakdown of the format of the Biology E/M Test:

Topics	Percentage of the Test
Common Core Questions:	
Cellular and Molecular Biology	12%
Ecology	12%
Classical Genetics	10%
Organismal Biology	30%
Evolution and Diversity	11%
Ecology/Evolution Section (Biology-E Test)	25%
Molecular/Evolution Section (Biology-M Test)	25%

As you can see, the SAT II: Biology E/M Test covers a broad range of topics. It requires you to think about those topics in ways that you may not have done before. As a result, it is likely that some of the questions on your subject test will explore topics that you did not cover in your biology class. If this is the case, do not be alarmed; there is so much to biology that you cannot possibly cover everything in a year, although it may seem like you've learned the entire field at the time. If you encounter this problem while you are taking the diagnostic test, plan to spend a little extra time on the chapters that cover your area(s) of weakness, so that on Test Day you are completely familiar with these topics.

While preparing for this exam, you should also make sure that you understand common algebraic concepts such as ratios and proportions and, more importantly, that you are able to apply these concepts to the word problems and data interpretation questions you will surely see on Test Day. You will not be allowed to use a calculator on this exam, but don't worry; the math should be nothing more complicated than simple calculations involving multiplication or division.

Test Strategy

First of all, you are obviously going to want to make sure that you are very familiar with either the ecology or the molecular biology covered on the exam, depending upon which option you select. And whether you select the Biology-E or the Biology-M section on Test Day, don't neglect classical genetics and evolution and diversity in your studies. These topics are likely to crop up in both of these sections.

The Biology E/M test stresses problem solving skills. You can prep yourself for this, by honing your mathematical skills and by getting as much experience as you can in the laboratory. The E/M Test will ask you to use simple algebraic concepts, and it assumes that you are comfortable with the metric system. It will also present a great deal of experimental data for you to synthesize, so the more familiarity you have with experimental situations, the better.

This book can help you both to decide whether you should choose the E-option or the M-option on your SAT II: Biology exam and to prepare appropriately for the option you've chosen. The first Biology E/M practice test in this book, Test One, is a diagnostic test. After you take this test, read through the instructions on how to score it that follow the answer key. These instructions identify which questions on the test are ecological and which are molecular in orientation. If you perform well on the ecological questions on the diagnostic test, it might be a good idea for you to select the Biology-E. Likewise, a good performance on the molecular questions should indicate that molecular biology might be your area of strength. If you've already decided which option to take, your score on the diagnostic test will help you to identify the areas that you need to brush up on. You can further hone your knowledge of ecology by taking a crack at Test Two (a Biology-E test), while Test Three will help you to prepare for the Biology-M test.

Test Tip

Our test strategies won't make up for a weakness in a given area, but they will help you to manage your time effectively and maximize points.

Scoring Information

The Biology E/M Test is scored in a range from 200–800, just like a section of the SAT I exam. Your raw score is calculated by subtracting $\frac{1}{4}$ of the number of questions you got wrong from the number of questions you got right. For example, if you answered 60 questions correctly and 20 incorrectly, your raw score would be:

Number correct	60.00
$\frac{1}{4}$ × Number incorrect	− 5.00
Raw Score	55.00

This raw score is then compared to the scores of all the other test takers to calculate a scaled score. This scaling accounts for any slight variations in difficulty between test administrations. Keep in mind that it is possible to miss a few questions and still receive a competitive score.

Question Types

You will encounter two main types of multiple choice questions on the Biology E/M exam: classification questions and five-choice completion questions. Make sure you feel comfortable with both types and their directions before test day. Don't waste time reading directions when you are being timed! We have also included a selection of sample ecology and molecular biology questions, to give you an idea of what to expect on these sections of the E/M Test.

Classification Questions

Classification questions consist of five lettered choices that are used in all of the questions that follow. Typically, the five choices will test your knowledge of ideas, organism names, graphs, or of some other type of data presentation.

Following the five choices will be three to five statements that can be functions of the choices, definitions, descriptive characteristics, or conditions that would favor the data set in question. Each of the five choices may be used more than once, so do not eliminate an answer just because you have already used it.

To familiarize yourself with this question type, read through the directions, and attempt to answer questions 1–4 below. Check your answers against the in-depth explanations that follow the question set.

Directions: Each set of lettered choices below refers to the numbered statements immediately following it. Select the one lettered choice that best fits each statement, and then fill in the corresponding oval on the answer sheet. A choice may be used once, more than once, or not at all in each set.

Questions 1–4:

(A) Prophase
(B) Metaphase
(C) Anaphase
(D) Interphase
(E) Anaphase I

Test Strategy

On Test Day, do classification questions first; they require less reading and will give you the most points for your time invested.

Next, do the Type 1 and Type 2 Five-Choice Completion questions. Again, you will get a lot of points for the amount of time you invest.

Test Strategy

Don't eliminate an answer choice just because you've used it. *Answer choices can be used more than once.*

1.   the stage during which a cell's DNA is replicated

2.   the stage during which homologous pairs of chromosomes are pulled to opposite poles of a cell

3.   a stage in meiosis

4.   the stage during which a cell's chromosomes condense

Explanations. The questions in this group deal with *mitosis*, the mechanism that a cell uses to replicate itself, and *meiosis*, the mechanism that produces gametes. A cell will spend roughly 90 percent of its time in Interphase. This phase may be broken down into a number of different stages. During the G_1 stage, the cell doubles in size, and new organelles such as mitochondria, ribosomes, the endoplasmic reticulum, and centrioles are produced. In the next stage, the S stage, all of the DNA is replicated (see question 1) so that during division, a complete copy of the genome can be distributed to both daughter cells. Following S stage is the G_2 stage, during which the cell continues to grow in size, preparing components for cell division.

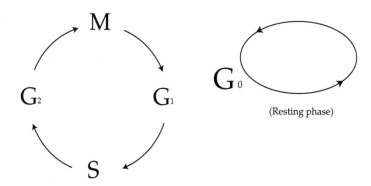

The Cell Cycle

During mitosis, the cell divides and distributes its DNA to its two daughter cells, such that each cell receives a complete copy of the original genome. Almost all cells, except for nerve cells and specialized muscle cells, can undergo mitosis. This process may be broken down into the following four stages:

- *Prophase.* The chromosomes condense (see question 4). The centriole pairs separate and move toward opposite poles of the cell, while the spindle apparatus forms between them. The nuclear membrane dissolves and the spindle fibers enter the nucleus.

- *Metaphase.* The chromosomes align at the metaphase plate and prepare to separate.

- *Anaphase*. The centromeres split so that each chromatid has its own centromere, and the sister chromatids are pulled toward the opposite poles of the cell (see question 2).

- *Telophase*. The spindle apparatus disappears, and new nuclear membranes are formed.

During meiosis, on the other hand, the gametocyte's chromosomes are replicated during the S phase of the cell cycle, just as in mitosis. In the first round of division, Meiosis I, the cell produces two intermediate daughter cells. In Meiosis II, the duplicated chromosomes split, resulting in four genetically distinct haploid gametes. In Anaphase I of Meiosis I, the homologous pairs of chromosomes are pulled to opposite poles of the cell. This process, called disjunction, accounts for the Mendelian law of independent assortment.

In light of all this, the correct answers for this question set are (D), (E), (E), and (A).

Five-Choice Completion Questions

These are common multiple-choice questions, and there are four types of them. The following directions apply to all four types.

<u>Directions:</u> **Each of the questions or incomplete statements below is followed by five suggested answers or completions. Select the one that is best in each case and then fill in the corresponding oval on the answer sheet.**

Type 1 Questions

These questions have a unique solution. The unique solution is often the only correct answer or the best answer. Sometimes, though, it will be the most inappropriate answer. These question types will have NOT, EXCEPT, or LEAST in capital letters somewhere in the stimulus.

Here are two examples:

Questions 5–6:

5. If the victim of an automobile accident suffered isolated damage to the cerebellum, which of the following would most likely occur?

   (A) loss of voluntary muscle contraction
   (B) loss of sensation in the extremities
   (C) loss of muscular coordination
   (D) loss of speech
   (E) loss of hearing

Four Types

Remember that there are four types of five-choice completion questions:

1. Unique solution
2. Roman numeral questions
3. Figure or diagram identification
4. Experimental data presentation

6.   Which of the following is NOT a type of genetic mutation?

  (A) point
  (B) silent
  (C) insertion
  (D) frameshift
  (E) malignant

Explanations. To answer question 5, you need to know that the cerebellum is located in the hindbrain, along with the pons and the medulla. All higher brain sensory neurons and motor neurons pass through the hindbrain. The main function of the cerebellum is to coordinate movement, such as hand-eye coordination, posture, and balance. Therefore, damage to the cerebellum would most likely affect (C) muscular coordination. As for the remaining choices, the loss of voluntary muscle contraction, sensation in the extremities, speech, or hearing may be caused by damage to specific areas of the cerebrum. The cerebrum is located in the forebrain and is divided into two hemispheres, the left and the right. This organ is responsible for the coordination of most voluntary activities, sensation, and "higher functions," including speech and cognition. Sensation of the extremities may also be controlled in part by the spinal cord.

Let's move on to question 6. Point mutations occur when a single nucleotide base is substituted for another nucleotide base. A silent mutation is a point mutation that occurs in a noncoding region, or when the mutation does not change the amino acid sequence due to the degeneracy of the genetic code. Meanwhile, a frame shift mutation is either an insertion or a deletion of a number of nucleotides. These mutations have serious effects on the protein coded for, since nucleotides are read as a series of triplets. The addition or subtraction of nucleotides (except in multiples of three) will change the reading frame of the mRNA. Finally, (E), "malignant," is not a type of mutation. A mutation may lead to a cell becoming malignant (cancerous), but it does not necessarily do so, and it is the cell that is malignant, not the mutation.

Test Strategy

Once you eliminate a roman numeral, make sure to eliminate all lettered answer choices with that roman numeral.

Test Strategy

Use the structure of a roman numeral question to your advantage. Eliminate choices as soon as you find them to be inconsistent with the truth or falsehood of a statement in the stimulus. Similarly, consider only those choices that include a statement that you've already determined to be true.

Type 2 Questions

Typically, these have three to five roman numerals following each question. One or more of these roman numerals may prove to be the correct answer(s). Following these roman numerals, you will encounter five lettered choices with various combinations of the roman numerals. You must select the combination that includes all of the correct answers and excludes all of the incorrect answers. Try your hand at the sample question below.

7. Which of the following are characteristic of animal cells, but NOT bacterial cells?

      I.   They are eukaryotic.
      II.   They possess ribosomes.
      III.   They possess cell walls.
      IV.   They reproduce asexually.

    (A) I only
    (B) I and II only
    (C) I, II, and III only
    (D) II, III, and IV only
    (E) none of the above

Explanation. This question requires you to know the differences between the basic animal cell, the fungal cell, and the bacterial cell. All animal (and plant) cells are eukaryotic, while all bacterial cells are prokaryotic. Bacteria have cell walls made of peptidoglycans, while animal cells do not have cell walls, and bacteria reproduce asexually via binary fission, while animal cells can replicate asexually by mitosis. Both cell types produce proteins with ribosomes, although they differ in size and composition. Therefore, the correct answer is (A).

Type 3 Questions

Organized in sets around a figure or a diagram, each question is nevertheless independent of the other questions in its set. These questions test your knowledge of morphology and function of a variety of biological structures. Typically, these questions are not as difficult as Type 4 questions, the experimental data questions. Here are three examples:

Questions 8–10 refer to the following diagram.

HUMAN HEART

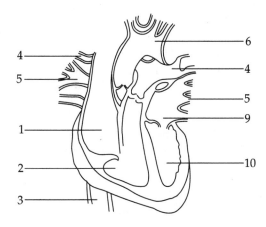

8.   Which part labeled in the diagram carries the most oxygenated blood?

(A)   4
(B)   5
(C)   1 and 2
(D)   6
(E)   10

9.   Where is the pacemaker of the heart?

(A)   1
(B)   2
(C)   9
(D)   10
(E)   8

10.   Which sequence is the correct flow of blood through the heart?

(A)   3 —> 9 —> 10 —> 4 —> 5 —> 1 —> 2
(B)   1 —> 2 —> 5 —> 4 —> 9 —> 10 —> 6
(C)   1 —> 2 —> 4 —> 5 —> 9 —> 10 —> 6
(D)   1 —> 2 —> 3 —> 4 —> 5 —> 9 —> 10
(E)   3 —> 2 —> 1 —> 4 —> 5 —> 9 —> 10

What's the Sinoatrial Node?

The sinoatrial node is another name for the pacemaker of the heart.

Test Strategy

Type 4 questions will often concentrate on trends or outliers in charts and graphs. Make an effort to pay close attention to these factors when skimming through the questions on Test Day.

Explanations. Let's begin with question 8. The heart is the driving force of the circulatory system. The right and left halves can be viewed as two separate pumps; the right side of the heart pumps deoxygenated blood into pulmonary circulation (toward the lungs), while the left side pumps oxygenated blood into systemic circulation (throughout the body). The two upper chambers are called atria, while the two lower chambers are called ventricles. The former are thin walled, while the latter are extremely muscular. The left ventricle is more muscular than the right ventricle because it is responsible for generating the force that propels systemic circulation, and because it pumps against high resistance. The most oxygenated blood is in the pulmonary veins. These veins return to the heart from the lungs, where the blood was oxygenated. This is the only type of vein in the body that carries oxygenated blood. In a similar fashion, the pulmonary artery carries deoxygenated blood to the lungs, and is the only artery in the body that carries deoxygenated blood. Therefore, the answer is (B).

As for question 9, an ordinary cardiac contraction originates in, and is regulated by, the sinoatrial node (SA node), which is also known as the pacemaker. This small mass of specialized tissue is located in the wall of the right atrium. It spreads impulses through both atria and ventricles, stimulating them to contract simultaneously. (A) is the answer here.

Finally, in question 10, the correct sequence of blood flow begins in the right atrium and then travels into the right ventricle. From there, the blood flows into the pulmonary arteries, and proceeds to the lungs to be oxygenated. It returns via the pulmonary veins and flows into the left atrium. From there, it flows into the left ventricle and is pumped throughout the body, starting at the aorta. The correct choice is (C).

Type 4 Questions

These are also organized in sets, but center on an experiment, chart, graph, or other experimental data presentation. These questions aim to assess how well you apply your scientific skills to unfamiliar situations. As in Type 3 questions, each question is independent of the others in its set. These questions are typically found during the latter part of the test, and are probably the most difficult ones you will encounter. They test your ability to identify a problem, evaluate experimental situations, suggest hypotheses, interpret data, make inferences and draw conclusions, check the logical consistency of hypotheses based on your observations, and select the appropriate procedure for further study. Four sample questions are presented below.

Questions 11–14 are based on the following pedigree:

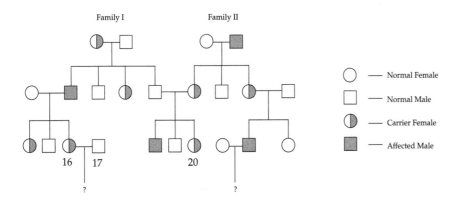

PEDIGREE

11. Based on the pedigree, what kind of trait is this?

    (A) autosomal recessive
    (B) sex-linked recessive
    (C) autosomal dominant
    (D) sex-linked dominant
    (E) heterozygous

12. What is the probability that 16 and 17 will have an unaffected daughter?

    (A) 50%
    (B) 0%
    (C) 100%
    (D) 25%
    (E) 75%

13. This trait is rare in females because

    (A) male-specific hormones trigger the disease
    (B) it is a dominant trait
    (C) the gene is on the Y chromosome
    (D) females must receive two faulty X chromosomes
    (E) none of the above

14. What is the genotype of individual 20?

(A) *XX*

(B) *Aa*

(C) $X_a X$

(D) *AA*

(E) *aa*

Pedigree Analysis

Note the following important facts about pedigrees:

- Autosomal recessive traits skip a generation.
- Autosomal dominant traits are found in every generation.
- Sex-linked traits show gender skewing.

Explanations. In answer to question 11, we can establish that this is a sex-linked recessive disorder (B). Sex-linked recessives typically affect only males, while females serve as carriers. They are differentiated from autosomal dominant traits that show up in every generation, with no gender skewing. Autosomal recessives also show no gender skewing, but skip generations.

As for questions 12 and 13, the pattern of inheritance for a sex-linked recessive is somewhat complicated. Since the gene is carried on the X chromosome, and males pass the X chromosome only to their daughters, affected males cannot pass the trait to their male offspring. Affected males will pass the gene to all of their daughters. However, unless the daughter also receives the genes from her mother, she will be a phenotypically normal carrier of the trait. Since all of the daughter's male children will receive their only X chromosome from her, half of her sons will receive the recessive sex-linked allele. Thus, sex-linked recessives generally affect only males; they cannot be passed from father to son, but can be passed from father to grandson via a daughter who is a carrier, thereby skipping a generation. All of the daughters will receive a functional X chromosome from their father, so they will all be outwardly normal. However, 50 percent of the daughters will receive an affected X chromosome from their mother, so they will be carriers and run the risk of their sons having the disorder. (C) is the correct choice for question 12, and (D) for question 13.

Individual 20 is a carrier female. Since the trait in these questions is sex-linked and recessive, this female must be $X_a X$. If you chose (C) in question 14, you were on the right track.

KAPLAN

Sample Questions for the Ecological and Molecular Sections

The sample questions below offer insights into the biology content that is explored in these sections.

Ecological Section

The first two sets of sample questions will focus on ecology as well as evolution and diversity.

Questions 1–3:

Within a particular species, six populations are either separated by a variety of geographical barriers or are able to interbreed. The diagram below shows which populations are able to interbreed and which are isolated.

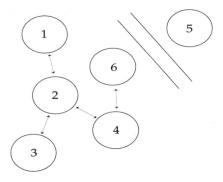

Use this information to answer questions 1–3.

1. Which of the following populations may be in Hardy-Weinberg equilibrium?

    (A) 1
    (B) 2
    (C) 3
    (D) 4
    (E) 5

2. The gain or loss of alleles that will occur between populations 4 and 6 is known as

    (A) gene flow
    (B) genetic drift
    (C) assortative mating
    (D) natural selection
    (E) mutation

Test Strategy

Look for opposing answers in the answer selections. If two answers are close in wording or if they contain opposite ideas, there is a strong possibility that one of them is the correct answer.

By the same token, if two answers mean basically the same thing, then they cannot both be correct—you can eliminate both answer choices.

3. Population 5 is wiped out by a flood. Only 10 percent of the population survives. This type of occurrence is known as

   (A) gene flow
   (B) founder effect
   (C) bottleneck effect
   (D) macroevolution
   (E) geographical variation

Questions 4–7:

Although goldfish have an optimal water temperature, they can tolerate higher or lower water temperatures if the change in water temperature to which they are subjected is gradual rather than abrupt. Swimming speed is used in the graph below to determine the general health of the fish in a variety of different water temperatures.

Test Strategy

Realize that the SAT II: Biology E/M Test emphasizes general trends and basic biology concepts. The test makers are probably not going to give you a question/graph that would take a rocket scientist 30 minutes to figure out. So, look for trends and outliers in graphs. If a value or a plot is vastly different from the others, it is likely that there will be a question about it.

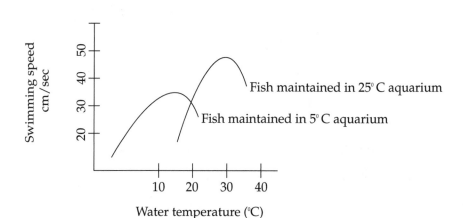

4. What is the optimal temperature for maintenance of goldfish?

   (A) 5° C
   (B) 10° C
   (C) 15° C
   (D) 25° C
   (E) 35° C

5. A fish that is maintained in a 5° C aquarium is placed in a 30° C aquarium. It would

    (A) swim at 20 cm/sec
    (B) swim at 50 cm/sec
    (C) swim at 40 cm/sec
    (D) be unable to adjust and suffer harm
    (E) be unable to remove oxygen from the water

6. Terrestrial organisms are often subjected to abrupt changes in the external temperature of their environment. As a result,

    (A) they suffer from a decrease in motility in warm weather
    (B) they are selected to tolerate such changes
    (C) they are able to control their external temperature
    (D) they suffer serious harm when abrupt temperature changes occur
    (E) they alter the morphology of their bodies

7. Goldfish and other aquatic animals generally require long periods of time to acclimatize themselves to different temperatures without suffering permanent damage. This reflects the fact that

    (A) water has a high heat capacity and changes temperature slowly
    (B) water is found in large bodies that maintain a constant temperature
    (C) aquatic organisms cannot utilize the sun as a heat source
    (D) the polar ice caps maintain the earth's water temperature
    (E) salt water is generally much warmer than fresh water

Explanations. The first few questions center around the topic of population genetics. In Hardy-Weinberg equilibrium, the allele frequencies, genotypes, and phenotypes are stable from one generation to the next. In order for Hardy-Weinberg equilibrium to be maintained, the following requirements must be met. There must be a large population; there must be no net mutations; no assortative mating may occur; and no migration may take place. Also, no natural selection can occur. In question 1, population 5 is the only isolated population in which there is no possibility that net migration of genes will occur. Therefore, (E) is the answer to this question.

The Power of Chance

Genetic drift is the term used to describe the changes in a gene pool that can be attributed to chance. The following are types of genetic drift:

- Bottleneck effect
- Founder effect

As for question 2, the gain or loss of alleles in a non-isolated population is known as gene flow (A). Gene flow occurs when fertile individuals migrate, or when gametes such as pollen grains are transferred. The migration between 4 and 6 makes A the correct answer. Genetic drift, meanwhile, refers to the changes in a gene pool that can be attributed to chance. Assortative mating, on the other hand, is the term assigned to the process by which individuals mate with partners who resemble themselves phenotypically, as when blister beetles select mates that are similar to themselves in size.

Examples of genetic drift include the bottleneck effect, the founder effect, gene flow, and mutations. In question 3, the bottleneck effect (C) occurs when a disaster (such as an earthquake, a flood, or a fire) reduces the size of a population in a drastic and unselective manner. The small surviving population is unlikely to be representative of the original population in terms of its genetic makeup.

Let's move on to the second question set. According to the chart, the goldfish at 25° C have the fastest swimming speed, an indication of general well-being. Of the fish described in this question set, these fish must therefore be the closest to their optimal temperature. The answer to question 4 is (D).

Fish are able to adjust to temperature changes without suffering serious harm only if these temperature changes occur gradually. For this reason, the fish in question 5 would most likely be unable to adjust to the new temperature, and would suffer harm (D).

As for question 6, terrestrial organisms live in environments that undergo abrupt changes in temperature. It is therefore logical to surmise that the environment selects for individuals who possess the capacity to tolerate rapid changes in temperature (B). Organisms that cannot tolerate such changes are unlikely to survive and reproduce on land. In aquatic environments, on the other hand, temperature changes occur slowly due to the high heat capacity of water. In contrast to terrestrial organisms, goldfish and other aquatic animals have not been selected to quickly acclimatize to temperature changes, and the answer to question 7 is (A).

KAPLAN

Molecular Section

Questions 8–11 refer to the diagram below.

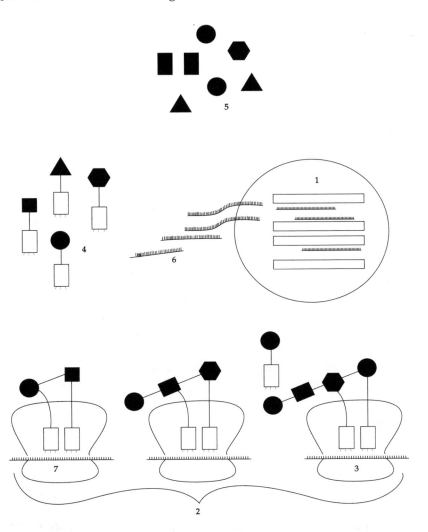

8. In the diagram above in the process labeled 2, what type of bond is being formed?

    (A) hydrogen
    (B) peptide
    (C) ionic
    (D) unbreakable
    (E) no bonds are being formed

Test Strategy

Predict your answer before you go to the answer choices so that you don't get persuaded by the wrong answers you will find there. This helps to boost your confidence and protects you from persuasive or tricky incorrect choices. Most wrong answer choices are logical twists on the correct choice.

9.   What process is occurring at 1?

   (A)   elongation
   (B)   translocation
   (C)   translation
   (D)   transcription
   (E)   termination

10.   Which of the following are made up of both protein and RNA?

   (A)   3
   (B)   4
   (C)   5
   (D)   3 and 6
   (E)   4 and 6

11.   A mutation in a gene that codes for a tRNA molecule may cause which of the following?

   (A)   autolysis of the cell
   (B)   premature ending of transcription
   (C)   a change in amino acid specificity for that tRNA
   (D)   mutated ribosomes
   (E)   a smaller number of free amino acids

Explanations.   Questions 8–11 explore the process of protein synthesis. In question 8, the type of bond being formed is (B) a peptide bond. This covalent bond is formed when tRNA (transport RNA) brings two amino acids into close proximity with each other. The tRNA's anticodons bind to the appropriate codon, or sequence, on the mRNA.

Transcription (D) is the process occurring in question 9. In transcription, information coded in the base sequence of DNA is transcribed onto a strand of mRNA (messenger RNA). mRNA leaves the nucleus through nuclear pores. The remaining events of protein synthesis take place in the cytoplasm.

As for question 10, structure 3 (A) is the correct answer. This ribosome is composed of two subunits, which in their turn consist of proteins and RNA that bind during protein synthesis. The ribosome has three binding sites—one for mRNA and two for tRNA. mRNA (structure 6) and tRNA (structure 4) are made up of RNA exclusively.

A mutation in a tRNA molecule may affect the anticodon, which would affect the specificity of that tRNA for its corresponding mRNA. Such a mutation may also affect the tRNA's site of amino acid attachment; this would affect the specificity of this tRNA for its amino acid. Hence the answer to question 11 is (C). Since stop codons are recognized by the ribosome and not the tRNA, a mutation in the tRNA could not result in a premature termination of translation. A premature ending of transcription (B) would be the result of a mutation in DNA, not tRNA.

Now that you've worked your way through our sample SAT II: Biology E/M questions, you should be ready to tackle our diagnostic test, Test One. This test will probe your knowledge of the various biology topics covered on both versions of the SAT II: Biology E/M exam. Use it to identify areas that you have not completely mastered, and plan to review the chapters that deal with these topics particularly carefully. Good luck!

Test Strategy

If you don't know what the answer is, eliminate obviously wrong answers and guess!

BIOLOGY REVIEW

ANSWER SHEET FOR BIOLOGY E/M
TEST ONE: DIAGNOSTIC TEST

1 (A) (B) (C) (D) (E) 26 (A) (B) (C) (D) (E) 51 (A) (B) (C) (D) (E) 76 (A) (B) (C) (D) (E)
2 (A) (B) (C) (D) (E) 27 (A) (B) (C) (D) (E) 52 (A) (B) (C) (D) (E) 77 (A) (B) (C) (D) (E)
3 (A) (B) (C) (D) (E) 28 (A) (B) (C) (D) (E) 53 (A) (B) (C) (D) (E) 78 (A) (B) (C) (D) (E)
4 (A) (B) (C) (D) (E) 29 (A) (B) (C) (D) (E) 54 (A) (B) (C) (D) (E) 79 (A) (B) (C) (D) (E)
5 (A) (B) (C) (D) (E) 30 (A) (B) (C) (D) (E) 55 (A) (B) (C) (D) (E) 80 (A) (B) (C) (D) (E)
6 (A) (B) (C) (D) (E) 31 (A) (B) (C) (D) (E) 56 (A) (B) (C) (D) (E) 81 (A) (B) (C) (D) (E)
7 (A) (B) (C) (D) (E) 32 (A) (B) (C) (D) (E) 57 (A) (B) (C) (D) (E) 82 (A) (B) (C) (D) (E)
8 (A) (B) (C) (D) (E) 33 (A) (B) (C) (D) (E) 58 (A) (B) (C) (D) (E) 83 (A) (B) (C) (D) (E)
9 (A) (B) (C) (D) (E) 34 (A) (B) (C) (D) (E) 59 (A) (B) (C) (D) (E) 84 (A) (B) (C) (D) (E)
10 (A) (B) (C) (D) (E) 35 (A) (B) (C) (D) (E) 60 (A) (B) (C) (D) (E) 85 (A) (B) (C) (D) (E)
11 (A) (B) (C) (D) (E) 36 (A) (B) (C) (D) (E) 61 (A) (B) (C) (D) (E) 86 (A) (B) (C) (D) (E)
12 (A) (B) (C) (D) (E) 37 (A) (B) (C) (D) (E) 62 (A) (B) (C) (D) (E) 87 (A) (B) (C) (D) (E)
13 (A) (B) (C) (D) (E) 38 (A) (B) (C) (D) (E) 63 (A) (B) (C) (D) (E) 88 (A) (B) (C) (D) (E)
14 (A) (B) (C) (D) (E) 39 (A) (B) (C) (D) (E) 64 (A) (B) (C) (D) (E) 89 (A) (B) (C) (D) (E)
15 (A) (B) (C) (D) (E) 40 (A) (B) (C) (D) (E) 65 (A) (B) (C) (D) (E) 90 (A) (B) (C) (D) (E)
16 (A) (B) (C) (D) (E) 41 (A) (B) (C) (D) (E) 66 (A) (B) (C) (D) (E) 91 (A) (B) (C) (D) (E)
17 (A) (B) (C) (D) (E) 42 (A) (B) (C) (D) (E) 67 (A) (B) (C) (D) (E) 92 (A) (B) (C) (D) (E)
18 (A) (B) (C) (D) (E) 43 (A) (B) (C) (D) (E) 68 (A) (B) (C) (D) (E) 93 (A) (B) (C) (D) (E)
19 (A) (B) (C) (D) (E) 44 (A) (B) (C) (D) (E) 69 (A) (B) (C) (D) (E) 94 (A) (B) (C) (D) (E)
20 (A) (B) (C) (D) (E) 45 (A) (B) (C) (D) (E) 70 (A) (B) (C) (D) (E) 95 (A) (B) (C) (D) (E)
21 (A) (B) (C) (D) (E) 46 (A) (B) (C) (D) (E) 71 (A) (B) (C) (D) (E)
22 (A) (B) (C) (D) (E) 47 (A) (B) (C) (D) (E) 72 (A) (B) (C) (D) (E)
23 (A) (B) (C) (D) (E) 48 (A) (B) (C) (D) (E) 73 (A) (B) (C) (D) (E)
24 (A) (B) (C) (D) (E) 49 (A) (B) (C) (D) (E) 74 (A) (B) (C) (D) (E)
25 (A) (B) (C) (D) (E) 50 (A) (B) (C) (D) (E) 75 (A) (B) (C) (D) (E)

Remove this answer sheet and use it to complete the Practice Test.

Use the answer key following the test to count up the number of questions you got right and the number you got wrong. (Remember not to count omitted questions as wrong.) The "Compute Your Score" section following the Answer Key will show you how to find your score.

Biology E/M Test One: Diagnostic Test

Part A

Directions: Each question or incomplete statement below is followed by five possible answers or completions, lettered A–E. Choose the answer that is the best in each case. Fill in the corresponding oval on your answer sheet.

1. In a climax community, which of the following will be observed?

   (A) The nitrogen cycle ceases to be important for primary producers.

   (B) There are no changes in seasonal population sizes in the community.

   (C) Predator-prey relationships between trophic levels in the food web remain constant from one generation to the next.

   (D) There is no loss of energy from one trophic level to the next.

   (E) There is only one species of primary producer.

2. Air entering the lungs of a tracheotomy patient through a tracheotomy (a tube inserted directly into the trachea) is colder and drier than normal, which often causes lung crusting and infection. This occurs primarily because the air

   (A) enters the respiratory system too rapidly to be filtered

   (B) is not properly humidified by the larynx

   (C) does not flow through the nasal passageways

   (D) does not flow past the mouth and tongue

   (E) none of the above

3. Smooth muscle develops from which of the following germ layers?

   (A) endoderm

   (B) mesoderm

   (C) epiderm

   (D) ectoderm

   (E) none of the above

GO ON TO THE NEXT PAGE

4.   What is the function of a lysosome's membrane?

(A)   It isolates an acidic environment for the lysosome's hydrolytic enzymes from the neutral pH of the cytoplasm.

(B)   It is continuous with the nuclear membrane, thereby linking the lysosome with the endoplasmic reticulum.

(C)   It is used as an alternative site of protein synthesis.

(D)   The cytochrome carriers of the electron transport chain are embedded within it.

(E)   It separates the nucleus from the cytoplasm.

5.   Which of the following statements regarding evolution is true?

(A)   Certain phenotypes are more fit in certain environments than others.

(B)   Natural selection creates new alleles.

(C)   Genotype, not phenotype, influences fitness.

(D)   Mutations always affect the fitness of an organism.

(E)   all of the above

6.   For the following organisms, which follows the correct sequence of evolution through time, starting with the organism that existed first?

(A)   sponge, flatworm, chordate, mollusk

(B)   flatworm, mollusk, sponge, chordate

(C)   sponge, flatworm, mollusk, chordate

(D)   mollusk, sponge, flatworm, chordate

(E)   flatworm, sponge, chordate, mollusk

7.   A culture of algae is inoculated with small numbers of two different species of protozoan ciliates, Protozoans A and C, that feed on the algae. Protozoan A replicates asexually once every hour and Protozoan C replicates asexually once every 1.5 hours under these conditions, as long as the algae is not limiting. Which of the following is most likely to be observed?

(A)   The protozoans will evolve to have a mutualistic relationship.

(B)   Both populations of protozoans will increase in size initially, but then Protozoan C will die off from the culture.

(C)   The algae will rapidly evolve to avoid predation.

(D)   The algae will die off from the culture due to overfeeding.

(E)   Protozoan C will evolve to replicate more rapidly.

GO ON TO THE NEXT PAGE

8. Oogenesis is the process by which

   (A) primary oocytes produce sperm
   (B) primary oocytes produce eggs
   (C) the egg implants in the uterus
   (D) the egg is released from the ovary
   (E) starfish regenerate limbs

9. An individual that has only one X chromosome is genotypically *XO*. This person

   (A) cannot survive
   (B) will have immature, ambiguous (both male and female) reproductive systems
   (C) will be phenotypically female but sterile
   (D) does not produce steroid hormones
   (E) none of the above

10. In the process of fat emulsification, bile salts make fats more susceptible to the action of lipases by

    (A) transporting fat globules to the region of lipase activity
    (B) increasing the surface area of the fat globules
    (C) functioning as a catalyst for the lipases
    (D) lowering the pH of the small intestine
    (E) none of the above

11. There is a recessive allele for a gene that made people more susceptible than normal to smallpox. Only homozygous recessive people display this trait; heterozygotes are indistinguishable from homozygous dominant people, with normal resistance to smallpox. After the point at which the smallpox virus was eliminated from the earth, which of the following occurred to the allele frequency for the allele that caused smallpox susceptibility?

    (A) The allele declined in frequency for several generations and then disappeared.
    (B) The allele remained at a constant frequency in the gene pool.
    (C) The allele increased in frequency, since it was no longer selected against.
    (D) The number of homozygous recessive people remained the same, but the number of heterozygous people increased for several generations.
    (E) none of the above

GO ON TO THE NEXT PAGE

12. Brain cells of the housefly *Musca domestica* have 6 pairs of chromosomes. Therefore, it can be concluded that

    (A) the fly's diploid number is 24
    (B) the fly's haploid number is 12
    (C) the fly's haploid number is 3
    (D) the fly's haploid number is 6
    (E) the fly's haploid number is 24

13. A snake eats frogs, which eat grasshoppers. The snake is an example of a

    (A) primary consumer
    (B) secondary consumer
    (C) producer
    (D) tertiary consumer
    (E) decomposer

14. Which of the following is an example of a scavenger?

    I. fungi
    II. vulture
    III. hyena

    (A) I only
    (B) II only
    (C) I and II
    (D) I and III
    (E) II and III

15. A black male mouse (I) is crossed with a black female mouse, and they produce 15 black and 5 white offspring. A different black male mouse (II) is crossed with the same female, and the offspring from this mating are 30 black mice. Which of the following must be true?

    (A) The female mouse is homozygous.
    (B) Male mouse II is heterozygous.
    (C) Two of the mice are heterozygous.
    (D) All the progeny of mouse II are homozygous.
    (E) All three mice are homozygous.

16. A difference between fats and carbohydrates is that

    (A) carbohydrates are always steroids
    (B) carbohydrates have a H:O ratio of 2:1
    (C) fats are known as starch
    (D) fats have a H:O ratio with much more oxygen than hydrogen
    (E) fats always contain nitrogen

17. Water travels into and out of cells via

    (A) carrier proteins
    (B) symport systems
    (C) ion channels
    (D) osmosis
    (E) active transport

GO ON TO THE NEXT PAGE

KAPLAN

18. Which of the following statements illustrates the principle of induction during vertebrate development?

   (A) The presence of a notochord beneath the ectoderm results in the formation of a neural tube.

   (B) A neuron synapses with another neuron via a neurotransmitter.

   (C) The neural tube develops into the brain, the spinal chord, and the rest of the nervous system.

   (D) Secretion of TSH stimulates the secretion of the hormone thyroxine.

   (E) none of the above

19. What is the correct sequence of events in the development of the embryo?

   (A) morula —> cleavage —> blastula —> gastrula

   (B) cleavage —> morula —> blastula —> gastrula

   (C) cleavage —> gastrula —> blastula —> morula

   (D) blastula —> cleavage —> gastrula —> morula

   (E) morula —> blastula —> cleavage —> gastrula

20. Members of a class are more alike than members of

   (A) an order
   (B) a phylum
   (C) a genus
   (D) a species
   (E) a family

21. Pancreatic lipase is involved in the digestion of

   (A) starch
   (B) protein
   (C) fat
   (D) cellulose
   (E) nucleic acids

GO ON TO THE NEXT PAGE

22. A population of horses is split into two populations by a new riverbed that forms after a flood. After many generations, the river changes course and the populations mix again. Which of the following indicates that the two populations have formed two separate species?

    (A) The populations refuse to cross the dry riverbed to interbreed.
    (B) The populations mix and mate and offspring are produced, although they are sterile.
    (C) One population has twice as many horses with white spots as the other.
    (D) Both populations are primary consumers.
    (E) The horses interbreed, but the offspring are shorter than either of the parents.

23. The mouse is known as *Mus musculus*. The *Mus* is the

    (A) phylum
    (B) class
    (C) order
    (D) genus
    (E) species

24. Albinos have a genotype of *aa*, while all other members of population are either *AA* or *Aa*. The offspring of a cross between a heterozygous male and an albino female would be

    (A) 100% albino
    (B) 100% normal
    (C) 50% normal, 50% albino
    (D) 25% normal, 75% albino
    (E) 75% normal, 25% albino

25. Which part of cellular respiration directly produces a pH gradient during the oxidative metabolism of glucose?

    (A) glycolysis
    (B) anaerobic respiration
    (C) Krebs cycle
    (D) electron transport chain
    (E) none of the above

GO ON TO THE NEXT PAGE

KAPLAN

26. Which of the following lions has the greatest fitness?

    (A) a male that dies young and leaves three cubs that are raised by an unrelated female
    (B) a female that raises four of her cousin's young
    (C) a male that is the leader of his social group, is the oldest and strongest, and has two cubs of his own
    (D) a female that raises five young of an unrelated female
    (E) a male that has one cub with two different females

27. The first organisms on Earth were thought to be

    (A) autotrophs
    (B) chemosynthetic
    (C) heterotrophs
    (D) oxygen producing
    (E) photosynthetic

28. Maple trees, apple trees, orchids, and palms are examples of

    (A) gymnosperms
    (B) bryophytes
    (C) angiosperms
    (D) chlorophytes
    (E) rhodophytes

29. Molds and yeast are classified as

    (A) rhodophytes
    (B) bryophytes
    (C) fungi
    (D) ciliates
    (E) flagellates

30. In humans, brown eyes are dominant over blue eyes. In the cross of $BB \times bb$, what percentage of the offspring will have brown eyes?

    (A) 75%
    (B) 50%
    (C) 0%
    (D) 100%
    (E) 25%

31. In humans, a normal sperm must contain

    I. an X chromosome
    II. 23 chromosomes
    III. a Y chromosome

    (A) I only
    (B) II only
    (C) III only
    (D) I and II
    (E) II and III

GO ON TO THE NEXT PAGE

32. Which of the following are bilaterally symmetrical?

    (A) planaria
    (B) roundworms
    (C) humans
    (D) arthropods
    (E) all of the above

33. In a small population of an endangered desert rodent, which of the following poses a concern for the future of the species?

    (A) The frequency of mutation increases in small populations.
    (B) Natural selection cannot occur in a small population.
    (C) Polyploidy is almost always lethal in higher vertebrates.
    (D) There is a lack of natural resources in the desert.
    (E) Recessive alleles become homozygous more frequently in small populations.

34. Examples of parasitism include all of the following EXCEPT

    (A) tick bird and rhinoceros
    (B) virus and host cell
    (C) tapeworm and man
    (D) tuberculosis bacteria and man
    (E) flukes and fish

35. Meiosis differs from mitosis in that

    I.   two cell divisions take place
    II. DNA replicates during Interphase
    III. haploid cells are produced from diploid cells

    (A) I only
    (B) II only
    (C) III only
    (D) I and III
    (E) I, II, and III

36. Stomata in plant leaves close at night to prevent the loss of

    (A) O_2
    (B) H_2O
    (C) CO_2
    (D) energy
    (E) chlorophyll

37. When calcium binds troponin in muscle cells, the binding site for which of the following is exposed?

    (A) tropomyosin
    (B) myosin
    (C) ATP
    (D) ADP
    (E) Pi

GO ON TO THE NEXT PAGE

KAPLAN

38. All organisms utilize

   (A) CO_2
   (B) a triplet genetic code to produce proteins
   (C) oxygen
   (D) ADP as cellular energy
   (E) membrane-bound organelles

39. In fruit flies, the gene for wing type is located on an autosomal chromosome. The allele for wild-type wings is dominant over the allele for vestigial wings. If a homozygous dominant male fly is crossed with a female with vestigial wings, what percentage of their female progeny are expected to have wild-type wings?

   (A) 0%
   (B) 25%
   (C) 50%
   (D) 75%
   (E) 100%

40. Scientists studied a pond and found water skimmers and minnows. Fifteen years later, the pond had filled in, resulting in swampy land; frogs and snakes were prevalent. This is a result of

   (A) predation
   (B) succession
   (C) speciation
   (D) natural disaster
   (E) global warming

41. Damsel flies and dragonflies can live in the same ecosystem because

   (A) they occupy different niches
   (B) they are commensal
   (C) they are both insects
   (D) dragonflies have stronger wing muscles than damsel flies
   (E) they mate at different times of the year

42. Legumes are good for the soil because

   (A) deer won't eat them
   (B) they are photosynthetic
   (C) they have nitrogen-fixing bacteria on their roots
   (D) animals convert them to energy
   (E) none of the above

43. If the DNA sequence is $5'TACAGA3'$, then the complementary mRNA sequence is

   (A) $3'UCUAUG5'$
   (B) $3'TACAGA5'$
   (C) $3'AUGUCU5'$
   (D) $3'UACAGA5'$
   (E) $3'ATGTCT5'$

GO ON TO THE NEXT PAGE

44. Sexually reproducing species can have a selective advantage over asexually reproducing species because sexual reproduction

    (A) is more energy efficient
    (B) allows for more genetic diversity
    (C) decreases the likelihood of mutations
    (D) always decreases an offspring's survival ability
    (E) can occur in any climate

45. The genetic code for eukaryotic protein sequences is passed from one generation to the next in

    (A) other proteins
    (B) rRNA
    (C) tRNA
    (D) mRNA
    (E) DNA

Questions 46–50 refer to the following figure:

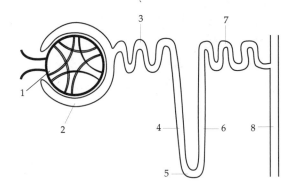

46. Structure 1 is known as the

    (A) glomerulus
    (B) Bowman's capsule
    (C) proximal convoluted tubule
    (D) loop of Henle
    (E) distal convoluted tubule

47. Structure 3 is where

    (A) urine is concentrated
    (B) almost all glucose and amino acids are reabsorbed
    (C) potassium is secreted
    (D) blood is oxygenated
    (E) the renal artery flows in

48. ADH acts on structure

    (A) 4
    (B) 5
    (C) 6
    (D) 7
    (E) 8

49. The organ the depicted structure is a part of is involved in

    (A) digestion
    (B) cellular respiration
    (C) homeostasis of extracellular ionic strength
    (D) digestion of lipids
    (E) maintenance of the heartbeat

GO ON TO THE NEXT PAGE

50. Which structure is part of the circulatory system?

(A) 1
(B) 2
(C) 3
(D) 4
(E) 5

Questions 51–54 refer to the following figure:

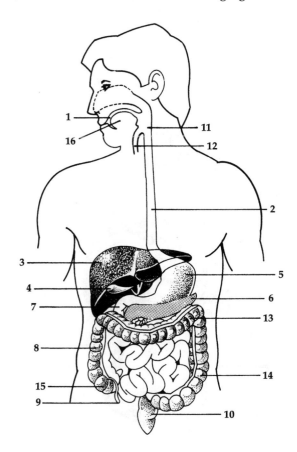

51. At which site does digestion of starches begin?

(A) 1
(B) 2
(C) 3
(D) 4
(E) 5

52. Structure 4

(A) produces bile
(B) stores bile
(C) secretes lipase
(D) secretes bicarbonate
(E) secretes HCl

53. Which structure is primarily responsible for water absorption during digestion?

(A) 5
(B) 6
(C) 7
(D) 8
(E) 9

54. Which structure has both exocrine and endocrine function?

(A) 5
(B) 6
(C) 7
(D) 8
(E) 9

GO ON TO THE NEXT PAGE

Part B

Each set of choices A–E below should be compared to the numbered statements that follow it. Choose the lettered choice that best matches each numbered statement. Fill in the correct oval on your answer sheet. Remember that a choice may be used once, more than once, or not at all in each set.

Questions 55–58:

(A) growth hormone

(B) oxytocin

(C) progesterone

(D) aldosterone

(E) glucagon

55. increases uterine contractions during child birth

56. stimulates the release of glucose to the blood

57. induces water resorption in the kidneys

58. prepares the uterus for implantation of the fertilized egg

Questions 59–61:

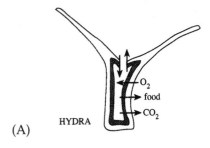

(A)

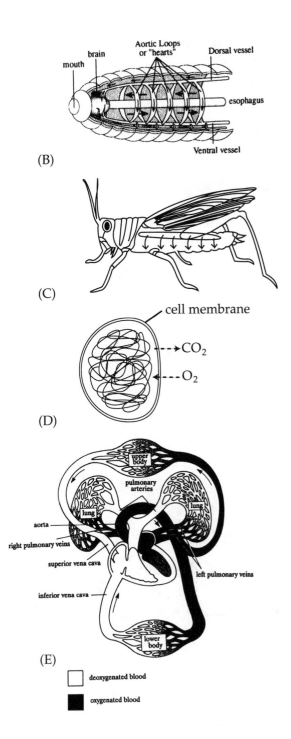

(B)

(C)

(D)

(E)

GO ON TO THE NEXT PAGE

KAPLAN

59. the transport system of a bacteria

60. the transport system of a human

61. the transport system of a segmented worm

Questions 62–65:

    (A) nucleus
    (B) endoplasmic reticulum
    (C) ribosomes
    (D) Golgi apparatus
    (E) lysosomes

62. membrane-bound convoluted organelle that is the site of synthesis of secreted proteins

63. membrane-bound organelle full of hydrolytic enzymes

64. membrane-bound organelle that contains the chromosomes

65. this consists of RNA and proteins and helps to translate mRNA during polypeptide synthesis

Questions 66–68:

    (A) mRNA
    (B) rRNA
    (C) tRNA
    (D) DNA
    (E) nucleolus

66. site of rRNA synthesis

67. product of transcription, directs translation

68. binds specific amino acids and carries them to the ribosomes during protein synthesis

Questions 69–71:

    (A) natural selection
    (B) adaptive radiation
    (C) vestigial structure
    (D) migration
    (E) parallel evolution

69. differential survival based on variations in phenotypes

70. production of several different species from a common ancestor

71. appears useless but had an ancestral function

GO ON TO THE NEXT PAGE

Questions 72–75:

(A)

(B)

(C)

(D)

(E)

72. building block of DNA or RNA

73. building block of proteins

74. lipid

75. carbohydrate

Part C

Each of the following sets of questions is based on a laboratory or experimental situation. Begin by studying the description of each situation. Next, choose the best answer to each of the questions that follow it. Fill in the corresponding oval on your answer form.

Questions 76–79 refer to the following paragraph:

All birds and mammals are able to maintain relatively constant body temperatures, despite fluctuations in external temperature. These animals have evolved thermoregulatory mechanisms that help them to adapt to their environments. One such mechanism is to regulate the metabolic rate. A plot of the rate of oxygen consumption versus body weight for various mammals reveals that metabolic rate is inversely proportional to body weight (see figure). However, metabolic rate and the transfer of heat to the environment are directly proportional to the surface area-to-volume ratio of the animal. For instance, a shrew has a higher metabolic rate and a greater surface area-to-volume ratio than a horse, which means that the shrew generates more internal heat per gram of body weight and loses more heat to the environment. This makes it especially difficult for small animals to maintain a constant body temperature in cold weather.

GO ON TO THE NEXT PAGE

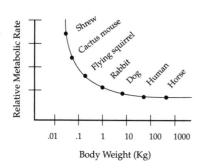

76. Small animals such as penguin chicks will huddle together when it is cold outside. The most likely explanation for this is that

(A) huddling decreases the effective surface area-to-volume ratio and decreases the loss of body heat
(B) huddling increases the effective surface area-to-volume ratio and increases the loss of body heat
(C) huddling increases the metabolic rate
(D) huddling is an instinct in penguin chicks
(E) baby penguins huddle following imprinting

77. A mechanism mammals have developed to dissipate excess heat in hot weather is

(A) huddling
(B) sitting in the shade
(C) sweating
(D) burrowing
(E) none of the above

78. Which of the following is most likely true of an animal that must keep all of its vital organs at approximately the same temperature?

(A) It could not survive in a desert environment.
(B) It sweats excessively and becomes dehydrated in hot weather.
(C) Its body temperature is determined by the most temperature-sensitive organ.
(D) It must always have a large supply of water.
(E) It will be found only on the tundra.

79. Which of the following animals has the highest metabolic rate?

(A) horse
(B) dog
(C) rabbit
(D) flying squirrel
(E) cactus mouse

GO ON TO THE NEXT PAGE

Questions 80–83 refer to the following experiment:

E. coli is a bacteria that can be used to study a variety of mechanisms. In this experiment, one wild type and four mutant strains were tested for their ability to grow in either minimal media or media that was supplemented with various amino acids.

Supplement	None	Arginine
Wild type	+	+
Strain 1	–	+
Strain 2	–	–
Strain 3	–	–
Strain 4	+	+

Supplement	Threonine	Histidine
Wild Type	+	+
Strain 1	–	–
Strain 2	+	–
Strain 3	–	+
Strain 4	+	+

A (+) sign indicates growth and a (–) sign indicates no growth.

80. Based on the table, strain 2

 (A) cannot produce arginine
 (B) cannot produce threonine
 (C) cannot produce histidine
 (D) cannot produce glycine
 (E) can produce all amino acids

81. Which amino acid(s) does strain 3 NOT need from the environment to grow?

 (A) arginine and histidine
 (B) threonine and histidine
 (C) histidine only
 (D) arginine and threonine
 (E) threonine only

82. The mutation in strain 1 that renders it incapable of growing without arginine occurs in

 (A) DNA
 (B) mRNA
 (C) protein
 (D) the anticodon region of the tRNA
 (E) rRNA

83. The mutation in strain 4

 (A) renders it dormant
 (B) does not affect its ability to synthesize amino acids
 (C) causes it to become haploid
 (D) is identical to that in strain 1
 (E) none of the above

GO ON TO THE NEXT PAGE

The following graphs depict growth in bacteria under a variety of different conditions with increasing numbers of bacteria on the *y*-axis and increasing time on the *x*-axis. In these conditions, the bacteria multiply at the fastest possible rate. The following graph depicts what occurs if bacteria are added to rich broth:

Use these graphs to answer questions 84–86.

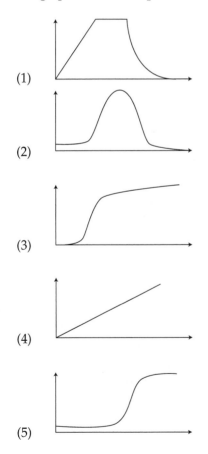

84. Which graph shows what happens when you introduce a bacteriophage that lyses the bacteria six hours after they enter the exponential growth phase?

    (A)   1
    (B)   2
    (C)   3
    (D)   4
    (E)   5

85. Which graph describes what happens when you introduce very minimal nutrients for the first six hours, then add surplus nutrients?

    (A)   1
    (B)   2
    (C)   3
    (D)   4
    (E)   5

86. Which graph shows growth in rich media with additional nutrients added?

    (A)   1
    (B)   2
    (C)   3
    (D)   4
    (E)   5

GO ON TO THE NEXT PAGE

In her experiment to determine where the C, H, and O come from when plants produce carbohydrates, a scientist grew plants in the presence of a variety of radioactive compounds and used a Geiger counter to determine whether the starch produced by the plants was radioactive.

	Molecule	Radioactive Starch
Expt. 1	$^{14}CO_2$	+
Expt. 2	$H_2^{18}O$	−
Expt. 3	$C^{18}O_2$	+

Use this data to answer questions 87–88.

87. The source of oxygen in carbohydrates produced by plants is

    (A) CO_2
    (B) H_2O
    (C) O_2
    (D) none of the above
    (E) all of the above

88. The oxygen released from the plant comes from

    (A) CO_2
    (B) H_2O
    (C) O_2
    (D) none of the above
    (E) all of the above

89. If a plant utilized H_2SO_4 instead of water, the plant would release

    (A) CO_2
    (B) H_2O
    (C) O_2
    (D) SO_4^{2-}
    (E) none of the above

90. Which compound captures light energy in plants?

    (A) O_2
    (B) CO_2
    (C) H_2O
    (D) chlorophyll
    (E) none of the above

91. The light reaction of photosynthesis occurs in the

    (A) stroma
    (B) thylakoid membranes
    (C) mitochondria
    (D) nucleus
    (E) ribosomes

92. The Calvin cycle

    (A) does not use light directly
    (B) occurs in the cytoplasm
    (C) releases CO_2
    (D) produces ATP
    (E) none of the above

GO ON TO THE NEXT PAGE

A study of a meadow yielded the following data:

Organisms	Number
Hawk	3
Various birds	50
Spiders	500
Insects	3,000
Shrubs and other plants	>5,000

Use this data to answer questions 93–95.

93. Which organisms are the primary consumers?

(A) hawks
(B) birds
(C) spiders
(D) insects
(E) shrubs

94. Which type of organism is not represented?

(A) primary producer
(B) secondary consumer
(C) tertiary consumer
(D) primary consumer
(E) decomposer

95. Which of the following is a correct pyramid of biomass?

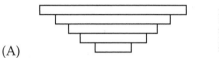

(A)

Hawks
Birds
Spiders
Insects
Shrubs

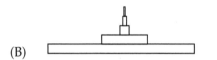

(B)

Hawks
Birds
Spiders
Insects
Shrubs

(C)

Hawks
Birds
Spiders
Insects
Shrubs

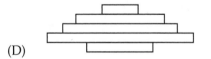

(D)

Hawks
Birds
Spiders
Insects
Shrubs

(E) none of the above

STOP

Test One: Diagnostic Test Answer Key

1.	C	20.	B	39.	E	58.	C	77.	C
2.	C	21.	C	40.	B	59.	D	78.	C
3.	B	22.	B	41.	A	60.	E	79.	E
4.	A	23.	D	42.	C	61.	B	80.	B
5.	A	24.	C	43.	C	62.	B	81.	D
6.	C	25.	D	44.	B	63.	E	82.	A
7.	B	26.	A	45.	E	64.	A	83.	B
8.	B	27.	C	46.	A	65.	C	84.	B
9.	C	28.	C	47.	B	66.	E	85.	E
10.	B	29.	C	48.	E	67.	A	86.	C
11.	B	30.	D	49.	C	68.	C	87.	A
12.	D	31.	B	50.	A	69.	A	88.	B
13.	D	32.	E	51.	A	70.	B	89.	D
14.	E	33.	E	52.	B	71.	C	90.	D
15.	C	34.	A	53.	D	72.	B	91.	B
16.	B	35.	D	54.	B	73.	C	92.	A
17.	D	36.	B	55.	B	74.	A	93.	D
18.	A	37.	B	56.	E	75.	D	94.	E
19.	B	38.	B	57.	D	76.	A	95.	B

Compute Your Practice Test Score

Step 1: Figure out your raw score. Refer to your answer sheet for the number right and the number wrong on the practice test you're scoring. (If you haven't checked your answers, do that now, using the answer key that follows the test.) You can use the chart below to figure out your raw score. Multiply the number wrong by 0.25 and subtract the result from the number right. Round the result to the nearest whole number. This is your raw score.

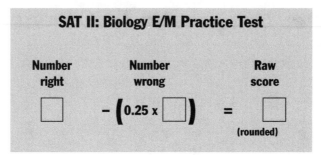

Step 2: Find your practice test score. Find your raw score in the left column of the table below. The score in the right column is an approximation of what your score would be on the SAT II: Biology E/M.

A note on your diagnostic practice test score: Don't take your score on the diagnostic test too literally. Practice test conditions cannot precisely mirror real test conditions. Your actual SAT II: Biology E/M Subject Test score will almost certainly vary from your practice test scores. Your scores on the practice tests will give you a rough idea of your range on the actual exam.

Find Your Practice Test Score

Raw	Scaled	Raw	Scaled	Raw	Scaled	Raw	Scaled	Raw	Scaled	Raw	Scaled
95	800	78	690	61	590	44	490	27	400	10	300
94	800	77	680	60	580	43	490	26	390	9	290
93	790	76	670	59	580	42	480	25	380	8	290
92	780	75	670	58	570	41	470	24	380	7	280
91	770	74	660	57	570	40	470	23	370	6	280
90	760	73	660	56	560	39	460	22	370	5	270
89	750	72	650	55	550	38	460	21	360	4	260
88	740	71	640	54	550	37	450	20	360	3	260
87	740	70	640	53	540	36	450	19	350	2	250
86	730	69	630	52	540	35	440	18	350	1	240
85	730	68	630	51	530	34	430	17	340	0	240
84	720	67	620	50	530	33	430	16	330	−1	230
83	720	66	620	49	520	32	420	15	330	−2	230
82	710	65	610	48	510	31	420	14	320	−3	220
81	710	64	610	47	510	30	410	13	320	−4	220
80	700	63	600	46	500	29	410	12	310	−5	210
79	690	62	590	45	500	28	400	11	300	−6 to −9	210

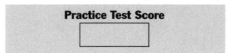

Practice Test Score

KAPLAN

How to Score the Biology E/M Diagnostic Test

This diagnostic test contains questions that you would find on both the E- and the M-options of the SAT II: Biology exam. Because it is designed to test your knowledge of the topics tested in both sections, it does not follow the exact format of a typical Biology E/M test: it has 95 questions instead of 80. Turn to Test Two for an in-format Biology-E practice test, and Test Three for a Biology-M test.

In the Diagnostic Test, the following questions are related to ecology and evolution. If you missed a large number of these, you may want to pay particular attention to the review sections on these topics, and consider taking the Molecular option of the test if you continue to experience difficulty with these topics. The ecological and evolution-oriented test questions are:

1–Ecology

5–Evolution

6–Evolution

7–Ecology

11–Population genetics

13–Ecology

14–Ecology

22–Evolution

26–Evolution

33–Population genetics

34–Ecology

40–Ecology

41–Ecology

69–Evolution

70–Evolution

71–Evolution

93–Ecology

94–Ecology

95–Ecology

Another set of questions is related to molecular biology, and may indicate your current strength or weakness in this area. If you missed several of these questions, you may want to review the relevant chapters in the text more closely, such as the chapter on molecular and cellular biology. If you did better on the ecological and evolution-oriented questions listed above, you should consider selecting the E-option when registering for the SAT II: Biology E/M exam. The molecular questions in the diagnostic test include:

16–Biological chemistry

17–Cell biology

25–Molecular biology

37–Molecular biology

38–Molecular biology

43–Molecular biology

45–Molecular biology

72–Biological chemistry

73–Biological chemistry

74–Biological chemistry

75–Biological chemistry

87–92: Molecular biology

Biology E/M Test One: Diagnostic Test Answers and Explanations

1. **(C)** A climax community is an ecological system in which the populations and the way they relate to each other remain the same from one generation to the next. The community is stable in its composition because the populations that live there affect the physical environment and each other in a way that promotes the continuance of the same populations. All living organisms need nitrogen, making the nitrogen cycle a part of any ecosystem ((A) is wrong). The fact that a climax community is stable does not mean there is no variation in population size between seasons, but that every generation replaces the existing one in the same pattern of ecological interactions ((B) is wrong). There is always loss of energy between trophic levels, since organisms need to burn energy to live ((D) is wrong). There might be a primary producer (plants, algae, etc.) that predominates in a climax community, but this is not necessarily true.

2. **(C)** When a patient breathes through a tracheotomy, the air entering the respiratory system bypasses a very important area—the nasal cavities. In a normally breathing individual, the extensive surfaces of the nasal passageways warm and almost completely humidify the air, and particles are filtered out by nasal air turbulence. Since the air reaching the lungs of a tracheotomy patient has not been warmed or humidified, lung crusting and infection often result. (B) and (D) are wrong because the mouth and larynx are much less effective at humidification.

3. **(B)** The endoderm develops into the lungs, the gastrointestinal tract, and the lining of the bladder. Ectoderm develops into the brain and nervous system, the lens of the eye, the inner ear, hair, nails, sweat glands, the lining of the mouth and nose, and the skin. The mesoderm becomes everything else, including the musculoskeletal system, the reproductive system, the circulatory system, and the kidneys.

4. **(A)** The lysosome acid has an acidic interior to enhance the activity of lysomal enzymes that degrade biomolecules. The lysosomal membrane separates the acidic interior of the lysome from the rest of the cell which has a neutral pH. The other responses do not involve functions of the lysomal membrane.

5. **(A)** In Darwin's theory of natural selection, some organisms in a species have variations in traits that give them an advantage over other members of the species. These adaptations enable these organisms and their offspring to survive in greater numbers than organisms that lack them, giving them greater fitness.

6. **(C)** On an evolutionary scale, sponges developed first, followed by flatworms, then mollusks, and finally chordates.

7. **(B)** When two populations are in direct competition for the same resource, one population will compete more effectively and will, with time, force out the other population. This is particularly true with the relatively short time spans and restricted conditions in laboratory experiments. This means that B is the best answer. While resources are not limiting, both populations will grow. When they compete, the protozoan that reproduces faster, Protozoan A, will out-compete Protozoan C and cause Protozoan C to die out. A is wrong—the relationship between these populations is a competitive one, and there is no reason to believe that this relationship will reverse itself to one in which both populations benefit. There is little an algae can do to escape predation ((C) is not the best choice). The populations of predator and prey are likely to reach an equilibrium state which is more or less stable. If the predator population increases and eats more of the prey, as the prey population decreases, the predator will decrease as well. Predators rarely hunt a prey to extinction ((D) is wrong). It is also unlikely that protozoans will be able to evolve so rapidly

BIOLOGY REVIEW

as to change a fundamental property in a few generations ((E) is wrong).

8. (B) Oogenesis is the process whereby primary oocytes undergo meiosis to produce one egg (or ovum) and two or three polar bodies.

9. (C) A Turner's female has the genotype *XO*; she carries only one *X* chromosome, has underdeveloped ovaries, and is sterile, but is female in appearance. These individuals are often shorter than normal and may have varying degrees of mental development problems.

10. (B) Bile salts act like detergents to break fat globules into smaller spheres. Emulsification, which occurs via the detergent action of bile salts, increases the surface area of fat globules exposed to the lipases. Since lipases can attack only these surfaces and not other parts of the fat globules, the action of bile salts greatly enhances the ability of lipase to act on fats.

11. (B) When an allele is harmful, as this one would be, it will be selected against. In this case, the allele is recessive and is only expressed in the homozygous recessive people, with heterozygotes as carriers. The homozygous recessive people will be selected against, and the allele frequency in the population will decrease slowly over time as long as the selection pressure is maintained. Once the selection pressure is removed by the absence of the virus, the allele frequencies will remain constant, unless some other selection pressure affects the allele ((B) is correct).

12. (D) If there are six pairs of chromosomes in a diploid cell, there are a total of 12 chromosomes in a $2n$ cell, and the haploid cell would have half that number; $n = 6$.

13. (D) Grasshoppers are primary consumers (they eat plants). The frog that eats these herbivorous insects is the secondary consumer, while the snake that eats the frogs is a tertiary consumer.

14. (E) A scavenger eats animals that have already been killed; typically, it is incapable of hunting. Vultures are scavengers, while fungi are saprophytic, decomposing dead plant material. Hyenas also act as scavengers, though they also hunt on occasion.

15. (C) In the cross of Mouse I with the female, the ratio of the offspring phenotypes is 3:1, indicating a *Bb* × *Bb* cross with *BB* and *Bb* animals black and *bb* animals white. Therefore, mouse I is *Bb*, while the female is *Bb*. In the second cross of mouse II and the female mouse, 100 percent of the offspring are black. Hence Mouse II must be homozygously dominant.

16. (B) Carbohydrates, including glucose, starches, or cellulose, have a H:O ratio of 2:1. Fats, in contrast, have a H:O ratio with much more hydrogen than oxygen.

17. (D) Osmosis is defined as the diffusion of water through semipermeable membranes from an area of lesser solute concentration to an area of greater solute concentration.

18. (A) In terms of vertebrate development, induction is defined as the process by which a particular group of cells causes the differentiation of another group of cells. (A) is an example of induction: The group of cells that forms the notochord induces the formation of the neural tube. Other examples of induction in vertebrate development include the formation of the eyes, in which the optic vesicles induce the ectoderm to thicken and form the lens placode, which induces the optic vesicle to form the optic cup. This in turn induces the lens placode to form the cornea and the lens.

19. (B) The zygote continues to divide from the cleavage stage until it becomes the morula, a solid

56 **KAPLAN**

ball of cells. This in turn forms the blastula, a hollow ball of cells. The blastula in its turn invaginates to become the gastrula.

20. (B) Remember the order of classification: kingdom, phylum, class, order, family, genus, and species. Members of a class are more alike than members of a kingdom or a phylum, but less alike than members of orders, families, genera, and species.

21. (C) Lipases are enzymes involved in dietary lipid digestion. Pancreatic lipase is an enzyme involved in the digestion of triglycerides in fat to release free fatty acids.

22. (B) The key to speciation is the inability of two populations to interbreed. If two populations can interbreed and produce fertile offspring, then they are not separate species ((B) is right). The unwillingness to cross the riverbed does not indicate a biological inability to interbreed, merely a geographical obstacle ((A) is wrong). The color and height of the offspring is irrelevant ((C) and (E) are wrong), as is the fact that the horses are primary consumers ((D) is wrong).

23. (D) The first half of the scientific name is the genus of the organism; the second half is the species of that organism.

24. (C) The trait is recessive since *Aa* people are normal. In an *Aa* × *aa* cross, 50% of the offspring will be *Aa* and the other 50% will be *aa*, and the phenotypes will have the same ratio.

25. (D) The electron transport chain directly produces the pH gradient by pumping protons out of the mitochondrial matrix. This proton gradient is used to make ATP.

26. (A) Fitness is defined as the ability of an organism to pass its traits and genes on to the next and future generations. This is determined, in part at least, by the number of young produced, as well as by the survival of the young. There is no reason to believe that any of the young are raised better or worse than the others. Choice (A) is best—this is the individual with the most direct descendants indicated, compared to (C) and (E). Lions are social animals and often care for the young of others, particularly of related individuals. Genes can be passed on by caring for the young of relatives in a process called kin selection. A cousin probably shares a relatively small percentage of genes, making the kin selection contribution to fitness weak ((B) is wrong). Choice (D) is raising unrelated young that do not share her genes, meaning that this does not improve her own fitness ((D) is wrong).

27. (C) According to the heterotroph hypothesis, the first forms of life lacked the ability to synthesize their own nutrients, requiring preformed molecules. Gradually, the molecules spontaneously formed by the environment began to prove inadequate to meet their energy needs, and autotrophs developed in response.

28. (C) Angiosperms are flowering plants that produce seeds enclosed in an ovary, e.g., the maple tree. Gymnosperms, on the other hand, do not have enclosed seeds; one example is the conifer.

29. (C) Fungi are eukaryotes, are typically saprophytic organisms, and are divided into true fungi—such as yeast, molds, and mushrooms—and slime molds.

30. (D) In a *BB* × *bb* cross, 100 percent of the offspring will be *Bb*. Genotypically, they will be heterozygous, and phenotypically, they will have brown eyes, since brown is dominant.

31. (B) A normal sperm will have a haploid (23) number of chromosomes. It could have either an X or Y chromosome.

32. (E) All of these organisms are bilaterally symmetrical, which implies that they have two identical mirror images. Radially symmetrical organisms, which are symmetrical from top to bottom but not from left to right, include hydra, starfish, and jellyfish. These organisms are often round.

33. (E) Small populations are subject to inbreeding. Due to the limited selection of mates available, it is much more likely in a small population that a mate will be related and will share recessive harmful alleles. These alleles can be homozygous and expressed at a much higher frequency than would have occurred in a larger population ((E) is correct). The rate of mutation is not affected by the population size, but by physical factors such as radiation and chemical exposure ((A) is wrong). There is no reason why natural selection will not occur in a small population ((B) is wrong). Polyploidy is irrelevant ((C) is wrong). If the animals live in the desert, then they are adapted for this environment and it cannot be said that this is a special hazard for this animal ((D) is wrong).

34. (A) The tick bird and rhinoceros engage in a symbiotic relationship known as mutualism. The rhinoceros provides the tick bird with insects that live on its hide, while the tick bird keeps the rhinoceros parasite-free.

35. (D) Meiosis has two divisions that create four haploid cells from one diploid cell, while mitosis results in two diploid cells from one diploid cell. Both mitosis and meiosis replicate DNA during interphase.

36. (B) Stomata are pores in leaves that allow gases in and out of the leaf. The stomata close at night, when there is no light energy to catalyze the light reaction, in order to prevent loss of water through transpiration in the stomata.

37. (B) When calcium binds troponin, it exposes a site that allows actin to bind to myosin, causing shortening of the sarcomere and muscle contraction.

38. (B) All organisms produce proteins through the triplet code. However, some of them do not need carbon dioxide or oxygen (and, in fact, some, such as tetanus, are poisoned by oxygen). All organisms use ATP as cellular energy. Only eukaryotes, however, have membrane-bound organelles.

39. (E) This is a basic cross using Drosophila melanogaster. You're told that the gene for wing type is located on an autosomal chromosome, a non–sex chromosome, which means that it is NOT inherited as a sex-linked trait. You're also told that the dominant allele codes for wild-type wings; "wild type" simply means that this is the phenotype that predominates in nature. The recessive allele codes for the vestigial wing type, which is a stumpy wing. The gender of the flies is of no relevance here; gender only comes into play for sex-linked traits. So, our male fly is homozygous dominant for wing type and our female fly is homozygous recessive. A cross between a homozygous dominant and a homozygous recessive yields 100% heterozygous individuals, which in this case means that 100% of the progeny will have wild-type wings.

40. (B) In an ecological succession, each stage causes changes in the ecosystem that allow for the next stage to develop. If you start with a pond, it will ultimately become filled in with earth. Populations change until the climax community, a woodland, is reached. Frogs and snakes would be found in an intermediate stage of this ecological succession, when the pond is transformed into moist land.

41. (A) In the same ecosystem, there can never be two species in the same niche; if two species are found together here, they MUST occupy different niches.

42. (C) The nitrogen-fixing bacteria form nodules on the roots of legumes so that elemental nitrogen (N_2) can be converted to usable nitrates, fertilizing the soil for other plants.

43. (C) Uracil is found in RNA instead of thymine. It bonds with adenine through two hydrogen bonds, while cytosine bonds with guanine through three hydrogen bonds.

44. (B) Asexual reproduction is more efficient than sexual reproduction in terms of the number of off-spring produced per reproduction, the amount of energy invested in this process, and the amount of time involved in the development of the young, both before and after birth. Overall, sexual repro-duction is a much more time-consuming, energy-consuming process than its asexual counterpart.

Asexual reproduction, however, must rely heavily on mutation to introduce phenotypic variability in future generations, since it almost exclusively pro-duces genetic clones of the parent organism. Sexual reproduction involves the process of meiosis. Recombination and independent assortment during meiosis allows new mixing of alleles that does not occur in asexual reproduction, as well as contribu-tions of alleles from each parent during fertilization. Two genetically unique nuclei, the sperm nucleus and the egg nucleus, fuse to form an equally genet-ically unique zygote. Phenotypic variability is in this way introduced into individuals in a popula-tion, and can allow the population to adapt to a wider variety of conditions.

45. (E) DNA is the genetic material for all prokary-otes and eukaryotes.

46. (A) The glomerulus is a network of capillaries in the Bowman's capsule of the kidney.

47. (B) In the proximal convoluted tubule, almost all of the amino acids, glucose, and important salts are actively resorbed from the glomerular filtrate back into the bloodstream.

48. (E) ADH acts to make the collecting duct more permeable to water, thereby increasing water reab-sorption from urine to make it more concentrated.

49. (C) The nephron is involved in maintaining fluid balance in the body and is crucial for mainte-nance of homeostasis. It also acts to excrete nitroge-nous wastes as part of the excretory system.

50. (A) As stated in the explanation to question 46 above, the glomerulus is a network of capillaries in the Bowman's capsule of the kidney. Capillaries are part of the circulatory system.

51. (A) Digestion of starches begins in the mouth as salivary amylase breaks them down into maltose. This is why if you chew bread for long enough, it begins to taste sweet.

52. (B) The gall bladder stores bile produced by the liver. Bile emulsifies (or increases the surface area of) fats. Hormones released in response to fat con-sumption cause the gall bladder to release bile into the small intestine.

53. (D) The large intestine is responsible for most of the water and Vitamin K absorption that our body needs.

54. (B) The pancreas both releases digestive enzymes and bicarbonate ions (in what is termed exocrine function) and secretes glucagon and insulin to control blood glucose levels (endocrine function). Exocrine secretions are released through a duct to act locally. Endocrine secretions enter the bloodstream and are carried to distant sites to act.

55. (B) Oxytocin is released by the posterior pituitary. It increases uterine contractions during childbirth.

56. (E) Glucagon is released by the pancreas. It causes the breakdown of glycogen into glucose, increasing the glucose concentration of the blood.

57. (D) Aldosterone increases sodium resorption in the nephron. Water and chloride ions follow passively, and water resorption is increased.

58. (C) Progesterone is secreted by the corpus luteum. Its function is to thicken the uterine lining, preparing it for implantation of the fertilized egg.

59. (D) Bacteria do not have a developed vascular system; instead, they exchange fluids, nutrients, and wastes with the environment through diffusion and transport across the cell membrane.

60. (E) Humans have a four-chambered heart that oxygenates blood at the lungs. It transports fluids, nutrients, and wastes throughout the body via a system of vessels known as arteries, capillaries, and veins.

61. (B) Segmented worms have a closed circulatory system, with dorsal and ventral vessels and five aortic arches (hearts) that force the blood through the body.

62. (B) The endoplasmic reticulum is a membranous network that is the site of synthesis of proteins that are secreted or are packaged for delivery to membranes.

63. (E) Lysosomes are membrane-bound sacs of digestive enzymes with very low pH. Degradation of proteins or other macromolecules can occur here.

64. (A) The nucleus contains the genome and is the site at which genes are used to make mRNA.

65. (C) The ribosome is either attached to the ER or floats freely in the cytoplasm. The site of protein synthesis, it is composed of proteins and rRNA.

66. (E) The nucleolus is a region of the nucleus where rRNA is synthesized.

67. (A) mRNA is coded directly from the DNA of the cell and directs protein translation.

68. (C) tRNA carries its specific amino acid to the ribosome, where it attaches to the growing polypeptide chain coded for by mRNA.

69. (A) Organisms in a population have differences in their phenotypes caused by their genes. These differences will cause some individuals to survive better and pass on their traits to more offspring. This is the essence of natural selection.

70. (B) Adaptive radiation may be defined as the development of a number of different species from a common ancestor as a result of differing environmental pressures.

71. (C) Vestigial structures appear useless, but in fact had a necessary ancestral function. One example is the appendix of man, which helped early humans to store and digest food, but now has no apparent function.

72. (B) A nucleotide is made up of a sugar, a phosphate group, and a nitrogenous base. This nucleotide is the building block of nucleic acids such as DNA and RNA.

73. (C) An amino acid contains an amino group, a carboxylic acids group, and one of 20 varying R groups. Amino acids serve as the building blocks of proteins.

74. (A) A trigyceride is a lipid that consists of three fatty acids (carboxyl groups) bound to one glycerol molecule.

75. (D) Carbohydrates are organic molecules with a H:O ratio of 2:1. Examples include fructose, glucose, sucrose, maltose, galactose, and other sugars.

76. (A) An animal's primary goal in cold weather is to keep warm. It wants to limit the amount of heat lost to the environment, because the more heat it loses, the colder the animal feels. The amount of heat transferred from a body to its external environment is directly proportional to the amount of surface area exposed to the external environment. If heat is lost through exposed surface area, then it follows that the greater the surface area that is exposed, the greater the amount of heat is that will be lost. Smaller animals have a tougher time than larger ones trying to maintain a constant internal body temperature, because smaller animals have a greater surface area-to-volume ratio and a higher metabolic rate than large ones. Therefore, smaller animals both generate more heat and lose more heat to the environment than larger ones. If a group of small animals can somehow decrease their collective surface area-to-volume ratio, then they will lose less heat as a unit. This is accomplished through huddling.

77. (C) Sweating dissipates heat produced by the animals. Sitting in the shade and burrowing, meanwhile, cool the external environment rather than dissipating heat.

78. (C) You're being asked to draw a conclusion about what it means for all organs, including the brain, to have a single body temperature. If an animal must keep ALL of its vital organs at the same temperature, and if one particular organ, such as the brain, is particularly sensitive to temperature, then the animal is obligated to maintain its body temperature at the temperature required by that organ.

79. (E) On the graph, the animal from the list with the highest relative metabolic rate is the cactus mouse. Only one animal has a higher metabolic rate—the shrew. The lowest metabolic rate included in this chart is the horse's.

80. (B) Because strain 2 cannot produce threonine on its own, it cannot grow unless the medium is supplemented with threonine.

81. (D) Strain 3 does not grow when it is supplemented with arginine and threonine; therefore, it obviously does not need them from the environment to grow. It does, however, need histidine to grow, and cannot produce it on its own, as evidenced by the fact that it will grow when histidine is supplemented.

82. (A) Mutations occur in the DNA and are passed on to the proteins that are translated via the mRNA.

83. (B) The mutation did not affect strain 4's ability to synthesize amino acids. This strain grows well regardless of whether or not its medium is supplemented. It is likely that its mutation is a silent mutation.

84. (B) Graph 2 shows normal growth with an abrupt cutoff when the bacteriophage is introduced and begins to lyse bacteria.

85. (E) Graph 5 shows a lag from normal growth, during which period the medium does not have enough nutrients, followed by an exponential surge in growth when nutrients are added.

86. (C) Graph 3 shows exponential growth—until the bacteria use up most of their nutrients and reach a plateau. The growth is the same as growth in rich media, since adding more nutrients cannot make the bacteria grow much faster.

87. (A) The labeled oxygen from the carbon dioxide shows up in the plant-formed carbohydrates; therefore, this must be the source of the oxygen for carbohydrate synthesis.

88. (B) The labeled oxygen from the water is not incorporated into the starch, and, therefore, must be released as molecular oxygen when water is split.

89. (D) Like water, the plant would split the H_2SO_4 into $2H^+$ and SO_4^{2-}.

90. (D) Chlorophyll captures light energy and passes it to the photosystems that create the NADPH and protein gradient used to make ATP and carbohydrates in the dark reactions.

91. (B) The light reaction occurs in the thylakoid membrane of the chloroplasts, while the dark reaction occurs in the stroma of the chloroplasts.

92. (A) The Calvin cycle is known as the dark reaction because it does not use light directly. Instead, it utilizes ATP and NADPH produced from the light reaction as energy.

93. (D) Primary consumers eat producers; in this population, insects eat shrubs.

94. (E) The decomposers that break down dead and decaying material are not presented in this pyramid. An example of a decomposer would be a fungus that feeds on fallen trees.

95. (B) Because organisms at upper levels derive their food energy from lower levels of the food chain, and because energy is wasted from one level to the next, smaller and smaller biomasses can be supported as the food chain is ascended. Biomass is continuously lost through wastes and other metabolic processes.

CHAPTER THREE

CELLULAR AND MOLECULAR BIOLOGY

One way to solve a puzzle is to put together the pieces in larger and larger assemblies until the entire puzzle is complete. Biologists try to gain understanding about living systems in a similar way, by studying life at many levels and then putting all of the pieces together in one complete picture. Looking at biology from this perspective, the behavior of molecules is observed to explain the workings of cells, which in turn explain the function of tissues, organs and organisms. From there, we can explain populations and ecosystems, and the changes in life through time called evolution that have created the great diversity of life on earth today.

In this chapter we will present the molecules of life and the workings of the cell. This will form the foundation for later chapters concerning organisms, genetics, ecology, and evolution. By the final chapter, it will be possible to view life not as a set of isolated facts, but as a rich interconnected network. This perspective should create a deep understanding that fosters improved problem solving and test-taking through the application of general concepts. It should also provide an appreciation for the unity of life in all its many facets, viewed from many different perspectives. Life is one thing and many things. It all depends on how you look at it.

Biological Chemistry

Life is, at the smallest level, an extremely sophisticated form of chemistry. Living organisms, whether they are rose plants or jellyfishes, are composed primarily of a few common types of molecules. The tissues within each organism and all organisms play many different roles, but contain the same chemical building blocks throughout.

At the elemental level, all life is composed primarily of carbon, hydrogen, oxygen, nitrogen, phosphorous and sulfur. Some other elements like iron, iodine, magnesium, and calcium are also essential for life. If all life is composed of the same elements, what produces such a broad variety of living organisms?

Salts like sodium chloride are significant and essential components of life, but since they do not contain carbon they are known as *inorganic compounds*.

Chemicals that contain carbon are called *organic compounds*, and include the major types of biological molecules found in all organisms. These primary types of molecules central to all life are proteins, lipids, carbohydrates, and nucleic acids. Before we explore these molecules, let's look at a vastly important and seldom appreciated molecule of life: water.

Water and Its Properties

Life is not possible without water. The presence of liquid water is probably one of the key factors that allowed life to evolve and to persist on earth. The unique properties of water that allow it to play this role are based on the way the water molecule is put together.

Each water molecule is composed of an atom of oxygen with two hydrogens attached at an angle. Oxygen draws the electrons in the molecule toward itself, giving itself a partial negative charge, while the hydrogens are partially positive. The water molecule as a whole is not charged but since the molecule is bent its unequal charge distribution makes one end positive and the other end negative. This unequal distribution of charge is called a *dipole moment*. When water molecules are together in a beaker, they interact with each other, with the partial positive and negative charges attracting each other. These interactions between water molecules are called *hydrogen bonds*, a particular type of dipole-dipole interaction.

The strong hydrogen bonds between water molecules give water its many special properties. These bonds between water molecules that hold the molecules together give water structure and take a lot of energy to break. One property of water caused by the bonds between molecules is that water has a great deal of cohesion and surface tension compared to other liquids. The cohesion of water allows trees to transport water from their roots all the way to their leaves in a single long column of water. Bonds between water molecules also mean that it takes a great deal of energy to heat water and to make it boil compared to other liquids. Remember, heat in a liquid or gas is carried in the movement of the molecules: More heat means more rapid movement of molecules. When you add heat energy to water, the energy must break bonds between molecules before it can increase their movement to increase the temperature of the water. Liquid hydrocarbons, like octane, for example, have very low boiling points because the molecules in the liquid are held together very weakly and when heat is added the molecules are able to move about rapidly (heat) and leave the liquid phase (boil). The great deal of energy that water requires to heat or boil means that our body temperature can be held stable and that we can cool ourselves through evaporation using sweat. Water's ability to absorb heat also means that water remains liquid over a range of temperatures that are common on our planet.

Another handy feature of water is that the solid form of water, ice, is less dense than its liquid form. The water molecules in ice are held apart from each other in a rigid matrix of hydrogen bonds, while in water they move

One of a Kind

Water is the only compound that exists in the earth's natural environment as a solid, a liquid, and a gas.

about more loosely and pack together more closely. As a result, ice floats on top of water. One of the many benefits derived from this property is that in winter, water freezes on top of lakes first and insulates the layers below from further cooling and freezing, allowing life to prosper in the water beneath the ice. If ice were denser than liquid water, then lakes would freeze from the bottom up and freeze solid.

One more feature of water that is important to life is its ability to dissolve many different things. Water molecules are polar, but they are not charged. Water is particularly good at dissolving other polar molecules, and at dissolving charged molecules like salts. Water is less able to dissolve non-polar molecules like hydrocarbons, which do not form hydrogen bonds or dipole interactions with water molecules. Life involves chemistry between molecules dissolved in water, so the ability of water to dissolve things is essential to life.

Solutions in Water

Life involves molecules in solution, but what does it mean if we say something is in solution? When we dissolve sugar in water, and the crystals disappear, what happens to the molecules in the sugar crystals? Solids like sugar crystals contain organic or inorganic molecules packed together and interacting together with one or more type of bond. If the molecules are polar, then they interact with each other in the crystal through polar bonds. Water can also form polar dipole interactions with these molecules, displacing their interactions with each other and allowing the molecules to leave the crystal to float surrounded by water molecules. As more and more molecules of sugar leave the crystal in this way, the crystal disappears, with the sugar molecules dissolved in water to form a solution.

In a solution, the substance that does the dissolving is called the *solvent*, and the molecules that are dissolved are called the *solute*. Since solutions and solutes are important in biology, we will talk a little more about measurements and calculations of solutions and about special classes of solutes called acids and bases.

Most biological reactions occur in water with solutes. Concentration is measured by *molarity*, which is the number of moles of solute in one liter. A mole is 6×10^{23} molecules of a substance, and has a specific weight called the molecular weight. For example, the molecular weight of salt, NaCl, is 28. Therefore, to obtain one mole of NaCl, 28 grams must be weighed out. One mole of one substance has the same number of molecules as a mole of another substance even if it does not have the same weight. The unit of concentration used in chemistry and biology is *molar* (M). A 1 molar (1M) solution contains 1 mole of solute in 1 liter of solution. To make a one-liter solution with one mole of NaCl, we would weigh out 28 grams of NaCl and completely dissolve it in water. Then we would bring the total volume to one liter and this would be a one-molar (1M) solution of NaCl.

Acid Rain

Acid rain forms when industrial wastes lower the pH of rain water. It has a pH that is twenty five times more acidic than normal rain. Acid rain causes considerable damage to the environment.

Acids and Bases: *Acids* and *bases* are particularly important types of solutes in biology. There are a few different ways that people define acids and bases. For our purposes, an acid is defined as a proton donor and a base is a proton acceptor. A proton (H^+) is a hydrogen atom stripped of its single electron leaving a positively charged proton. One famous example of a strong acid is hydrochloric acid: HCl. Chlorine atoms have a relatively weak affinity for hydrogen atoms, and in HCl they are held together not covalently but by ionic bonds. Water can easily dissolve HCl, to form H^+ and Cl^-. In this case, since HCl donates its proton, it is an acid. In the reverse reaction, Cl^- would accept a proton to form HCl again, making Cl^- a base in this reverse reaction. However, Cl^- has very little affinity for H^+ ions, making it a weak base, and meaning that the reverse reaction is not favored.

Equation: $HCl \longrightarrow H^+ + Cl^-$

Another example of an acid is water itself. H_2O can dissociate to donate a proton, forming H^+ and OH^- (see equation). Not only can water be an acid, but water can also be a base, acting as a proton acceptor. The H^+ donated by one water molecule can be accepted by another one to form H_3O^+.

Water as an acid: $H_2O \longrightarrow H^+ + OH^-$

Water as a base: $H^+ + H_2O \longrightarrow H_3O^+$

A substance that reduces the hydrogen ion concentration in a solution is known as a base. This disassociation of the base results in more OH^- ions than H^+ ions. An example of this is NaOH, a strong base that favors dissociation into sodium and hydroxide ions. The hydroxide ions in solution react with protons to reduce the acidity of the solution.

Equation: $NaOH \longrightarrow Na^+ + OH^-$

Other bases reduce H^+ ion concentration directly by accepting H^+ ions into themselves. Ammonia is an example of a base that will bind a hydrogen ion from the solution.

Equation: $NH_3 + H_2O \longrightarrow NH_4^+ + OH^-$

Either case will result in a reduction of the H^+ concentration, a decrease in acidity, and an increase in basicity. Solutions with a relatively high concentration of OH^- are called basic solutions.

In pure water, the spontaneous dissociation of water molecules, and the spontaneous reassociation of H^+ and OH^- to form water again, results in a constant and equal concentration of H^+ and OH^- at 10^{-7} M each. Adding a different acid or base as a solute will change the concentration of protons and hydroxide ions. As an acid is added, and the H^+ concentration increases to become greater than 10^{-7} M, the hydroxide ion concentration decreases in a proportional manner. If the concentrations of protons and hydroxide ions are multiplied by each other, the product is always 10^{-14}. If there are 10^{-5}M H^+ ions, then there are 10^{-9} M hydroxide ions.

pH Scale: The concentration of hydrogen ions can be expressed as molar, but it can be hard to compare the acidity of solutions this way. Another common way to express the concentration of protons in a solution is by expressing them as a pH. The $pH = -\log[H^+]$, where $[H^+]$ is the concentration of hydrogen ions given in units of molarity. Thus, if the concentration of protons in solution is 10^{-8} M, then the pH = 8. The pH for acidic solutions is less than 7 and the pH of basic solutions is greater than 7. In the human body, the pH in the blood and tissues is about 7.4. This pH is carefully maintained and controlled since large changes in pH can harm cells and tissues.

Buffers are substances that help to maintain a relatively constant pH in the lab or the body. In pure water, with no buffer present, adding a very small quantity of acid can have a large affect on pH. To change the pH from pH 7 to pH 5 requires only .00001 M acid. Normal metabolic activity would have large harmful affect on pH in the body if buffers were not present. Buffers absorb protons by acting as proton acceptors and donors. The more protons they absorb, the fewer the protons that are available in solution, and the smaller the change in pH.

Functional Groups

Functional groups are attached to carbon skeletons in organic molecules (including biological molecules) and give compounds their functionality. The basic backbone of organic molecules is the *hydrocarbon chain*, upon which functional groups are found. The functional groups, together with their hydrocarbon backbone, provide the physical characteristics, structure, and function of biological molecules.

- *Hydroxyl* (OH). Hydroxyl groups are contained in polar compounds, such as ethanol or glucose. The hydroxyl group is polar, increases water solubility, and is involved in hydrogen bonds.

- *Carbonyl* (C=O). Carbonyl groups are polar groups with a double-bond between C and O contained in aldehydes and ketones, including formaldehyde.

- *Carboxyl* (COOH). Carboxylic groups are polar groups contained in carboxylic acids. They lose their H^+ ions to form acids like acetic acid (vinegar) and have a negative charge when they become deprotonated. Fatty acids and amino acids have carboxyl groups.

- *Amino* (NH_2). Amino groups are polar groups found in molecules such as methylamine and amino acids. Amines can be primary, secondary or tertiary depending on where they are found in the hydrocarbon backbone. Amines can act as bases, with primary amines becoming positively charged when protonated in molecules like amino acids.

- *Sulfhydryl* (SH). Sulfhydryl groups help stabilize proteins by reacting with each other to form cross bridges in the protein structure. They are contained in thiols, such as mercaptoethanol and the amino acid cysteine.

- *Phosphates* (PO_4^{2-}). Found in organic phosphates like glycerol phosphate, these groups store energy that can be passed from one molecule to another by the transfer of a phosphate group. DNA and ATP are important biological molecules that contain phosphate.

Carbohydrates

One of the main classes of biological molecules is *carbohydrates*, or sugars. Another name for carbohydrates is *saccharides*. The functions of carbohydrates include important roles in energy metabolism and storage, and structure of the cell and organisms. One carbohydrate, *cellulose*, provides the cell wall of plants, and is the singularly most abundant biological molecule on earth.

Carbohydrates are built from simple building blocks, starting with simple sugars that have only a single sugar unit. Simple sugars are called *monosaccharides*, sugars with two subunits are called *disaccharides*, and sugars with lots of subunits are called *polysaccharides*. All carbohydrates are composed of carbon, hydrogen and oxygen. Simple sugars have the general molecular formula $C_nH_{2n}O_n$. The name carbohydrate was given to sugars since the molecular formula suggests that these molecules contain water (hydrates) combined with carbon, thus carbo-hydrates. The molecular formula does not indicate, however, that there is water hidden in the structure of these molecules.

Monosaccharides—Simple Sugars

Monosaccharides are simple sugars, with only one sugar building block. Examples of monosaccharides include *glucose* and *fructose* (see figure). The number of carbons in a monosaccharide can vary, including sugars with as few as three carbons. Some of the more common simple sugars in metabolism have five or six carbons. Glucose and fructose have six carbons and ribose has five carbons. Although the number of carbons may vary, the ratio of carbon: hydrogen: oxygen is always the same in simple sugars, 1:2:1. Sugars always have one carbonyl group, either on the carbon on the end of the chain (an aldehyde), or on one of the middle carbons (a ketone). The remaining carbons that do not have a carbonyl group have hydroxyl groups.

One important aspect of the structure of sugars is their *stereochemistry*. Sugars commonly have one or more chiral carbons that make sugars optically active. The sugar molecules found in living organisms almost always occur with a D stereochemistry on the last carbon in the chain.

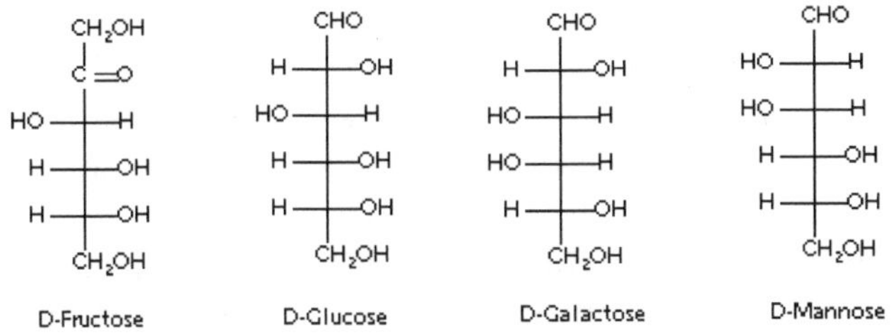

Monosaccharides

Although simple sugars are often drawn as stick structures, they do not necessarily exist in the straight chain stick structure when the sugar is placed into solution. When placed into water, the hydroxyl group on one end of the sugar chain in five and six carbon sugars can bend around to attack the carbonyl on the other end. Since the two components in this reaction are in the same molecule, this reaction occurs quite readily. The formation of a new bond between the two ends of the sugar molecule causes the sugar to go from a straight chain form to a ring form spontaneously in solution.

A single sugar can yield two different version of the ring form called anomers, depending on the whether the carbonyl oxygen is pointed one way or the other when it is attacked by the other side of the molecule. In solution the chain form and the two ring forms are in equilibrium, switching continually between one state and the next.

Sugar—Sugar Bonds

Simple sugars are only one form of carbohydrates. Larger sugars are created when covalent bonds called *glycosidic linkages* are formed between simple sugars, joining them together. Maltose is a disaccharide formed by joining two glucose molecules together (see figure). The reaction to form the disaccharide is called a *dehydration reaction* since a water molecule is removed in the process. The reverse reaction to form two glucose molecules from a single maltose is called *hydrolysis*, with the addition of a water molecule. Another very common disaccharide is *sucrose*, common table sugar, which consists of a glucose molecule and a fructose molecule joined together.

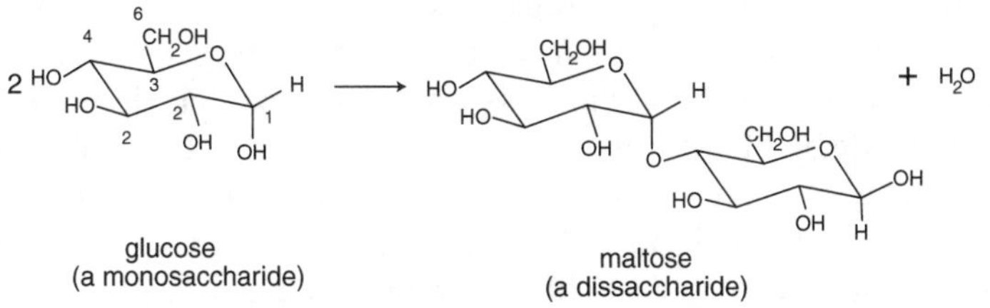

glucose
(a monosaccharide)

maltose
(a dissaccharide)

Disaccharides

Polysaccharides

Polysaccharides are carbohydrates formed by joining many monosaccharides together into large polymers. Being large, polysaccharides are not generally soluble in water, and do not pass through membranes, unlike monosaccharides and disaccharides, which are very water-soluble. *Glycogen, starch* and *cellulose* are three very common examples of polysaccharides, formed by joining many glucose monomers together. The glucose monomers in these polysaccharides are in the ring form and do not equilibrate with the straight chain form. Glycogen and starch are both polysaccharides that store energy. Glycogen stores energy in animals, mostly in muscle and the liver in humans. It occurs as large branching polymers of glucose that form granules in cells. When energy is required, hormones signal liver enzymes to release glucose from glycogen and transport it through the blood to be burned as fuel in cells. When energy is abundant, hormones cause some of the available glucose to be directed to rebuild glycogen reserves. Glycogen is only a short-term energy store, however. The glycogen store in the body is a rapid energy source that is a small fraction of the total energy store and that is rapidly consumed during strenuous exercise. When the glycogen is depleted, the body converts to burning fats, a much more efficient form of energy storage.

Cellulose, a 1,4'-β-D-Glucose polymer

Starch, a 1,4'-α-D-Glucose polymer

Polysaccharides

Starches are a form of energy storage used by plants. Like glycogen, starch consists of many glucose units joined together by glycosidic bonds into large branching polymers. *Cellulose* is also a polysaccharide made of glucose that is made by plants but is not used as an energy store. Cellulose provides structure for plants, forming the wall between cells. Wood is composed of cellulose left behind as the cell wall from many cells of the growing tree. Cellulose also consists of glucose molecules joined together by a different type of bond. While humans produce enzymes that can break down glycogen and starch for energy, releasing their constituent glucose molecules, human enzymes cannot break down cellulose. Microorganisms, however, can break down cellulose and make it possible for termites or ruminant mammals like cows to digest grasses and plant material.

Lipids (Fats and Oils)

Structure and Function

Lipids are very nonpolar, or hydrophobic, molecules. They are composed mostly of nonpolar bonds, and tend to repel water. The structure of lipids allows them to play important roles in energy metabolism and in the structure of cells and organisms. Like carbohydrates, lipids are composed of carbon, hydrogen, and oxygen. Lipids, however, are very distinct from carbohydrates in their structure and function. Lipids have much lower oxygen content than carbohydrates. With less oxygen, they are less oxidized and can store more energy than carbohydrates. The low degree of oxygen they carry also contributes to their nonpolar, water-repelling nature.

Energy Storage of Lipids

Lipids are the chief means of energy storage in animals, since lipids store and release more energy for their weight than any other class of biological compounds. A key trait for most animals is motility, which requires efficient energy storage. A potato can afford to store its energy as starch, since it does not need to get around too much, but if a deer stored all of its energy as carbohydrates rather than fat it would be a slow-moving snack for any predator it encountered. The lipids used by animals to store energy are *triglycerides*, which contain three *fatty acids* joined by ester linkages to one *glycerol* molecule. Fatty acids contain long carbon chains that are very hydrophobic, so triglycerides gather together to form globules inside fat cells that contain very little water. A major component of fatty adipose tissue, triglycerides also provide insulation and protection against injury.

Synthesis of Triglycerides

During periods of low glucose or extended exercise, hormones can induce the mobilization of triglycerides for energy production in cells. Cells that store the triglycerides can convert some of them to release free fatty acids into the blood to be transported to cells that absorb the fatty acids to burn them for energy.

Lipids and Cellular Membranes

Phosholipids are lipid molecules with phosphate-containing groups on one end and fatty acid side chains on the other end. The fatty acid chains in phospholipids are very hydrophobic (water-excluding), while the phosphate-alcohol end is very hydrophilic (water-loving). As a result, phospholipids are like detergents, and when mixed with water will spontaneously form structures with the fatty acids gathered together to keep out water and the phosphate group pointing out toward water. This is the basic structure of the lipid bilayer formed by phospholipids in cell membranes. Phospholipids contain glycerol, two fatty acids, a phosphate group, and a nitrogen-containing alcohol like *lecithin*. Lecithin is a major constituent of lipid bilayer cell membranes.

Other Lipid Functions

Waxes are esters of fatty acids and alcohols. They form protective coatings to keep water out of skin, fur, and leaves of higher plants, and are found on the exoskeleton of many insects in the form of lanolin. *Steroids* have three fused cyclohexane rings and one fused cyclopentane ring. Examples of steroidal molecules include cholesterol in membranes and steroid hormones like estrogen, testosterone, and cortisol. *Bile* acids from the liver are steroids as well. The *carotenoids* contain conjugated double bonds and carry six-membered carbon rings at each end. As pigments, they produce red, yellow, orange, and brown colors in plants and animals. Two carotenoid subgroups are the carotenes and the xanthophylls. Many vitamins including vitamin A and vitamin D are very hydrophobic and are often associated with lipids.

Proteins

Carbohydrates and lipids provide energy and structure for cells. There is much more to life, however, than these functions. One of the characteristics of life is that it is very active, with cells continually carrying out a broad range of functions in order to grow, reproduce and survive. *Proteins* provide cells with the ability to carry out these functions, including the following:

Type of Protein	Function	Examples
Hormonal	Chemical messengers	Insulin, glucagon
Transport	Transport of other substances	Hemoglobin, carrier proteins
Structural	Physical support	Collagen

Type of Protein	Function	Examples
Contractile	Movement	Actin/myosin
Antibodies	Immune defense	Immunoglobulins, interferons
Enzymes	Biological catalysts	Amylase, lipase, ATPase

One of the key functions of proteins is their role as biological catalysts called enzymes. Enzymes allow cells to capture and use energy, to replicate their DNA, and to carry out other activities that are essential to life.

The basic structure of proteins is simple. Proteins are polymers formed by joining simple building blocks called *amino acids* together in a process called *translation*. There are 20 common amino acids that are incorporated into proteins during translation. Every amino acid has an amino portion and a carboxylic acid portion that is the same in all of the amino acids. The portion that makes each of the 20 amino acids distinct from the others is a variable functional group generically called the "R" group in the middle of the amino acid molecule (see figure).

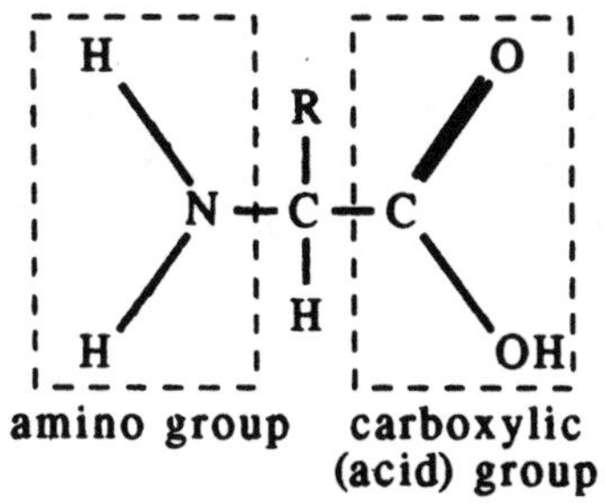

amino group   carboxylic
(acid) group

Amino Acid

The basic structure of proteins is formed by joined amino acids together end to end, like beads on a string. The bonds between amino acids in the chain are called *peptide bonds*, and the resulting string of amino acids is called a *polypeptide* (See figure).

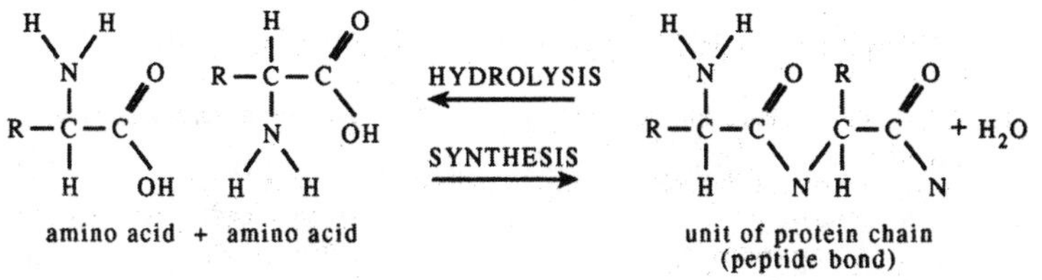

amino acid + amino acid

unit of protein chain
(peptide bond)

Polypeptide

The side groups of amino acids vary in size, shape, charge (positive, negative or neutral), hydrophobicity, and their reactivity with other molecules. Glycine has the smallest side chain, with only a hydrogen. Tryptophan, tyrosine, leucine, isoleucine, alanine and valine have hydrophobic side chains. Aspartate and glutamate have carboxylic acid side chains that are deprotonated to have a negative charge at normal body pH, while lysine and arginine have basic amine groups that accept protons at normal body pH levels. Serine, threonine and tyrosine have hydroxyl groups that can be modified with phosphate groups to regulate the activity of proteins. Cysteine has a sulfhydryl group that will react with other cysteines in a protein to create disulfide bridges that stabilize protein structure. The variety of amino acids provides a great variety of chemical abilities in proteins.

BIOLOGY REVIEW

In a Nutshell

Protein Levels of Structure:

- *Primary*: a linear sequence of amino acids

- *Secondary*: regular coils or folds of the protein with nearby amino acids in the polypeptide chain interacting

- *Tertiary*: the interactions between different areas of the protein to form the overall 3D shape of a polypeptide

- *Quaternary*: the interactions between polypeptides in proteins with more than one subunit

The Relationship Between Protein Structure and Function

If the basic structure of proteins is so simple, formed by joining a few different building blocks together in a string, how can proteins play so many different roles in the cell? It is the variety of the amino acids and the order in which they are joined that allows proteins to do so many different things.

The polypeptide is a chain of amino acids joined together in a linear string. After the polypeptide is made it does not remain a straight chain like a stick, however. The bonds in the polypeptide are flexible and can rotate, allowing the amino acids in the polypeptide chain to move about and the chain to fold flexibly in many different ways. Proteins can fold to form ball-like globular structures, or long extended fibrous structures, and countless structures in between. Protein folding does not occur randomly, though. The order and the identity of the amino acids in the chain determine how the protein folds. For example, hydrophobic, water-repelling amino acids tend to fold into the interior of proteins, away from water, while charged and polar amino acids tend to fold on the outside of proteins.

Protein structure is described in four levels, *primary*, *secondary*, *tertiary*, and *quaternary*, with each level describing more and more complex levels of protein structure.

The *primary level of protein structure* is the sequence of the amino acids joined together in the linear polypeptide chain. With 20 different amino acids in proteins, even a very short polypeptide can be assembled in many different ways. Imagine a short peptide, with four different amino acids joined together. For this peptide there are 20 X 20 X 20 X 20 possible combinations (20 to the fourth power) = 160,000 possible peptides that could be made. The primary level does not describe any of the folding that occurs in the polypeptide chain, but does determine how the protein will fold.

The *secondary level of protein structure* includes regular repeating elements of folding that involve interactions between local neighbors in the polypeptide chain. The most common types of secondary structure are called *alpha helix* and *beta-pleated sheet*. In both of these, the interactions and identity of amino acids are not as important as hydrogen bond interactions between the polypeptide backbone of amino acids that are a few units away from the interacting partner in the chain. The alpha helix is a spiral-shaped structure common in many proteins, with different parts of the backbone forming hydrogen bonds in the spiral. The beta-pleated sheet contains straight lengths of the polypeptide chain that lie one on top of the other to form a flat layer of structure.

The *tertiary level of structure* can be characterized as folding due to bonding between side chains of the various amino acids. These interactions can be hydrophobic interactions between amino acids, hydrogen bonds, ionic interactions or even covalent bonds. Tertiary structure can be reinforced by strong covalent bonds, disulfide bridges between cysteines that anchor the overall tertiary structure in place, making it difficult to unfold. Tertiary structure can be modular, with two or more globular regions called domains that fold independently and are connected by relatively flexible regions of the polypeptide chain. A given domain usually has a specific function within the overall protein. Secondary structure involves bonds between nearby amino acids in the linear sequence. Tertiary structure can bring amino acids near to each other in space even if they are far apart in the linear stretched-out polymer.

The final *quaternary level of protein structure* results from interactions between different polypeptides to form a single functional unit. Many proteins are made of more multiple polypeptides that interact in quaternary structures to create large protein complexes. These subunits are separate polypeptides, not part of the same chain, and can be synthesized separately, then joined later. The different subunits of the mature protein can in many cases regulate the activity of other subunits by communicating information through the interface they share with each other. The information is carried as slight changes in the overall polypeptide folding, which are transmitted to create small changes in the other polypeptides in the quaternary structure. Hemoglobin, for example, the protein in red blood cells that carries oxygen, has four polypeptide subunits in its quaternary structure. Interactions between the polypeptide subunits allow the subunits to communicate with each other to bind oxygen cooperatively. This cooperation between subunits allows hemoglobin to pick up oxygen in lungs and deliver it to tissues more efficiently.

Some proteins, particularly small simple globular enzymes, fold correctly without any help. This indicates that the information for protein folding is contained in the protein itself, in the linear amino acid sequence of the polypeptide. The polypeptide probably folds first into secondary structure, followed by tertiary and quaternary structure. More complicated proteins and those that are headed for special locations in the cell appear to need other proteins to help them to fold correctly.

The structure of proteins is not just a physical property, but determines and is required for their function. One way scientists study the importance of protein structure is by forcing a protein to unfold, also called *denaturation*. Boiling a protein or exposing it to harsh chemicals can disrupt its structure, and in cases where its structure is denatured, the protein usually loses its function as well. An enzyme that is denatured by boiling will not catalyze its reaction when unfolded. Some proteins will refold themselves after cooling however, supporting the idea that proteins contain the information to fold themselves.

Spontaneous folding of proteins is one aspect of a larger phenomena displayed by biological systems—spontaneous self-assembly. If you mix phospholipids, they spontaneously assume their lowest energy state, the lipid bilayer membrane. Proteins in many cases will fold and form complex structures simply if you mix them together in solution in the right conditions. Viruses will even assemble themselves from their component proteins and nucleic acids. The spontaneous self-assembly of biological systems hints at the robust nature of structure in these complex systems.

Enzymes

Life uses chemistry to build, to generate energy, to move, and to perform other functions. Many of these reactions would either occur very slowly or not at all on their own if the chemicals involved were simply mixed in a tube on the countertop. If you put some sugar on the countertop, it will eventually become oxidized and release energy, but on its own it will happen so slowly that the sugar might outlive the countertop. *Enzymes* act as biological catalysts to speed up these reactions and make them useful for living organisms.

Every chemical reaction begins with reactants, and proceeds to products. The reactants have a certain amount of energy contained in their bonds and movement, and the products contain a unique amount of energy as well. In enzymatic reactions, the reactants are called *substrates*. To go from being reactants (substrates) to products, chemicals must first pass through an unstable high-energy state called the *transition state*. The amount of energy required to reach the transition state is called the *activation energy*. Enzymes increase the rate of reactions by lowering the activation energy required for a reaction (see figure). They do not change the energy of reactants or products and do not change the equilibrium of the reaction, but they can increase the rate at which equilibrium is reached to a great degree. In this way they are like other catalysts, even if they are made of protein. Also, like all good catalysts, enzymes are not consumed or changed as a result of the reaction.

How do enzymes decrease the activation energy? Like other proteins, enzymes must fold into specific 3D structures to do their work. In a properly folded enzyme molecule, the folded polypeptide contains a three-dimensional binding pocket called the *active site* in which the reaction is catalyzed. The active site contains specific amino acids precisely oriented in the protein's structure to form a substrate binding pocket. In the active site, the substrates are brought together in close proximity and in the correct orientation for the reaction to proceed, lowering the activation energy. The active site also stabilizes the transition state intermediate, making it easier for the transition state to form, and further lowering the activation energy.

In a Nutshell

Enzymes:

- Lower the activation energy rate of a reaction
- Increase the rate of the reaction
- Do not affect the overall energy change of the reaction
- Are not changed or consumed in the reaction

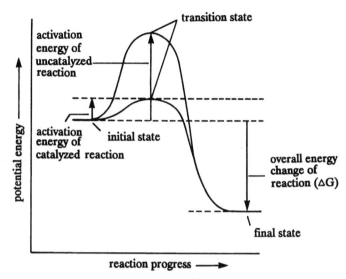

Reactions With and Without Enzymes

Almost all enzymes are proteins. Proteins make versatile enzymes due to the great variety of amino acids that can be employed. Any biological molecule that can catalyze a reaction would be an enzyme, however. In fact, several enzymes made from RNA are now known in which the active site is formed by nucleotides in a folded RNA molecule rather than amino acids in a folded protein.

The Model for Enzyme Activity

Enzymes are said to be active if they are able to catalyze their reaction. The amount of catalysis by an enzyme is called *enzyme activity*, and is commonly studied in the laboratory in a variety of conditions. One of the early observations of enzyme activity was that as you add more substrate (reactant) to an enzyme reaction, product is generated more rapidly. This makes sense, and is what you would expect for any chemical reaction. At high concentrations of substrate, though, adding more substrate to an enzymatic reaction does not increase the reaction rate further. This property of enzyme-catalyzed reactions is called *saturability*, and is a hallmark of enzyme-catalyzed reactions that distinguishes them from ordinary chemical reactions. Ordinary chemical reactions will keep on increasing in rate as you add more and more of the starting materials, and will not display saturability.

The maximal reaction rate observed with the saturated enzyme reaction is called V_{max}. Why are enzymatic reactions saturable? The key is the active site. Enzymatic reactions are only catalyzed at active sites, and the number of active sites in a reaction is limited, depending on how much enzyme you add. Usually, the concentration of enzyme in a reaction is far lower than the concentration of the compounds involved in the reaction. Also, each active site can only work so fast to catalyze a reaction. As the enzyme works, it will bind substrate, catalyze the reaction, and release the products, ready to begin again. If there is not much substrate around, then the active site may be empty for a time before it happens upon another substrate molecule. If there is a lot of substrate, however, the active sites may be full all of time. When the active sites are full 100% of the time, the enzyme cannot work any faster and adding more substrate does not increase the reaction rate further.

Another hallmark of enzymatic reactions is their *specificity*. With organic chemistry in the lab, the reaction products can be unpredictable and depend on the chance orientation of molecules flying about in solution. Enzymes bind molecules in the active site in very precise ways, however. To fit in the active site and be used as a substrate, a chemical must have the specific shape of the substrate that the enzyme recognizes, determined by the precise orientation of amino acids in the binding pocket. Other molecules will not fit in the site and will not act as substrates. Changing an amino acid in the active site can change the specificity of an enzyme from one substrate to another. The high degree of enzyme specificity even extends to specific synthesis of one stereoisomer in a reaction and not the other mirror image molecule. This type of synthesis is still very difficult to achieve in the lab but is carried out routinely by all living things. Sugars and amino acids are all very stereospecific in living systems and the enzymes that work with these molecules can distinguish isomers that different only in the orientation of bonds, with all of the same atoms and connections present.

Enzymes and the Environment

The nature of an enzyme's environment affects its activity. Since the active site is a complex structure and depends on the correct folding of proteins at all levels, anything that disturbs protein folding disturbs the active site structure and reduces the ability of enzymes to catalyze reactions. Important factors in the environment affecting enzymes include the temperature, the presence of coenzymes, the location of an enzyme in the cell, the pH of the surrounding solution, factors that affect enzyme folding, and the presence of molecules that inhibit or activate enzyme activity.

Temperature will affect enzyme activity for two main reasons. At low temperatures, most enzymes have very little activity. As the molecules in solution move more slowly at lower temperatures, they take longer to find active sites, slowing the reaction rate. When they do find the active site, they may not have sufficient kinetic energy to form the transition state. As the temperature increases, and molecules start bouncing about more energetically, substrate molecules bounce into the active site more often and with more energy. The more energy that molecules have, the more likely they are to have the energy to form the transition state, and the more molecules that reach the transition state, the faster the reaction rate. At high temperatures, though, heat energy makes the enzyme molecule vibrate, unfold and denature, losing activity.

Reaction Rates

The rate of an enzyme-related reaction may be changed by:

- Concentration
- Temperature
- pH

KAPLAN

Most enzymes in the human body have evolved to have an optimal reaction rate at 37° C, the temperature of the human body, but there is nothing magical about this temperature. Other organisms adapted to different environments have evolved enzymes with much different temperature sensitivity. Microorganisms that live in the arctic have enzymes with optimal activity at freezing temperatures while microbes in boiling hot springs have enzymes with optimal performance in boiling water. The unique properties of these enzymes have been exploited by molecular biologists to make new technologies like polymerase chain reaction (PCR) possible.

pH also affects enzyme activity markedly. Most enzymes have optimal activity within a narrow range of pH. Acidity can affect enzyme activity through alteration of the charge of key amino acids in the enzyme, and through changing the folding of the protein, including unfolding of the enzyme at extremes of acid or base conditions. If you try to mix orange juice and milk, the curdled mixture that you see is the result of proteins in milk being denatured, unfolded by the acid in the orange juice. Most human enzymes have optimal activity at about pH 7, the pH of blood and most parts of the body. One exception to this is the interior of the stomach, in which the digestive enzyme pepsin that breaks down proteins resides and has a pH optimum at the acidic pH of 2. Another exception includes enzymes that reside in the lysosome organelle in the cell, which also has an acidic pH.

Many enzymes require components called *coenzymes* that are not a part of the polypeptide itself to perform their activities. Since the coenzyme is not part of the encoded polypeptide, it must join the polypeptide after synthesis. Specialized nucleotides are often used as coenzymes in reactions that reduce or oxidize material. Coenzymes can either be bound to an enzyme or can be in solution and join the enzyme with each reaction cycle. The importance of coenzymes is illustrated by their importance in the diet. Many coenzymes are not synthesized by the body and are important nutrients required from food, vitamins.

Regulation of Enzyme Activity

A key to the role of enzymes in life is not just that enzymes can speed up the rate of reactions, but that their activity is controlled and regulated to speed up reactions at the right time in the right place. Most enzymes are not active at full blast all of the time. Life must react to its environment, and tailor the activity of its enzymes to match what it needs, or else squander energy and resources and perhaps damage itself. An enzyme from the pancreas that breaks proteins down into small pieces may be quite useful in the intestine as part of digestion but quite harmful if it is active inside the cell that produces it.

Enzyme activity is regulated in many ways. One way is by making more enzyme when it is needed. This is very effective and is a common mechanism, although not very fast since it requires turning genes on and off and

Make Sure You Get Your Vitamins

Many coenzymes are vitamins. This is why vitamin deficiencies often cause severe diseases like scurvy. Scurvy, caused by a lack of vitamin C, used to be commonplace on board ships, where no fresh food was available.

waiting for protein synthesis to occur. Another very simple way to regulate enzyme activity is increasing the amount of substrate. When more substrate is present, the reaction rate will increase and use up the substrate until its concentration decreases again.

Since enzymes are often part of complex metabolic pathways, more sophisticated and rapid ways to regulate activity are sometimes needed. Rather than make more enzyme, cells can regulate enzymes with compounds that bind to enzymes to increase or decrease their activity. Compounds that decrease enzyme activity are called *inhibitors*. There are three common varieties of enzyme inhibitors: *irreversible inhibitors*, *competitive inhibitors*, and *noncompetitive inhibitors*. *Irreversible inhibitors* bind to an enzyme active site and covalently modify it to permanently cripple the enzyme molecule. More enzyme must be made to overcome the effects of an irreversible inhibitor. *Competitive inhibitors* resemble the substrate and are able to displace the substrate from the active site. *Noncompetitive inhibitors* do not bind at the active site but at a regulatory part of the enzyme molecule that transmits a change in structure through the enzyme molecule to reduce activity.

A very common mechanism used to regulate metabolic pathways is called *feedback inhibition*. In a pathway, the final product will often act as an inhibitor of the first step in the pathway. As the final product accumulates and its concentration increases, the cell needs less to be made. If the product inhibits the first step in the pathway, the pathway slows down, and the cell gets exactly the amount of product it needs. Feedback inhibitors are often competitive inhibitors since they resemble the initial starting material for the pathway that they inhibit. Such automated self-regulation of metabolic pathways in the cell is what enables cells and the body to maintain constant conditions in untold different ways all of the time.

Enzymes are often regulated by other means that alter a protein's structure. Proteins are sometimes activated by cleavage of one or more peptide bonds. The cell can make precursor protein that is larger and lacks activity until after it is cut into smaller pieces by another enzyme called a *protease* (enzymes that hydrolyze peptide bonds). This form of regulation is not reversible, but is used to produce digestive enzymes in a safe inactive state until after secretion when they are activated. Modification of the amino acids with phosphate by enzymes called *protein kinases* is often used to regulate the activity of enzymes, particularly in response to extracellular signals that require a rapid response. The phosphate is charged and negative and can alter the protein folding to either decrease or increase activity. Other proteins called *phosphorylases* can remove the phosphate to reverse the change in enzyme activity. Sometimes proteases or kinases act as part of a network of enzymes that modify each other in response to a signal. Clotting in the blood involves a cascade of protease enzymes that cut and activate each other, resulting ultimately in the rapid formation of blood clots to stop bleeding. Converting glycogen into glucose involves a series of kinases that phosphorylate each other. *Signal transduction cascades* with

CELLULAR AND MOLECULAR BIOLOGY

enzymes such as protein kinases or proteases linked together in a series are able rapidly to achieve a great amplification of signal.

Nucleic Acids

Nucleic acids are another class of the essential biological molecules found in all living organisms. Nucleic acids are informational molecules and include DNA and RNA. All organisms (except for some viruses, which most people do not classify as truly living) use DNA as their genome. The genome is the information that an organism must carry to conduct all of its activity of living and must transmit to its offspring if they are to do the same. The genome does not directly do these things, but carries information in a code called the genetic code that conducts these activities. The structure and function of nucleic acids will be addressed in more detail in the section describing the genome and gene expression.

How Cells Get Energy to Make ATP

Life is energy. One of the essential features of life is the ability to capture and harness energy from the environment and use this energy to build, move, grow, replicate and even think. All of these activities require energy. What energy is used and where does it come from?

Organisms eat carbohydrates and fats that contain chemical energy. In digesting these molecules, organisms trap some of their chemical energy, then release it as glucose to the rest of the body. Cells use glucose to make a molecule called ATP. ATP is then used in the cells to do most activities that require energy input to occur.

The physics behind this is that processes that require energy input will not occur on their own, catalyzed or not. This includes most of the things that life wants to do. In fact, without energy input, most of the molecules of life would tend to move in the other direction, toward oxidation and a loss of structure. By capturing food energy and converting it into ATP, life uses the energy of ATP to drive forward all of the energetically uphill reactions it needs to perform.

The common currency used by cells of all living things is a molecule called ATP, a nucleotide, adenosine triphosphate. ATP consists of adenosine, and three phosphate groups, a triphosphate. Each phosphate is negatively charged, and negative charges repel each other. Having three negative charges close together in the triphosphate stores a lot of energy, and when the triphosphate is split the energy is released and can be used to carry out important cellular functions.

Where does the ATP come from? Cells in humans and other organisms use a common set of biochemical reactions to make ATP. These reaction pathways are *glycolysis*, the *Krebs cycle,* and *electron transport*. It all starts with glucose. In humans, glucose is always present in the blood as a fuel for all cells. Cells take in glucose, which then enters the glycolytic pathway as the first step in the path to ATP.

Glycolysis

A metabolic pathway is a linked series of biochemical reactions that have a common purpose. *Glycolysis* is probably a very ancient pathway in the evolution of life, given its presence in all of the kingdoms of life, from bacteria to humans. The reason for its importance is that glycolysis is the first biochemical pathway in the capture of energy from glucose to make ATP. The glycolytic pathway consists of ten steps, each catalyzed by an enzyme uniquely evolved to catalyze that reaction. We will not go into all of the individual reactions or the individual enzymes. Being familiar with the idea of metabolic pathways and the function of glycolysis is a good idea, though. Glycolysis takes glucose, a sugar molecule with six carbon atoms, and breaks it into two pyruvate molecules, each with three carbons, capturing energy in two different ways.

One way to capture energy is to directly produce ATP as part of the glycolytic pathway. At two different steps in the pathway, ATP is put into the reaction to drive the splitting of glucose, consuming 2 ATP molecules for each glucose molecule that enters the top of the pathway. Further along in the pathway 4 ATP are generated, for a net production of 2 ATP per glucose molecule.

This is not where most of the energy of glucose is extracted, however. In the burning of fuels to extract energy, the fuel molecules become more and more oxidized, and have less energy, releasing their energy to their surroundings. If you take a sugar cube and burn it, the sugar molecules are oxidized directly to carbon dioxide and release their energy rapidly as heat. In the body, the oxidation of glucose is more controlled, and the energy is transferred to other molecules that carry the energy and transfer it later in the process to make ATP. In glycolysis, glucose is oxidized to form pyruvate, and some of the energy is captured to make NADH. NADH is an energy carrier that the cell uses to make ATP through electron transport, as we will see.

If you add up all of the reactions of glycolysis, the overall net reaction describing this whole reaction pathway becomes:

$$\text{Glucose} + 2Pi + 2ADP + 2NAD^+ \longrightarrow$$

$$2 \text{ pyruvate} + 2ATP + 2NADH + 2H^+ + 2H_2O$$

In eukaryotic cells, glycolysis occurs in the cytoplasm. In prokaryotes, where there are no organelles, all reactions occur in the same compartment.

Fermentation

Since NAD^+ is required for glycolysis to occur, and it is converted to NADH as part of glycolysis, NAD^+ must be regenerated or glycolysis would run out of it and stop, halting ATP production as well (and probably the life of the cell or organism involved). The NAD^+ is regenerated in one of two ways. In the first, NADH goes on to the electron transport chain and is used to produce more ATP, as described in the sections that follow, being converted back to NAD^+ in the process. This process is termed oxidative metabolism because it requires oxygen. The second way to regenerate NAD^+ occurs in the absence of oxygen or in anaerobic organisms that do not use oxidative metabolism. This alternate pathway is called *fermentation*.

If NAD^+ runs out in a cell or organism, then glycolysis cannot proceed and ATP generation will halt. ATP is essential for life; no ATP, no life, so ATP generation must be continued whether oxygen is around or not. In the presence of oxygen, NAD^+ is regenerated allowing glycolysis and oxidative respiration to continue ATP production. Fermentation allows glycolysis to continue even in the absence of oxygen. In fermentation, NADH is regenerated back to NAD^+ in the absence of oxygen to allow glycolysis to continue to produce ATP, producing either ethanol or lactic acid as byproducts. *Ethanol fermentation* is carried out by yeast in the absence of oxygen and is used to make beer and wine. The first step in ethanol fermentation occurs when pyruvate is decarboxylated (loses a CO_2) to become the two-carbon molecule acetaldehyde. NADH from glycolysis then reduces acetaldehyde to ethanol and is itself oxidized back to NAD^+. The regenerated NAD^+ allows glycolysis to proceed and to continue ATP production. This process can continue until the alcohol level rises so high that it kills the organisms producing it.

Lactic acid fermentation occurs in some bacteria and fungi and, closer to home, in human muscles during strenuous exercise. Fermentation in muscle allows ATP production to continue when oxygen in the muscle is limited because it is consumed more rapidly than the blood stream can supply it. In lactic acid fermentation, pyruvate is reduced to lactic acid by NADH that is oxidized to regenerate NAD^+. The NAD^+ then allows glycolysis and ATP production to proceed. During exercise, the lactic acid produced can accumulate in muscle and cause pain as well as a drop in blood pH. After the exercise is complete, the lactic acid will be oxidized back to pyruvate to reenter oxidative metabolism pathways.

Aerobic Respiration

Although glycolysis produces two ATP and two NADH for every molecule of glucose, this is not where the eukaryotic cell extracts most of its energy from glucose. Glycolysis is only the beginning. *Aerobic respiration* is the rest of the story. In aerobic respiration, glucose is fully burned by the cell as an energy source, going through the Krebs cycle and electron transport to trap

Ouch!

Lactic acid buildup, a product of lactic acid fermentation, is what makes your muscles sore after a good workout.

energy that ultimately is used to make ATP. The role of oxygen in the process is to help complete the electron transport process.

The pyruvate left at the end of glycolysis still contains a great deal of energy that is extracted in oxidative metabolism. Oxidative metabolism of glucose produces a maximum total of 38 ATP in the complete oxidation of one glucose molecule, compared to two from glycolysis (or fermentation) alone (note that the actual total is 36 ATP/glucose in eukaryotes due to a slight energy cost for transport into the mitochondria). Oxidative respiration is a far more efficient energy production system than glycolysis or fermentation, so the ability to perform oxidative respiration gives organisms a competitive advantage in conditions where oxygen is available.

To accomplish this more efficient energy production, pyruvate from glycolysis is oxidized all the way to carbon dioxide in a pathway called the *Krebs cycle*. The Krebs cycle and the other steps of oxidative metabolism occur in mitochondria. As pyruvate is oxidized in the Krebs cycle, NADH and another high-energy electron carrier called $FADH_2$ are produced. In electron transport, the energy of these high-energy electron carriers creates a pH gradient by pumping protons (H^+ ions) out of the mitochondria. The energy of this pH gradient drives ATP synthesis, and this is the ultimate source of most of the ATP produced in the oxidative metabolism of glucose.

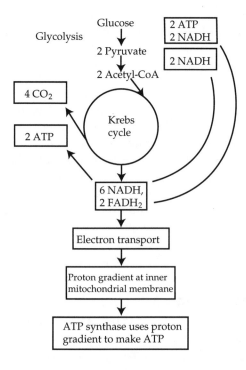

Aerobic Respiration

The Krebs Cycle

The Krebs cycle is a series of reactions linked in a circle that extracts energy from the products of glycolysis to make the high-energy electron carriers NADH and $FADH_2$ (see figure). This pathway is also sometimes called the citric acid cycle or tricarboxylic acid cycle. In aerobic conditions, when oxygen is present, the pyruvate produced in the cytoplasm in glycolysis is transported through both mitochondrial membranes into the interior of the mitochondria. Here, pyruvate is converted into a two-carbon molecule called acetate, with the release of CO_2. The acetate group is linked to a large carrier molecule called coenzyme A, to make acetyl-coenzyme A. The oxidation of pyruvate to acetyl-CoA also produces more NADH that will go on to make ATP via the electron transport chain.

$$\text{Pyruvate} + \text{CoA} + \text{NAD}^+ \longrightarrow \text{acetyl-CoA} + CO_2 + \text{NADH}$$

Acetyl-CoA is an important metabolic junction in the cell, and is involved in many different metabolic pathways. Acetyl-CoA enters the Krebs cycle by combining with a four-carbon intermediate to make the six-carbon citrate (or citric acid). In every complete loop of the Krebs cycle, two CO_2 molecules are produced and leave the cycle for every atom of acetyl-CoA that enters the cycle. This leaves the net level of intermediates in the cycle constant and regenerates the four-carbon intermediate that combines with another acetyl-CoA in the next round of the cycle. Every cycle also produces three NADH, one $FADH_2$ and one ATP. Since two acetyl-CoAs are produced from every glucose, the net result of the Krebs cycle is (for each glucose that enters glycolysis):

$$2 \text{ Acetyl-CoA} + 6\text{NAD}^+ + 2\text{FAD} + 2\text{ADP} + 2\text{Pi} + 4H_2O \longrightarrow$$

$$4CO_2 + 6\text{NADH} + 2\text{FADH}_2 + 2\text{ATP} + 4H^+ + 2\text{CoA}$$

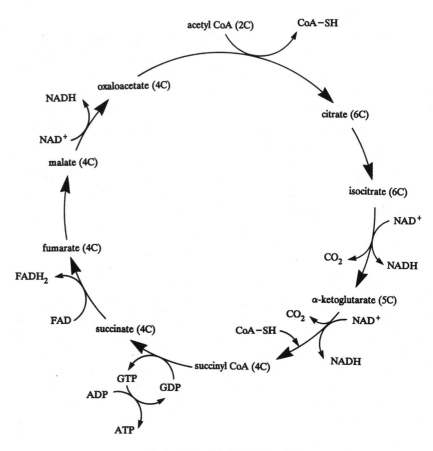

Krebs Cycle (Citric Acid Cycle)

It is not necessary to memorize all of the intermediate steps of the Krebs cycle for the SAT II test, but it is a good idea to understand how it fits into the overall metabolism of glucose to make ATP.

Electron Transport

Glycolysis directly produces two ATP for every glucose, and the Krebs cycle directly produces two more ATP for every glucose. However, most of the energy of glucose is not extracted directly into ATP. The high-energy electron carriers NADH and $FADH_2$ contain much more of the energy that is extracted by the oxidation of glucose in glycolysis and the Krebs cycle. *Electron transport* is the mechanism used to convert the energy held by these carriers into a more useful form that ultimately results in ATP production.

The electron transport chain is a series of proteins and electron carriers located in the inner mitochondrial membrane. NADH and $FADH_2$ transfer the high-energy electrons they carry to the chain and are oxidized back to NAD^+ and $FADH^+$, respectively. The high-energy electrons that enter the electron transport chain are transferred from one carrier to another, transferring energy to the carriers along the way. The carriers alternate between oxidized and reduced forms as they transfer electrons through the chain to the final electron acceptor at the end of the chain, oxygen, which is reduced to water. This oxygen is the oxygen needed for aerobic respiration and the oxygen we breathe and transport throughout the body. As electrons move through the series of oxidation-reduction reactions in the electron transport chain, their energy is used at three points in the chain to pump protons, hydrogen ions, out of the mitochondrial interior. The pumping of these H^+ ions out of mitochondria creates a pH gradient across the inner mitochondrial membrane. This is yet another form of energy conversion in the process of getting the energy from food into the ultimate state. We still have not made ATP, but we are getting close.

In glycolysis and the Krebs cycle, a small amount of the energy stored in the chemical bonds of glucose is converted to form a few high-energy ATP bonds directly, and more energy is used to reduce the high-energy electron carriers NADH and $FADH_2$. In electron transport, the energy trapped in NADH and $FADH_2$ is converted again, this time to the energy of the *proton gradient* between the inside and the outside of the mitochondria. Since these protons cannot diffuse through the membrane, this energy is stored and the cell can harness it to make ATP.

This proton gradient created by the electron transport system is another form of energy. When anything is pumped against a concentration gradient, that concentration gradient contains energy. The inner mitochondrial membrane forms a tight seal that protons cannot leak through so once the protons are pumped out of mitochondria, their energy is stored. The next step is to take the energy of the mitochondrial proton gradient and use it to make ATP.

A protein called *ATP synthase* found in the inner mitochondrial membrane harvests the energy of the pH gradient to produce the bulk of the ATP from oxidative glucose metabolism. ATP synthase allows the protons on the outside of the mitochondria to flow down their concentration gradient back into the mitochondria. Protons do not freely diffuse through the membrane, however. ATP synthase harnesses the energy of proton movement down the gradient through ATP synthase itself to make ATP, making three ATP for every NADH that enters electron transport and two ATP for every $FADH_2$ that enters electron transport.

The Krebs cycle, electron transport, production of the proton gradient, and ATP production are all linked. If oxygen is removed, all of these processes stop, since they are all dependent on each other. If metabolic poisons allow protons to flow into mitochondria through the membrane, without going through ATP synthase, then the proton gradient will be destroyed. In this case, ATP production is said to be uncoupled from the Krebs cycle and electron transport, since ATP production would cease although the Krebs cycle and electron transport would increase their rate of activity. If metabolic needs of the cell increase, then all of these processes increase in rate. In this way the cell regulates ATP production and the associated pathways to meet its own needs.

Photosynthesis

To survive, all organisms need energy. Herbivores get energy by eating plants and carnivores by eating herbivores. The foundation of all ecosystems and the source of the energy in these ecosystems and on planet earth as a whole is *photosynthesis*. Plants are *autotrophs*, or self-feeders, that generate their own chemical energy from the energy of the sun through photosynthesis. There are also many prokaryotic photosynthetic organisms such as algae that contribute to global productivity significantly. The chemical energy that plants get from the sun is used to produce glucose that can be burned in mitochondria to make ATP, which is then used to drive all of the energy-requiring processes in the plant, including the production of proteins, lipids, carbohydrates, and nucleic acids. Animals eat plants to extract this energy for their own metabolic needs. In this way, photosynthesis supports almost all living systems.

Photosynthesis occurs in plants in the *chloroplast*, an organelle that is specific to plants. In algae, a prokaryote, there are no chloroplasts, and photosynthesis occurs throughout the cytoplasm. Chloroplasts are found mainly in the cells of the *mesophyl*, the green tissue in the interior of the leaf. The leaf contains pores in its surface called *stomata* that allow carbon dioxide in and oxygen out to facilitate photosynthesis in the leaf. The chloroplast has an inner and outer membrane and within the inner membrane a fluid called the *stroma*. In addition, the interior of the chloroplast contains a series of membranes called the *thylakoid membranes* that form stacks called *grana*.

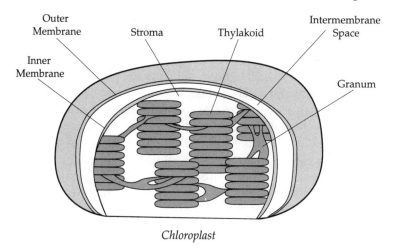

Chloroplast

Photosynthesis can be summarized with this equation:

$$6CO_2 + 12H_2O + uv \text{ light} \longrightarrow C_6H_{12}O_6 + 6O_2 + 6H_2O$$

Photosynthesis involves the reduction of CO_2 to a carbohydrate. It can be characterized as the reverse of respiration, in that reduction of CO_2 to produce glucose occurs instead of oxidation of glucose to make CO_2. One of the by-products of photosynthesis, oxygen, is of keen interest to all of us air-breathers since it is the source of the oxygen that we need to survive.

Photosythesis has two main parts, the *light reaction* and the *Calvin cycle*. The light reaction occurs in the interior of the thylakoid while the Calvin cycle occurs in the stroma. Plants are green because they reflect green color the most. The pigments involved in photosynthesis absorb the most strongly in the red and blue wavelengths.

Light Reactions

The first part of photosynthesis is made up of *light reactions*, in which light energy is used to generate ATP, oxygen, and the reducing molecule NADPH. The molecule that captures light energy to start photosynthesis is a pigment called *chlorophyll* found within the thylakoid membranes of the chloroplast. Chlorophyll is used by two complex systems called *Photosystem I* and *Photosystem II* in the thylakoid membrane. When photons strike chlorophyll, electrons are excited and transferred through the photosystems to a reaction center. As the electrons work their way through this system, their energy pumps protons out of the stroma and into the interior of the thylakoid membranes, creating a proton gradient similar to the gradient created in mitochondria during aerobic respiration. The reactions of the light reactions can be called *photophosphorylation* are grouped into two types of reactions using the two photosystems: *cyclic* and *noncyclic photophosphorylation*.

Cyclic reactions are conducted by Photosystem I and contribute to ATP production but do not produce NADPH. When chlorophyll in Photosystem I is excited, the excited electrons pass through a system of electron carriers that pump protons to build a proton gradient. This proton gradient is used to make ATP (sound familiar?). At the end of the chain, the electron is given back to chlorophyll to be excited once again, making this reaction cyclic.

Noncyclic photophosphorylation requires both photosystems and produces ATP, NADPH and O_2, molecular oxygen. The non-cyclic reactions begin with excitation of chlorophyll in Photosystem II. This excited chlorophyll uses water as an electron donor, producing protons and oxygen. Photosystem II then sends these excited electrons through a chain of redox factors that use the energy to pump protons that will be used to make ATP. At the bottom of the redox chain, the electrons are donated to Photosystem I, which will excite the electrons and use their energy to make NADPH.

A proton gradient is produced by both the cyclic and non-cyclic reactions. The proton gradient in photosynthesis involves pumping protons into the interior of the thylakoids. This proton gradient is used to make ATP similar to the way that a proton gradient is used to make ATP in mitochondria. Protons flow down this chloroplast proton gradient back out into the stroma through an ATP synthase to produce ATP. The NADPH and ATP produced during the light reactions are used to complete photosynthesis in the Calvin cycle, using carbon from carbon dioxide to make sugars. The oxygen produced in the light reactions is released from the plant as a byproduct of photosynthesis. This oxygen helps to maintain the oxygen atmosphere of earth that organisms need for aerobic respiration. Photosynthesis almost certainly helped to create the atmosphere rich in oxygen found today.

Calvin Cycle

The Calvin cycle, also known as the "dark cycle," creates carbohydrates using the energy of ATP and the reducing power of NADPH produced in the light reactions. The carbon used in the creation of carbohydrates comes from atmospheric carbon dioxide, CO_2, so the process is sometimes called carbon fixation. CO_2 first combines with, or "is fixed to," ribulose bisphosphate, a five-carbon sugar with two phosphate groups attached. The resulting six-carbon compound is promptly split, resulting in the formation of two molecules of 3-phosphoglycerate, a three-carbon compound. The 3-phosphoglycerate is then phosphorylated by ATP and reduced by NADPH, which leads to the formation of glyceraldehyde 3-phosphate. This molecule can then be utilized as a starting point for the synthesis of glucose.

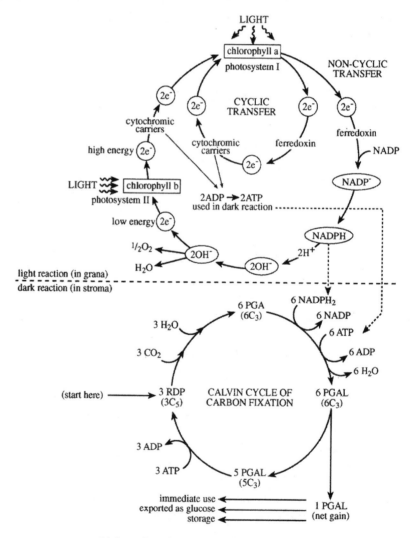

Light and Dark Reactions of Photosynthesis

Plants, animals, and bacteria may differ in their form, biochemistry, and lifestyle, but they all share a common molecular biology that underlies the inheritance and expression of traits. All living organisms inherit traits from their parents and these traits are encoded by the molecule called DNA. Understanding the structure and function of DNA is an important part of understanding all forms of life. In the following section, we'll discuss DNA's role in the genome and gene expression.

The Genome and Gene Expression

By comparing the features of parents with their children, people have long known intuitively that animals transfer genetic traits from one generation to another. Many years ago Mendel pioneered studies of the genetic behavior of traits passed between generations of pea plants. The discovery of the identity of the molecules that store and transfer genetic information is relatively recent, however. *Genes* encode these physical traits, and many scientists once believed that proteins had to be the genetic material, since nucleic acids such as DNA had such simple components it was difficult to see how they DNA could carry such complex information. Through many elegant experiments, however, it was proven that DNA is the genetic material except in certain viruses, and with the elucidation of the structure of DNA by Watson and Crick in 1953, it became clear how DNA could play this role.

The basic outline of information flow in living organisms is sometimes called the *Central Dogma*. The Central Dogma includes the following concepts that are the foundation of modern molecular biology:

1. DNA is the genetic material, containing the genes that are responsible for the physical traits (phenotype) observed in all living organisms.

2. DNA is replicated from existing DNA to produce new genomes.

3. RNA is produced by reading DNA in a process called *transcription*.

4. This RNA serves as the message used to decode and transmit the genetic information and synthesize proteins according to the encoded information. This process of protein synthesis is called *translation*.

DNA Structure

DNA is a polymer built from simple building blocks called *nucleotides*, of which there are four types: *adenine* (A), *guanine* (G), *thymine* (T), and *cytosine* (C). Each nucleotide contains three parts, a five-carbon sugar (deoxyribose), a phosphate group, and a nitrogenous base that distinguishes each of the four nucleotides (A, G, T or C). There are two types of bases: *purines* and *pyrimidines*. The purines are larger, with two rings in each base, and include adenine and guanine. The pyrimidines have one ring and in DNA are

Now You Know Who to Blame

DNA, the molecule of heritability, is what actually causes you to end up with your mother's nose or your father's ears. DNA encodes the genes responsible for these genetic traits.

thymine and cytosine. The size of the bases is important, since this affects the way the bases fit together to make DNA.

To make DNA, nucleotides are polymerized, joined together in long regular strands of nucleotide building blocks. The phosphate group on one nucleotide forms a covalent bond to the sugar group on the next nucleotide to make a phosphate-sugar backbone in the polymer with the base groups projecting to one side, exposed. When a chain of nucleotides is polymerized, it always proceeds from the 5' toward the 3' end with new nucleotides added onto the 3' end, where 5' and 3' (pronounced "5 prime" and "3 prime") refers to the numbering of the sugar ring. On the 5' end of a DNA strand, the 5' OH group on the last deoxyribose sugar in the chain is free, while the 3' OH group is free on the last sugar at the other end of the strand. One polymer strand alone forms half of a DNA double helix—the other half is another strand oriented in the opposite direction (antiparallel). The two strands bind together to form the familiar DNA *double helix* with two strands wrapped around each other.

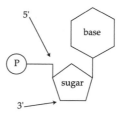

Nucleotide

The binding of two DNA strands to each other requires that the base pairs in each strand hydrogen bond, or *base pair*, to the other strand in a very specific and restricted manner. An A in one strand can only bind to a T in the other.

(A-T) and G in one strand only bind to C in the other strand (G-C). This specificity is determined by the way that the base pairs hydrogen bond with each other, with A and T forming two hydrogen bonds and G-C forming three. Also, each base pair must include one purine (big—two rings) and one pyrimidine (small—one ring) to fit in the space allowed inside the double helix. When the bases in two strands match correctly, the bases stack like plates one on top of the other on the inside of the double helix, with each strand wrapped around the other, and the phosphate-sugar backbones facing outward. The two complementary strands of DNA are always oriented in opposite directions in the double helix, with one strand oriented 5' to 3' and the complementary strand pointing in the other direction. The two strands for this reason are said to be *antiparallel*.

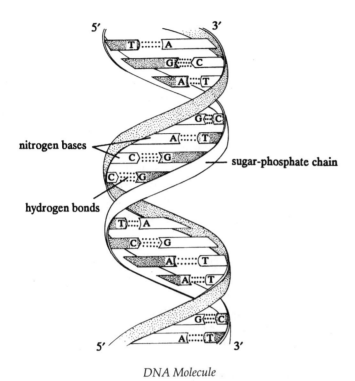

DNA Molecule

DNA Replication

When Watson and Crick built their model for DNA, they felt it must be correct because of the elegance of its design and the solution for DNA replication that the structure immediately suggested. The precise base-pairing of the DNA double helix means that each of the two strands contains a complete and complementary copy of the encoded information found on the other strand. A on one strand means that the opposite strand must contain T in the same position, and so on, throughout the length of each chromosome in the genome. If the two strands are separated, breaking apart the hydrogen bonds between bases that hold the strands together, then a single strand has all of the information needed for a new matching strand. As the new strand of DNA is made, all that must be done is to match up the correct bases on the new strand to have the same base-pairing as the old strand and then link the new nucleotides together. The existing DNA strand that is read to make a new complementary copy is called the *template*.

During cell division, this is exactly what happens. Each daughter cell must get a complete and accurate copy of the genome after mitosis. To replicate the genome, the two strands of DNA are separated one region at a time, and a new DNA strand is synthesized using each of the separated old strands as a template. When the process is complete, two complete copies of the genome have been generated from one copy, with one complete copy of the genome for each daughter cell. This type of DNA replication is called *semi-conservative* since one old strand is maintained (conserved) while a new strand is made from the template of the old strand.

Snowball Effect

Mutations in the DNA repair mechanism enzymes lead to a much higher mutation rate in the organism as a whole.

DNA Repair

The DNA in the genome encodes all of the information cells and organisms need to function. Every enzyme in the cell is encoded in the genome, each with its own coding gene. If there are mistakes in the genome, then defective enzymes will be made and the cell or organism may not be able to function normally. DNA replication and the synthesis of new DNA would introduce errors in the genome if the process were not tremendously accurate. Exposure to certain chemicals, UV light, and radiation can all alter DNA and potentially introduce these sort of harmful mistakes into the genome. During the growth and life of an organism, cells will go through many rounds of cell division, making it all the more important that mistakes are not created in the genome during DNA replication.

The structure of DNA provides a way to keep the genome free of mistakes during DNA replication. If the genome is altered, or a mistake is made during DNA replication, then the base pairs will not fit properly into the normal double helical structure of DNA. There are enzymes that can detect and fix these mistakes, proofreading DNA and correcting the mistakes to form the correct base pairing once again. Occasional mistakes do occur still, resulting in changes in the genome called mutations, but these mistakes are much more unusual than they would be in the absence of DNA repair and proofreading. Individuals lacking effective proofreading mechanisms are much more susceptible to cancer as a result.

The Genetic Code

As part of the Central Dogma, it is stated that DNA contains genes that are transcribed to create messenger RNA, which is in turn translated to make proteins. How do the four base pairs in DNA encode the twenty amino acids found in a protein polypeptide chain? The order of the four base pairs in DNA is the basis of this encoded information, and is called the *genetic code*. Within a gene, every set of three base pairs encodes one amino acid in a polypeptide chain. This three base pair unit of the genetic code is called a *codon*. Since there are four possible base pairs at each of the three positions in a codon, there are a total of $4 \times 4 \times 4$ possible codons = 64 codons. This is in fact what is observed, with 64 different codons used in genes, more than enough to encode the twenty amino acids. Since there are more codons than amino acids, more than one codon encodes a given amino acid in many cases. For example, the codons GAA and GAG both encode the amino acid glutamate in a protein. This use of multiple codons to encode a single amino acid is called *degeneracy*. There is also one codon that indicates the start of a protein and there are three codons that indicate the end of a protein chain (stop codons). The degeneracy of the code means that in many cases mutations can occur in a gene and still not change the resulting protein if the new codon still codes for the same amino acid.

Mutations

The sequence of nucleotides in a gene determines the sequence of amino acids in the resulting protein. Since the function of a protein depends on the sequence of its amino acids, a change in the sequence can change or harm a protein's function. In a gene *mutation*, nucleotides in a gene are added, deleted, or substituted to change the sequence of nucleotides and change the gene. In some cases, inappropriate amino acids are created in a polypeptide chain, and a mutated protein is produced. Hence, a mutation is a genetic "error."

One example of a genetic error in human heredity is phenylketonuria (PKU), a molecular disease that involves the inability to produce the proper enzyme for the metabolism of phenylalanine, resulting in the accumulation of a toxic degradation product (phenylpyruvic acid). Sickle-cell anemia is another condition resulting from an inborn genetic error. Widespread in Africa, this disease cripples red blood cells and is the result of a change in a single amino acid in the hemoglobin gene. In people who carry two copies of the mutant hemoglobin gene, the mutant hemoglobin protein tends to aggregate and disrupt the structure of the cells that carry it. The disorder can be traced to the presence of valine (GUA or GUG) instead of glutamic acid (GAA or GAG) in the hemoglobin in these cells. Curiously, individuals who carry only one copy of the sickle-cell gene are less likely to contract malaria, explaining the prevalence of this gene in regions where malaria is common.

Point Mutations

A *point mutation* occurs when a single nucleotide in DNA is substituted by another nucleotide. If the substitution occurs in a noncoding region, or if the substitution changes a codon into another codon for the same amino acid, there will be no change in the resulting protein's amino acid sequence and the mutation is said to be silent. If the substitution changes the amino acid sequence in a polypeptide, the result can range from insignificant to lethal, depending on the effect the substitution has on the protein. Some changes in a protein's sequence are called *conservative* if an amino is changed to an amino acid that is chemically similar. Conservative changes are less likely to change a protein's function than nonconservative changes.

Frameshift Mutations

Insertions and deletions are two additional types of mutations. Insertions involve the addition of base pairs into the DNA sequence of a gene, while deletions involve a loss of nucleotides. Such mutations usually have serious effects on the protein coded for, since nucleotides are read as a series of triplets. The addition or loss of nucleotide(s) (except in multiples of three) will change the reading frame of the mRNA, and are known as *frameshift mutations*. All of the codons after the frame shift will be read wrong, unless

Hello, Dolly

Recombinant DNA technology forms the scientific basis for cloning, as in the recent cloning of Dolly the sheep.

the frame shifts by three nucleotides. If this occurs, one amino acid will be added or lost, but the rest of the amino acids will be normal.

Recombinant DNA Technology

Until recently, biology was mostly a descriptive science, devoted to observing the many facets of life and attempting to deduce functions and mechanisms. Soon after the structure of DNA was elucidated, however, scientists found tools to manipulate DNA to change genes and organisms in highly specific ways. Not only has this helped scientists to understand life, it has led to a greatly improved understanding of disease and to medicines derived from genes and the proteins they encode.

One thing scientists do to DNA is to take pieces containing genes and move the genes from one DNA to another. To move genes, scientists must have a way to cut out a specific section of DNA that has a gene or a piece of a gene in it. To manipulate DNA and cut out genes, scientists use *restriction enzymes*. Restriction enzymes are proteins that cut the DNA double helix at specific sequences of nucleotides. If two cuts are made, then a piece of DNA can be removed from the chromosome it usually is found in. Restriction enzymes often cut DNA in a way that leaves small segments of single-stranded DNA on the end. This small single-stranded overhang is often called a "sticky end." If the sticky end of one DNA matches the sticky end of another, then the two sticky ends can match and bind to each other and hold the DNA pieces together in a new location. Another enzyme called *ligase* can then seal the DNAs together by covalently joining the strands of DNA in their new location.

If a gene from a human is cut out using restriction enzymes, it can be inserted and ligated into a small circular bacterial DNA called a *plasmid*. If the plasmid is reintroduced into bacteria, the bacteria may then produce the human protein encoded by the human gene. The plasmid serves as a vehicle that carries the desired gene into a new organism and allows the gene to be expressed. A DNA that plays this role of carrying a gene into a new organism is called a *vector*. The same idea is employed in gene therapy. If a human lacks an essential gene due to a genetic inborn condition, then supplying them with the gene might alleviate the illness caused by the absence of the gene. To perform gene therapy, the necessary gene is engineered into a virus that infects human cells, like the adenovirus. If given to patients, the virus can then carry the gene back into humans that lack the gene, curing the disease.

Sequencing DNA allows the entire nucleotide sequence of genes to be known and studied. Today, the genomes of entire organisms are being sequenced, including the sequencing of the human genome in the Human Genome Project. The information from the Genome Project is likely to have profound consequences in the future for medical research.

RNA

In the central dogma, RNA is produced by reading genes from DNA. Like DNA, RNA is a polymer of nucleotides. Both DNA and RNA are nucleic acids and the structure of RNA is very similar to single-stranded DNA. During RNA synthesis, the nucleotides in RNA are matched to base pair with the DNA template similar to the base-pairing of DNA with DNA during DNA replication. There are, however, a number of important differences. These differences include the use of the sugar ribose in the RNA backbone rather than deoxyribose, the presence of the base uracil in RNA rather than thymine, and the fact that RNA is usually single-stranded, while DNA is usually double-stranded. Also, RNA is not proofread when it is made, unlike DNA.

DNA-Unique Features:

- Double-stranded except when replicating

- Deoxyribose sugar in the nucleotides

- Thymine base forms a thymine-adenine base pair (T-A)

- Replicates DNA $\longrightarrow$ DNA

- Only one type of DNA per organism. This DNA acts as the original source of information, acting like a master record. Its information is copied onto RNA molecules.

RNA-Unique Features:

- Nearly always single-stranded

- Ribose sugar in nucleotides

- Uracil base instead of thymine. The The base pair is uracil-adenine (U-A)

- Does not normally replicate (except in the case of some viruses)

- Three types of RNA (mRNA, tRNA, rRNA)

Don't Mix These Up on Test Day

There are three types of RNA, each with its own function:

- *mRNA*, or messenger RNA, carries messages that encode proteins.

- *tRNA*, or transfer RNA, carries amino acids to make proteins.

- *rRNA*, or ribosomal RNA, is a structural component of ribosomes.

There are three types of RNA with distinct functions: *messenger* RNA (mRNA), *ribosomal* RNA (rRNA) and *transfer* RNA (tRNA). mRNA encodes gene messages that are to be decoded in protein synthesis to form proteins. rRNA is a part of the structure of ribosomes and is involved in translation (protein synthesis). tRNAs also play a role in protein synthesis, with an anticodon that recognizes one of the three base pair codons in mRNA and brings the amino acid that matches that codon to the translation process. tRNAs are relatively short, do not encode any proteins, have a compact, complex 3D structure, including base pairing within the molecule, and have one end specialized to be bound to amino acids. The central role of RNAs in key cellular processes is believed by some to support the idea that life originated as an RNA-centered form that later evolved to use protein enzymes and DNA genomes.

Quick Quiz

Match the numbered stages with the correct lettered descriptions below.

1. splicing

2. translation

3. transcription

(A) the stage during which the information coded in the base sequence of DNA is read to produce a strand of mRNA

(B) the stage during which mRNA are shortened to remove parts that do not code for proteins

(C) the process in which the genetic code in mRNA is used to assemble amino acids in the correct sequences to make a protein

Answers:
1. = (B)
2. = (C)
3. = (A)

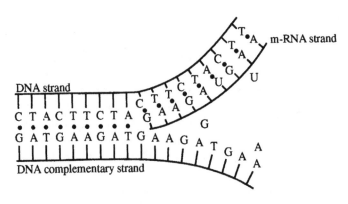

Messenger RNA (mRNA)

Transcription and RNA Processing

Each gene in DNA has the information to make a protein, but DNA does not do this directly. First genes are read to make RNA. *Transcription* is the process in which genes in the DNA genome are used as templates to produce mRNA messages for proteins. The enzyme that synthesizes RNA in transcription, RNA polymerase, uses single-stranded DNA as a template to read the gene, matching base pairs as it synthesizes new RNA from the DNA template (G matching with C and Uracil matching with A). RNA is synthesized like DNA in one direction only, from the 5′ end of the polymer to the 3′ end. The messenger RNA is not proofread as it is produced, however, unlike DNA synthesis.

There are probably about 100,000 genes in the human genome, but not all of these are expressed in every tissue. Genes can be turned on or off by regulating gene transcription according to the needs of the cell and the organism. Transcription is turned on or off by regions of DNA near the start of genes called *promoters*. Promoters are short sequences of DNA that bind proteins called *transcription factors* that regulate transcription of genes by RNA polymerase. Transcription factors bind to specific sequences of DNA in promoters and turn genes on and off in response to hormones or other signaling mechanisms perceived by the cell.

In eukaryotic cells, mRNA is produced in the nucleus and is translated into proteins in the cytoplasm. Before the mRNA is translated, however, it is usually modified in the nucleus. The modifications include the addition of a special cap to the 5′ end of the mRNA, the addition of poly A tail to the 3′ end, and the removal of RNA sections that do not encode a protein, a process called *splicing*. The part of the RNA molecule that encodes the protein message is called an *exon* and the part between coding blocks, the part that is removed, is called the *intron*. Splicing removes introns from mRNA and connects the exons together. Once splicing and processing are complete, the mRNA can be exported from the nucleus through the nuclear pores and is ready for translation to make proteins from the message.

Prokaryotic genes do not have introns and do not go through splicing. In fact, since there is no nucleus in prokaryotes, transcription and translation occur in the same compartment at the same time, with ribosomes translating an RNA before transcription is even complete. Prokaryotic genes are often found with several related genes next to each other in the genome and even transcribed together in the same RNA molecule. These prokaryotic genes are *polycistronic* while eukaryotic genes are *monocistronic*, with only one gene per RNA message.

Translation

Protein translation is the process in which the genetic code in mRNA is used to assemble amino acids in the correct sequence to make a protein. Translation occurs in the cytoplasm. After mRNA is processed and spliced in the nucleus and is ready to be translated, it is exported through a nuclear pore to the cytoplasm. To initiate translation, the mRNA is bound by a ribosome at the site on the mRNA where protein synthesis will begin. The start site of translation is the start codon AUG in the mRNA module, which codes for the amino acid methionine. Since all proteins have AUG as the start codon, all proteins have methionine as the first (or N-terminal) amino acid.

In the processed mRNA, each three-base pair codon codes for a specific amino acid that will be included in the protein amino acid chain. How do ribosomes match amino acids up to the correct codons? There are intermediary molecules, molecular "middle-men," that match each amino acid up to its codons in mRNA. These middle-men are tRNAs. Each tRNA is activated to have a specific amino acid bound covalently at one end. At the other end, the tRNA has a three-base pair region called the *anticodon* that will match up and hybridize to the correct codon in mRNA, base-pairing with the mRNA during translation.

Enzymes called aminoacyl-tRNA synthetases attach amino acids to the correct tRNAs in a very accurate manner. If a tRNA has the wrong amino acid attached, the wrong amino acid will be built into a protein and the wrong protein sequence will be made. This could make the resulting protein unable to do its normal job, in the same manner as a mutation in a gene. There is proofreading in the production of activated tRNAs, but not in protein synthesis once an amino acid is built into a protein chain.

After the ribosome recognizes the first codon, the start codon methionine, it matches up the next codon in the mRNA to the tRNA with the correct anticodon. If the mRNA has the three base pair code GAU for aspartate, then the tRNA for aspartate, with CUA in its anticodon and aspartate bound to the other end, will match up to the codon on the ribosome. The ribosome will then join aspartate to the end of the growing protein chain, move down the mRNA one codon, and start again to match up another tRNA to the next codon in the mRNA. Once the next tRNA is bound to the ribosome, with its anticodon matching the mRNA codon, the next amino acid will be transferred from the bound tRNA to the end of the protein chain.

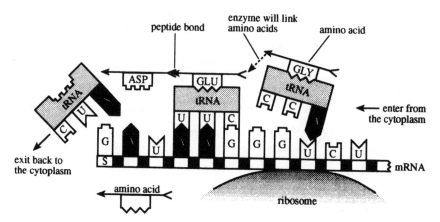

Translation

With each step in protein synthesis, the ribosome matches another tRNA to the correct mRNA codon, adds the next amino acid to the end of the protein chain, forms a peptide bond in the growing polymer, and releases the used tRNA to go back to the cytoplasm. The used tRNA will be recycled by the addition of the correct amino acid once again. When the ribosome reaches a stop codon in the mRNA, it stops translation and releases the mRNA. The mRNA can then be translated again or chewed into pieces by enzymes that degrade mRNA in the cell.

In eukaryotes, translation can occur either on ribosomes in the cytoplasm or on ribosomes bound to the rough endoplasmic reticulum, depending on where signals on the newly forming protein direct it to go. Proteins that will live in the cytoplasm are translated by ribosomes in the cytoplasm. Proteins that are destined for the endoplasmic reticulum (ER), Golgi, or the plasma membrane or that are to be secreted from the cell are synthesized by ribosomes bound to the rough ER. When the protein is synthesized, it is inserted into the ER through the ER membrane. The protein sequence tells the ER where to send the protein. From the ER, the newly synthesized proteins are packaged into small spheres of membrane called vesicles. These vesicles move from the ER to the Golgi, where the proteins are further modified, then on to the plasma membrane where they are either secreted or remain as transmembrane proteins.

Cell Structure and Organization

Cell Theory

The role of the cell in modern biology is so inherent in the way we view life that is easy to overlook the importance of cell theory. Cells were unknown until the development of the microscope in the seventeenth century allowed scientists to see cells for the first time. Matthias Schleiden and Theodor Schwann proposed that all life was composed of cells in 1838, while Rudolph Virchow proposed in 1855 that cells arise only from other cells. The *cell theory* based on these ideas unifies all biology at the cellular level and may be summarized as follows:

- All living things are composed of cells.

- All chemical reactions of life occur in cells or in association with cells.

- Cells arise only from preexisting cells.

- Cells carry genetic information in the form of DNA. This genetic material is passed from parent cell to daughter cell.

Prokaryotic versus Eukaryotic Cells

Prokaryotic Cells

Prokaryotes include bacteria and cyanobacteria (blue-green algae), unicellular organisms with a simple cell structure. These organisms have an outer lipid bilayer cell membrane, but do not contain any membrane-bound organelles, unlike their cousins the eukaryotes. Prokaryotes have no true nucleus and their genetic material consists of a single circular molecule of DNA concentrated in an area of the cell called the nucleoid region.

Prokaryotes may also contain plasmids, small circular extrachromasomal DNAs containing few genes. Plasmids replicate independently from the rest of the genome and often incorporate genes that allow the prokaryotes to survive adverse conditions. Bacteria also have a cell wall, cell membrane, cytoplasm, ribosomes, and, sometimes, flagella that are used for locomotion. Respiration in prokaryotes occurs at the cell membrane, since there are no other membranes present at which a proton gradient could be created for ATP synthesis to take place.

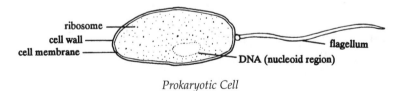

Prokaryotic Cell

Eukaryotic Cells

All multicellular organisms (you, a tree, or a mushroom) and all protists (amoeba or paramecia) are composed of *eukaryotic cells*. A eukaryotic cell is enclosed within a lipid bilayer cell membrane, as are prokaryotic cells. Unlike prokaryotes however, eukaryotic cells contain organelles, membrane-bound structures within the cell that have a specific function. The inside of organelles and the outside are separate compartments. The separation of the organelle membrane and interior from the rest of the cell

Don't Mix These Up on Test Day

Prokaryotes have no nucleus and no membrane-bound organelles, but do have ribosomes and cell walls made up of peptidoglycans.

Eukaryotes have a nucleus, membrane-bound organelles, and ribosomes. Examples include protists, fungi, plants, and animals. Fungi and plant eukaryotic cells have cell walls made of cellulose.

allows organelles to perform distinct functions isolated from other activities, which is not possible in prokaryotes. This prevents incompatible processes from mixing together, allows step-wise processes to be more strictly regulated, and can make processes more efficient by making them happen in a single constrained place. The *cytoplasm* is the liquid inside the cell surrounding organelles.

Although both animal and plant cells are eukaryotic, they differ in a number of ways. For example, plant cells have a cell wall and chloroplasts, while animal cells do not. Centrioles, located in the centrosome area, are found in animal cells but not in plant cells.

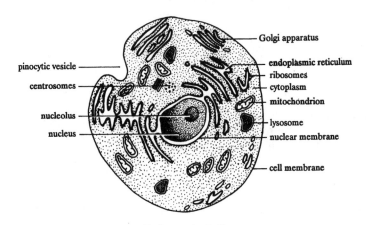

Eukaryotic Cell

Summary of Cell Properties:

Structure	Nucleus?	Genetic Material?	Cell Wall?	Cell Membrane?
Eukaryote	Yes	DNA	Yes/No	Yes
Prokaryote	No	DNA	Yes	Yes

Structure	Membrane Organelles?	Ribosomes?
Eukaryote	Yes	Yes
Prokaryote	No	Yes*

*Ribosomes in prokaryotes are smaller and have a different subunit composition than those in eukaryotes.

Plasma Membrane

The *plasma membrane* is not an organelle but it is an important component of cellular structure. The plasma membrane (also called the *cell membrane*) encloses the cell and exhibits *selective permeability*; it regulates the passage of materials into and out of the cell. To carry out the biochemical activities of life, life must retain some molecules inside the cell and keep other material out of the cell. This is what the selective permeability of the membrane provides.

According to the *fluid mosaic model*, the cell membrane consists of a *phospholipid bilayer*. Phospholipids have both a (water-loving) polar phosphoric acid region and a hydrophobic fatty acid region. If they are mixed with water, phospholipids will spontaneously organize themselves into a lipid bilayer structure, in which the hydrophilic regions are found on the exterior surfaces of the membrane facing water, while the hydrophobic regions are found on the interior of the membrane, facing each other.

The lipid bilayer membrane contains phospholipids, cholesterol and proteins as major components. In a lipid bilayer membrane, lipids and many proteins can move freely sideways within the plane of the membrane. Because of this fluid motion of material in the membrane, this model of membrane structure is called the fluid mosaic model. Cholesterol molecules embedded in the hydrophobic interior contribute to the membrane's fluidity. Proteins interspersed throughout the membrane may be partially or completely embedded in the bilayer; one or both ends of the protein may extend beyond the membrane on either side.

How does the membrane create *selective permeability*, restricting the flow of material into and out of the cell? The lipid membrane itself is one factor responsible for the control of material into and out of the cell, in addition to proteins in the membrane. With the membrane itself, the hydrophobic interior of the membrane prevents charged or very polar molecules from diffusing across the membrane, although noncharged small molecules such as water, oxygen, and carbon dioxide diffuse freely through the membrane. The proteins within the membrane also allow material to pass in and out of the cell. Cell membrane proteins contain both ion channels that act as selective pores for ions and receptors that bind signaling molecules outside of the cell and send signals into the cell. They also carry out the functions of cell adhesion and nutrient transport.

History of the Cell Membrane

As early as 1895, scientists began to create models of cell membranes. But it wasn't until 1972 that Singer and Nicolson proposed that the cell membrane was a phospholipid bilayer with fluid movement within the plane of the membrane.

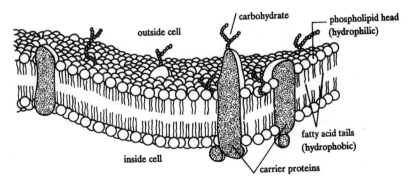

The Plasma Membrane

Organelles

Eukaryotic cells have specialized membrane-bound structures called *organelles* that carry out particular functions for the cell. Organelles include the nucleus, endoplasmic reticulum, Golgi apparatus, lysosomes, microbodies, vacuoles, mitochondria, and chloroplasts. The lipid bilayer membranes that surround organelles also regulate and partition the flow of material into and out of these compartments, just as the plasma membrane does for the cell with its exterior environment.

Nucleus

One of the largest organelles of the cell is the *nucleus*. The nucleus is the site in which genes in DNA are read to produce messenger RNA (transcription), RNA is spliced, and the DNA genome is replicated when the cell divides. Other activities like glycolysis and protein synthesis are excluded from the nucleus. The nucleus is surrounded by a two-layer membrane *nuclear membrane* (or nuclear envelope) that maintains a nuclear environment distinct from that of the cytoplasm. Nuclear pores in this membrane allow selective two-way exchange of materials between the nucleus and cytoplasm, importing some proteins into the nucleus that are involved in transcription, mRNA splicing, and DNA replication, and keeping out other factors like those involved in glycolysis and translation. The nucleus contains the DNA genome complexed with proteins called *histones* involved in packaging DNA and regulating access to genes. DNA packaged with histones is called *chromatin* and forms chromosomes, the highest level of structure in the genome, in which each chromosome contains a fully packaged and immensely long molecule of DNA containing many different genes. The activity of chromosomes in cell division and the role this plays in heredity will be discussed in chapter 5, Classical Genetics.

A dense structure within the nucleus in which ribosomal RNA (rRNA) synthesis occurs is known as the *nucleolus*.

Ribosomes

Ribosomes are not organelles but are relatively large complex structures that are the sites of protein production and are synthesized by the nucleolus. They consist of two subunits, one large and one small; each subunit is composed of rRNA and many proteins. Free ribosomes are found in the cytoplasm, while bound ribosomes line the outer membrane of the endoplasmic reticulum. Prokaryotes have ribosomes that are similar in function to eukaryotic ribosomes, although they are smaller.

Endoplasmic Reticulum

The *endoplasmic reticulum* (ER) is a network of membrane-enclosed spaces connected at points with the nuclear membrane. The network extends in sheets and tubes through the cytoplasm. If this network has ribosomes lining its outer surface, it is termed *rough endoplasmic reticulum* (RER); without ribosomes, it is known as *smooth endoplasmic reticulum* (SER). The ER is involved in the transport of proteins in cells, especially proteins destined to be secreted from the cell. SER is involved in lipid synthesis and the detoxification of drugs and poisons, while RER is involved in protein synthesis. Proteins that are found in the cytoplasm are made by free ribosomes. Proteins that are secreted, found in the cell membrane, the ER, or the Golgi are made by ribosomes on the rough ER. Proteins synthesized by the bound ribosomes cross into the *cisternae* (the interior) of the RER. Small regions of ER membrane bud off to form small round membrane-bound vesicles that contain newly-synthesized proteins. These cytoplasmic vesicles are transported next to the Golgi apparatus.

Golgi Apparatus

The *Golgi* is a stack of membrane-enclosed sacs. It receives vesicles and their contents from the ER and modifies proteins (through glycosylation, the process of adding sugar, for example). Next, it repackages them into vesicles and ships the vesicles to their next stop, such as lysosomes, or the plasma membrane. In cells that are very active in the secretion of proteins, the Golgi is particularly active in the distribution of newly synthesized material to the cell surface. Secretory vesicles, produced by the Golgi, release their contents to the cell's exterior by the process of exocytosis.

Lysosomes

Lysosomes contain hydrolytic enzymes involved in intracellular digestion, degrading proteins and structures that are worn out or not in use. Maximally effective at a pH of 5, these enzymes are enclosed within the lysosome, which has an acidic environment distinct from the neutral pH of the cytosol (the fluid portion of the cytoplasm). Lysosomes fuse with endocytic vacuoles, breaking down material ingested by the cells. They also aid in renewing a cell's own components by breaking them down and releasing their molecular building blocks into the cytosol for reuse.

A cell in injured or dying tissue may rupture the lysosome membrane and release its hydrolytic enzymes to digest its own cellular contents. This is referred to as *autolysis*, and is not common in adult organisms.

Microbodies

Microbodies can be characterized as specialized containers for metabolic reactions. The two most common types of microbodies are *peroxisomes* and *glyoxysomes*. Peroxisomes contain oxidative enzymes that catalyze a class of reactions in which hydrogen peroxide is produced through the transfer of

A Closer Look

Endoplasmic refers to something that is within the cytoplasm, while *reticulum* is derived from a latin word that means "network."

Think of It This Way

It may help to visualize the *Golgi apparatus* as the warehouse of the cell, a place where proteins are packaged for shipment.

The *lysosome* serves as the "stomach" of the cell—both the lysosome and the stomach have acidic pHs.

hydrogen from a substrate to oxygen. These microbodies break fats down into small molecules that can be used for fuel; they are also used in the liver to detoxify compounds, such as alcohol, that may be harmful to the body. Glyoxysomes, on the other hand, are usually found in the fat tissue of germinating seedlings. They are used by the seedling to convert fats into sugars until the seedling is mature enough to produce its own supply of sugars through photosynthesis.

Vacuoles

Vacuoles are membrane-enclosed sacs within the cell. They are formed after endocytosis and can fuse with a lysosome to digest their contents. Contractile vacuoles in freshwater protists pump excess water out of the cell. Plant cells have a large central vacuole called the tonoplast that is part of their endomembrane system. In plants, the tonoplast functions as a place to store organic compounds, such as proteins, and inorganic ions, such as potassium and chloride.

Mitochondria

Mitochondria are sites of aerobic respiration within the cell and are important suppliers of energy. Each mitochondrion has an outer and inner phospholipid bilayer membrane. The outer membrane has many pores and acts as a sieve, allowing molecules through on the basis of their size. The area between the inner and outer membranes is known as the intermembrane space. The inner membrane has many convolutions called cristae, as well as a high protein content that includes the proteins of the electron transport chain. The area bounded by the inner membrane is known as the *mitochondrial matrix*, and is the site of many of the reactions in cell respiration, including electron transport, the Krebs cycle, and ATP production. For more information about the role of mitochondria in energy metabolism, see the earlier portion of this chapter.

Mitochondria are somewhat unusual in that they are semiautonomous; that is, they contain their own circular DNA and ribosomes, which enable them to produce some of their own proteins. In addition, they are able to self-replicate through binary fission. They are believed to have developed from early prokaryotic cells that evolved a symbiotic relationship with the ancestors of eukaryotes and still retain vestiges of this earlier independent life. This hypothesis for the evolution of the eukaryotic cell is called the endosymbiont theory.

Chloroplasts

Chloroplasts are found only in algal and plant cells. With the help of one of their primary components, chlorophyll, they function as the site of photosynthesis. They contain their own DNA and ribosomes and exhibit the same semiautonomy as mitochondria. They are also believed to have evolved via symbiosis. For more information, see the earlier portion of this chapter.

Quick Quiz

What do *mitochondria* and *chloroplasts* do?

Answer:

Chloroplasts capture light energy and use it to make glucose. Mitochondria extract energy from glucose to make ATP.

Cytoskeleton

The cell is not a blob of gelatin enclosed by a membrane bag. The cell has shape, and in some cases actively moves and changes its shape. The cell gains mechanical support, maintains its shape, and carries out cell motility functions with the help of the *cytoskeleton*. The cytoskeleton is composed of *microtubules*, *microfilaments*, *intermediate fibers*, and chains and rods of proteins each with distinct functions and activities.

Microtubules. *Microtubules* are hollow rods made of polymerized tubulin proteins. When polymerized, microtubules radiate throughout the cells and provide it with support. They also provide a framework for organelle movement within the cell. *Centrioles*, which direct the separation of chromosomes during cell division, are composed of microtubules.

Cilia and Flagella. *Cilia* and *flagella* are specialized arrangements of microtubules that extend from certain cells and are involved in cell motility. Prokaryotic flagella are entirely distinct in structure from eukaryotic flagella.

Microfilaments. Cell movement and support are maintained in part through the action of solid rods composed of actin subunits; these are termed *microfilaments*. Muscle contraction, for example, is based on the interaction of actin with myosin in muscle cells. Microfilaments move materials across the plasma membrane; they are active, for instance, in the contraction phase of cell division and in amoeboid movement.

Intermediate Fibers. These structures are a collection of fibers involved in the maintenance of cytoskeletal integrity. Their diameters fall between those of microtubules and microfilaments.

Membrane Transport across the Plasma

It is crucial for a cell to control what enters and exits it. In order to preserve this control, cells have developed the mechanisms described below.

Permeability-Diffusion Through the Membrane

Traffic through the membrane is extensive, but the membrane is selectively permeable; substances do not cross its barrier indiscriminately. A cell is able to retain many small molecules and exclude others. The sum total of movement across the membrane is determined by *passive diffusion* of material directly through the membrane and selective transport processes through the membrane that require proteins.

Most molecules cannot passively diffuse through the plasma membrane. The hydrophobic core of the membrane impedes diffusion of charged and polar molecules. Hydrophobic molecules such as hydrocarbons and oxygen can readily diffuse through the membrane, however. The ability of cells to get oxygen to fuel electron transport depends on the ability of oxygen to diffuse through membranes into the cell and for carbon dioxide to passively diffuse back out again through the cell membrane and into the blood stream. Although it is polar, water is also able to readily diffuse through the membrane. If two molecules are equally soluble, then the smaller molecule will diffuse through the plasma membrane faster. Small, polar, uncharged molecules can pass through easily, but the lipid bilayer is not very permeable to large, uncharged polar molecules like glucose. It is also relatively impermeable to all ions, even small ones such as H+ and Na+, in part because of the large sphere of water molecules around the ion that hydrate it.

Transport Proteins

Molecules that do not diffuse through the membrane can often get in or out of the cell with the aid of proteins in the membrane. Hydrophilic substances avoid contact with the lipid bilayer and still traverse the membrane by passing through *transport proteins*. There are three types of transport proteins: uniport, symport, and antiport. Uniport proteins carry a single solute across the membrane. Symport proteins translocate two different solutes simultaneously in the same direction; transport occurs only if both solutes bind to the proteins. Antiport proteins exchange two solutes by transporting one into the cell and the other out of the cell.

Diffusion/Passive Transport

Diffusion is the net movement of dissolved particles down their concentration gradients, from a region of higher concentration to a region of lower concentration. *Passive diffusion* does not require proteins since it occurs directly through the membrane. Since molecules are moving down a concentration gradient, no external energy is required.

Facilitated Diffusion

The net movement of dissolved particles down their concentration gradient—with the help of carrier proteins in the membrane—is known as *facilitated diffusion*. This process does not require energy. Ion channels are one example of membrane proteins involved in facilitated diffusion, in which the channel creates a passage for ions to flow through the membrane down their concentration gradient. These ions will not flow through the membrane on their own. Some ion channels are always open for ions to flow through them. Other ion channels open only in response to some stimuli. The stimuli that open ion channels might include a change in the voltage across the membrane or the presence of a molecule like a neurotransmitter that opens the channel.

Active Transport

Active transport is the net movement of dissolved particles against their concentration gradient with the help of transport proteins. This process requires energy, and is necessary to maintain membrane potentials in specialized cells such as neurons. The most common forms of energy to drive

Don't Mix These Up on Test Day

Passive transport:
- Moves with the gradient
- Requires no protein
- Requires no energy

Facilitated diffusion:
- Moves with the gradient
- Requires a protein
- Requires no energy

Active transport:
- Moves against the gradient
- Requires a protein
- Requires energy

active transport are ATP or a concentration gradient of another molecule. Active transport is used for uptake of nutrients against a gradient.

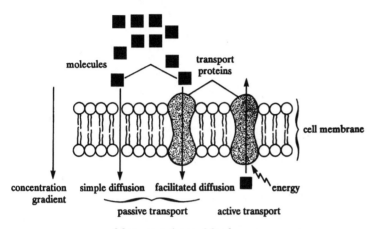

Movement Across Membranes

Osmosis

The process known as *osmosis* is the simple diffusion of water from a region of lower solute concentration to a region of higher solute concentration. Water flows to equalize the solute concentrations. If a membrane is impermeable to a particular solute, then water will flow across the membrane until the differences in the solute concentration have been equilibrated. Differences in the concentration of substances to which the membrane is impermeable affect the direction of osmosis.

Water diffuses freely across the plasma membrane. When the cytoplasm of the cell has a lower solute concentration than the extracellular medium, the medium is said to be *hypertonic* to the cell; water will flow out, causing the cell to shrink. On the other hand, when the cytoplasm of a cell has a higher solute concentration than the extracellular medium, the medium is *hypotonic* to the cell, and water will flow in, causing the cell to swell. If too much water flows in, the cell may lyse. Red blood cells, for example, lyse when put into distilled water. Finally, when solute concentrations are equal inside and outside, the cell and the medium are said to be *isotonic*. There is no net flow of water in either direction.

Don't Mix These Up on Test Day

In a *hyp**o**tonic* solution, a cell will swell until it looks like an O, and eventually burst.

In a *hyp**er**tonic* solution, a cell will shrivel.

In an *isotonic* solution, a cell will remain the same size.

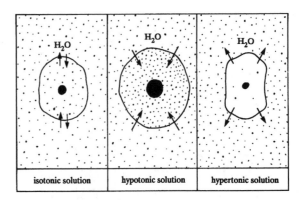

Osmosis

Endocytosis/Exocytosis

Endocytosis is a process in which the cell membrane is invaginated, forming a vesicle that contains extracellular medium. Meanwhile, *pinocytosis* is the ingestion of liquids or small particles, while *phagocytosis* is the term assigned to the engulfing of large particles. In the latter, articles may first bind to receptors on the cell membrane before being engulfed.

These processes differ from *exocytosis*, which occurs when a vesicle with the cell fuses with the cell membrane and releases its contents to the outside. This fusion of the vesicle with the cell membrane can play an important role in cell growth and intercellular signaling. Exocytosis is used for cells to release secreted proteins into their exterior, for example. In both endocytosis and exocytosis, the material inside the vesicle never actually crosses the cell membrane but is enclosed within the membrane of the vesicle and kept separate from the cytoplasm. Endocytosis is usually mediated by cell-surface receptors that are internalized along with the membrane vesicle.

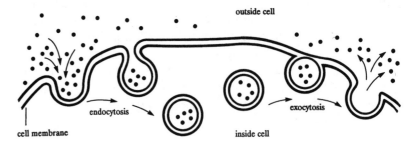

Endocytosis and Exocytosis

Now that you've worked through this chapter, you shold have a good understanding of important concepts in cellular and molecular biology. Test yourself with the short quiz that follows.

Cellular and Molecular Biology Quiz

1. Which of the following processes utilizes proofreading to increase its accuracy?

   (A) Transcription
   (B) DNA replication
   (C) Recombination
   (D) Peptide bond formation
   (E) Electron transport

2. Which of the following is NOT an organelle?

   (A) nucleus
   (B) Golgi apparatus
   (C) lysosome
   (D) chlorophyll
   (E) chloroplast

3. Which is correctly associated?

   (A) RNA: thymine
   (B) DNA: uracil
   (C) RNA: replication
   (D) mRNA: picks up amino acids
   (E) RNA: ribose sugars

4. What is the best evidence that genes encode the amino acid sequence of proteins?

   (A) Proteins are macromolecules.
   (B) RNA directs amino acid synthesis.
   (C) The amino acid sequence of polypeptides is changed by gene mutation.
   (D) DNA serves as a template for RNA.
   (E) mRNA is found in the ribosome.

5. The tRNA anticodon for the amino acid valine is CAA (reading $3'$ to $5'$). What is the mRNA codon for valine (reading $5'$ to $3'$)?

   (A) TTG
   (B) GGU
   (C) CCA
   (D) CCG
   (E) GUU

6. The source of oxygen given off in photosynthesis is

   (A) water
   (B) carbon dioxide
   (C) glucose
   (D) starch
   (E) chlorophyll

7. Which of the following is NOT a lipid derivative?

   (A) waxes
   (B) steroids
   (C) carotenoids
   (D) albumins
   (E) lecithin

8. Which is NOT a characteristic of proteins?

   (A) They contain genetic information.
   (B) They can act as hormones.
   (C) They can catalyze chemical reactions.
   (D) They act in cell membrane trafficking.
   (E) They can bind foreign materials.

GO ON TO THE NEXT PAGE

9. The rough endoplasmic reticulum differs from the smooth endoplasmic reticulum due to the presence of

   (A) lysosomes
   (B) ribosomes
   (C) mitochondria
   (D) Golgi apparati
   (E) histones

10. Which of the following is found in eukaryotes but not in prokaryotes?

    (A) ribosomal RNA
    (B) plasma membrane
    (C) nuclear membrane
    (D) ribosomes
    (E) none of the above

11. Which of the following statements regarding photosynthesis is NOT true?

    (A) The light cycle occurs only during exposure to light.
    (B) The dark cycle occurs only in the absence of light.
    (C) ATP is produced during the light cycle.
    (D) During the dark cycle, sugars are produced.
    (E) Red and blue light are optimal for photosynthetic function.

12. Which of the following statements about the Krebs cycle is NOT true?

    (A) The Krebs cycle occurs in the matrix of the mitochondrion.
    (B) The Krebs cycle is linked to glycolysis by pyruvate.
    (C) The Krebs cycle is the single greatest direct source of ATP in the cell.
    (D) Citrate is an intermediate in the Krebs cycle.
    (E) The Krebs cycle produces nucleotides such as NADH and $FADH_2$.

13. Cells that are involved in active transport, such as cells of the intestinal epithelium, utilize large quantities of ATP. In such cells, there are

    (A) high levels of adenylate cyclase activity
    (B) many polyribosomes
    (C) many mitochondria
    (D) high levels of DNA synthesis
    (E) many lysosomes

14. The process by which a cell engulfs large particulate matter is called

    (A) pinocytosis
    (B) exocytosis
    (C) cytokinesis
    (D) phagocytosis
    (E) osmosis

GO ON TO THE NEXT PAGE

KAPLAN

15. The basis for the pairing of the two strands of DNA in the double helix is

    (A)   covalent bonding

    (B)   ionic bonding

    (C)   hydrogen bonding

    (D)   hydrophobic interactions

    (E)   tertiary structure

16. Which of the following statements about enzymes is NOT true?

    (A)   The activity of enzymes is unaffected by genetic mutation.

    (B)   Enzymes may interact with nonprotein molecules in order to engage in biological activity.

    (C)   Enzymes optimally operate at a particular pH.

    (D)   Enzymes optimally operate at a particular temperature.

    (E)   Enzymes are almost always proteins.

17. Which of the following is a correct association?

    (A)   mitochondria: transports materials from the nucleus to the cytoplasm

    (B)   lysosome: digestive enzymes for intracellular use

    (C)   endoplasmic reticulum: selective barrier for the cell

    (D)   ribosome: electron transport chain

    (E)   polysome: group of Golgi complexes

18. Which of the following is in a different chemical category than the others?

    (A)   cytosine

    (B)   thymine

    (C)   arginine

    (D)   guanine

    (E)   uracil

STOP

Answers and Explanations to the Cellular and Molecular Biology Quiz

1. **(B)** DNA replication is a highly accurate process in part because of the proofreading of newly synthesized DNA that occurs, remopving and fixing base mismatches. This is not true of RNA synthesis during transcription or any of the other processes.

2. **(D)** Chlorophyll is an essential component of an organelle, the chloroplast, but it is a chemical, not an organelle.

3. **(E)** RNA is made up of a ribose sugar bound to a phosphate group, which is then bound to one of the four bases. It is also characterized by having uracil as one of its bases rather than thymine (as in (A)). DNA, meanwhile, utilizes thymine rather than uracil (B). As for (C), RNA does not replicate in eukaryotic cells. Although RNA will be used to synthesize DNA in some retroviruses, this is DNA replication. In (D), tRNA carries the amino acids to the ribosome, while the mRNA carries the message of the protein that is to be produced to the ribosome. Neither are associated with ribose sugars.

4. **(C)** A point or frameshift mutation in a gene will be evidenced by a corresponding mutation in the protein, such as a change in the amino acid sequence of polypeptides. (A), on the other hand, has nothing to do with how DNA might control the synthesis of a protein, and (B) is not the most direct connection between genes and peptides. (D) and (E) are both true, but are not the best answers; they could be nothing more than coincidental occurrences and do not fully support the statement.

5. **(E)** mRNA includes a coded base sequence called a codon. tRNA, which carries the amino acids to the mRNA and the ribosome, has a complementary strand called the anticodon. Therefore, a tRNA with an anticodon of CAA should match the mRNA codon GUU. Remember that in RNA, adenine bonds with uracil, and guanine bonds with cytosine. Also, nucleic acid strands always bond with one strand 5′ to 3′ and the other in the opposite direction of 3′ to 5′.

6. **(A)** In the light reaction, light splits H_2O into excited electrons, H^+ and O_2. The excited electrons go on to form ATP and the H^+ electrons are incorporated into the carbohydrates produced during the dark reaction. O_2 is released into the environment as a waste product of this reaction. In (B), CO_2 donates the carbon and the oxygen required for carbohydrate formation in the dark reaction. (C) and (D) are end products of photosynthesis, and (E) chlorophyll is involved in the initial capture of sunlight.

7. **(D)** Albumins are not lipid derivatives; they are globular proteins. Waxes (A) are esters of fatty acids and monohydroxylic alcohols. They are found as protective coatings on skin, fur, leaves of higher plants, and on the cuticle of the exoskeleton of many insects. Steroids (B) such as cholesterol and estrogen have three fused cyclohexane rings and one fused cyclopentane ring. Carotenoids (C) are fatty, acidlike carbon chains containing conjugated double bonds and carrying six-membered carbon rings at each end. These compounds are pigments, and produce red, yellow, orange, and brown colors in plants and animals. Finally, lecithin (E) is an example of a phospholipid. Phospholipids contain glycerol, two fatty acids, a phosphate group, and nitrogen-containing alcohol.

8. **(A)** DNA is the only molecule in eukaryotes that contains genes. Proteins may function as hormones (chemical messengers), enzymes (catalysts of chemical reactions), structural proteins (providers of physical support), transport proteins (carriers of important materials), and antibodies (binders of foreign particles).

9. **(B)** Ribosomes, the site of protein production, give the rough ER its characteristic appearance. Sections of the ER are lined with ribosomes, where proteins are produced and then transported to the appropriate areas. (A) Lysosomes are membrane-bound organelles in the cytoplasm with very low pHs. Filled with proteolytic enzymes, they are the site of macromolecule degradation in the cell. As for (C), mitochondria, another type of membrane-bound organelle, serve as the site of cellular respiration. The Golgi apparatus (D) is also a membrane-bound organelle, and follows the ER in the production of proteins. It is here that proteins are glycosylated, post-translationally modified, and packaged. In the final choice, (E), histones are proteins in the nucleus which bind to DNA like "beads on a string."

10. **(C)** One basic difference between eukaryotes and prokaryotes is that eukaryotes possess membrane-bound organelles, while prokaryotes do not. There is no prokaryotic nucleus. Both of the cells types in (A) and (B) have ribosomal RNA (although it differs in composition and size) and both have plasma membranes.

11. **(B)** Photosynthesis occurs in two steps. Step one is the light reaction in which visible light, especially that in the red/blue wavelengths, produces ATP and $NADPH_2$ through the splitting of CO_2 and H_2O. O_2 is produced during the splitting of water. During the dark cycle, the second step, carbohydrates such as glucose are synthesized whenever $NADPH_2$, ATP, and CO_2 are present, and this will occur regardless of the presence or absence of light.

12. **(C)** The single greatest direct source of ATP in the cell is the proton gradient created by the electron transport chain, not the Krebs cycle. The Krebs cycle does occur in the matrix of the mitochondria (A), however, and oxidative phosphorylation (the electron transport chain) does occur in the inner membrane of the mitochondria known as the cristae. Citrate is an intermediate in this cycle (D). The Krebs cycle only forms two ATP directly; all of the other ATP that form during this cycle are pro-

duced when NADH and $FADH_2$ donate their electrons to the electron transport chain, which pump protons so that ATP synthase can use the proton gradient to make ATP.

13. **(C)** Cells that are involved in active transport, such as the epithelial cells of the intestine, will require large amounts of ATP. If a cell utilizes large amounts of ATP, it must produce it in many mitochondria. In (A), high levels of adenylate cyclase activity are found in cells that are the target cells for hormone activation. Many polyribosomes (B), meanwhile, are found in cells that have a high level of protein synthesis. As for (D), high levels of DNA synthesis are found in cells that undergo rapid reproduction and mitosis, and many lysosomes would be found in phagocytic cells, enabling them to digest the foreign material they have endocytosed (E).

14. **(D)** Phagocytosis is the process of engulfing large matter, such as a bacterium. Meanwhile, pinocytosis (A) is the process of taking in small amounts of liquid, and exocytosis (B) is the term given to the release of proteins from the cell. In (C), cytokinesis is the division of the cytoplasm during mitosis, while osmosis is the movement of water from an area of lower solute to an area of higher solute concentration (E).

15. **(C)** DNA is a double-stranded helix, composed of the purines adenine and guanine and the pyrimidines cytosine and thymidine. Adenine binds with thymidine, while guanine binds cytosine via weak hydrogen bonds. These weak bonds enable the helices to separate easily to facilitate DNA replication. (A) Covalent bonding is characterized by shared electron pairs, while (B) ionic bonding is characterized by electron transfer. These are both strong forms of intermolecular bonds. In the last two choices, (D) hydrophobic interactions are attractive forces between nonpolar molecules, while (E) tertiary structure is found in proteins.

16. (A) Enzymes are biological catalysts usually composed of proteins. They work at an optimal temperature and pH, typically around the physiological temperature of 37° C and a pH of 7.2. At higher temperatures, the proteins will denature and lose their function. They will often have to interact with cofactors such as vitamins or ions for optimal activity. Mutation will also affect the DNA sequence coding for these proteins, leading to an altered polypeptide and, often, a change in spatial configuration, which leads in its turn to a change or loss in function.

17. (B) Lysosomes are membrane-bound organelles containing digestive enzymes. Typically, they have a low pH. (A) Mitochondria, on the other hand, are involved in cellular respiration, while (C) the ER transports polypeptides around the cell and to the Golgi apparatus for packaging. Finally, (D) the ribosome is the site of protein synthesis, and (E) polysomes are groups of ribosomes that make large quantities of a particular polypeptide.

18. (C) Of the compounds listed, (A), (B), (D), and (E) are all nitrogenous bases. They are present in RNA and DNA. The nitrogenous bases guanine and adenine are purines, and the nitrogenous bases cytosine, thymine, and uracil are pyrimidines. Arginine (C) is an amino acid and not a nitrogenous base. Amino acids are the building blocks for proteins.

CHAPTER FOUR

ORGANISMAL BIOLOGY

Living organisms must maintain constant interior conditions in a changing environment. The interior environment that cells must maintain includes water volume, salt concentration, and appropriate levels of oxygen, carbon dioxide, toxic metabolic waste products, and essential nutrients. Organisms must also respond to their environment to avoid harm and seek out beneficial conditions. In addition, organisms must reproduce themselves. Single-cell organisms like prokaryotes or protists have relatively simple ways to meet these needs. Multicellular organisms have evolved more complex body plans that provide a variety of solutions to the common problems that all organisms face.

As multicellular organisms have evolved into larger and more complex forms over time, their cells have become more removed from the external environment and more specialized toward one specific need. These specialized cells form *tissues*, cells with a common function and often a similar form. Cells from different tissues come together to form *organs*, large anatomical structures made from several tissues working together toward a common goal. Organs in turn are part of organ systems, including systems for digestion, respiration, circulation, immune reactions, excretion, reproduction, the nervous system and the endocrine system.

The following chapter will give you the basics of topics including reproduction, physiology, and animal behavior. Along the way, organisms such as protozoans, cnidarians (hydra), annelids (earthworms), and arthropods (insects) will be described as prototypes to illustrate the range of solutions evolution has provided in different types of organisms. The unique solutions of plants will be presented separately later in the chapter.

Reproduction

One of the essential functions for all living things is the ability to reproduce, to produce offspring that continue a species. An individual organism can survive without reproduction but a species without reproduction will not survive past a single generation. Reproduction in eukaryotes can occur as

Believe It or Not

It's hard to believe, but true—all the nucleated cells of your body, regardless of structure or function, have *exactly* the same number of chromosomes (including two sex chromosomes). The only exceptions are gametes, which have half the normal number of chromosomes. This means that different cell types have different structures and functions not because their DNA is different, but because what is *expressed* by that DNA is different.

either asexual or sexual reproduction. Prokaryotes have a different mechanism called binary fission for reproduction.

Mechanisms of Cell Division

One of the inherent features in reproduction is cell division. Prokaryotic cells divide, and reproduce themselves, through the relatively simple process of binary fission. Eukaryotic cells divide by one of two mechanisms: mitosis and meiosis. *Mitosis* is a process in which cells divide to produce two daughter cells with the same genomic complement as the parent cell; in the case of humans there are two copies of the genome in each cell. Mitotic cell division can be a means of asexual reproduction, and also is the mechanism for growth, development, and replacement of tissues. *Meiosis* is a specialized form of cell division involved in sexual reproduction that produces male and female gametes (sperm and ova, respectively). Meiotic cell division creates cells with a single copy of the genome in preparation for sexual reproduction in which gametes join to create a new organism with two copies of the genome, one from each parent.

Prokaryotic Cell Division and Reproduction

Prokaryotes are single-celled organisms and their mechanism for cell division, *binary fission*, is also their means of reproduction. As with all forms of cell division, one of the key steps is DNA replication. Prokaryotes have no organelles and only one chromosome in a single long circular DNA. The single prokaryotic chromosome is attached to the cell membrane, and replicated as the cell grows. With two copies of the genome attached to the membrane after DNA replication, the DNAs are drawn apart from each other as the cell grows in size and add more membrane between the DNAs. When the cell is as big as two cells, the cell wall and membrane close off to create two independent cells. The simplicity of prokaryotic cells and the small size of their genome compared to eukaryotes may be a factor that assists in their rapid rate of reproduction, dividing as rapidly as once every thirty minutes under ideal conditions.

Bacteria and other prokaryotes do not reproduce sexually, but they do exchange genetic material with each other in some cases. *Conjugation* is one mechanism used by bacteria to move genes between cells by exchanging a circular extrachromosomal DNA with each other. In a process called *transduction*, viruses that infect bacteria can accidentally carry bacterial genes with them into a new cell that they infect. These processes can introduce new genes into bacteria, but do not involve the union of gametes from two parents that is involved in sexual reproduction.

Mitosis

Eukaryotic cells use *mitosis* to divide into two new daughter cells with the same genome as the parent cell. The growth and division of cells to make new cells occurs in what is known as the *cell cycle*. The cell cycle is a highly regulated process, linked to the growth and differentiation of tissues. Growth factors can stimulate cells to move through the cell cycle more rapidly, and other factors can induce cells to differentiate and stop moving forward through the cell cycle. Failure to control the cell cycle properly can result in uncontrolled progression through the cell cycle and cancer. Cancer cells contain mutations in genes that regulate the cell cycle.

The four stages of the cell cycle are designated as G_1, S, G_2, and M. The first three stages of this cell cycle are interphase stages—that is, they occur between cell divisions. The fourth stage, mitosis, includes the actual division of the cell.

A Typical Cell's Day

Typically, cells spend 90 percent of their time in interphase, and only 10 percent in the mitotic phase.

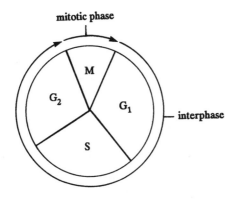

The Cell Cycle

Stage G_1. G_1 is characterized by intense biochemical and biosynthetic activity and growth. The cell doubles in size, and new organelles such as mitochondria, ribosomes, and centrioles are produced.

Stage S. This is the stage during which synthesis of DNA takes place (S is for synthesis). Each chromosome is replicated so that during division, a complete copy of the genome can be distributed to both daughter cells. After replication, the chromosomes each consist of two identical sister *chromatids* held together at a central region called the *centromere*. The ends of the chromosomes are called *telomeres*. Cells entering G_2 actually contain twice as much DNA as cells in G_1, since a single cell holds both copies of the replicated genome.

Stage G_2. The cell prepares for mitosis, making any of the components still needed to complete cell division.

Stage M (Mitosis and Cytokinesis). Mitosis, the stage in which the cell divides to create two similar but smaller daughter cells.

Mitosis is further broken down into four stages: *prophase, metaphase, anaphase,* and *telophase.* Upon completion of mitosis, the cell completes its split into daughter cells via *cytokinesis.*

In a Nutshell

Mitosis proceeds in the following stages:

- *Prophase*: chromosomes condense, spindles form
- *Metaphase*: chromosomes align
- *Anaphase*: sister chromatids separate
- *Telophase*: new nuclear membranes form

Prophase. During interphase, the chromosomes in the nucleus are very extended despite their packaging into chromatin and it is not possible to see the chromosomes even under a microscope. *Prophase* begins with the condensation and separation of chromosomes so that they are visible in a microscope as dark solid bands if stained. Cell division cannot occur without this condensation or the chromosomes become tangled with each other. Another step in prophase is the separation of the *centrioles*, centers of microtubule formation. As the centrioles separate, they form the *spindle apparatus*, a specialized system of microtubules that spans between the centrioles during mitosis. As the centrioles separate toward opposite poles of the cell, the spindle apparatus elongates between them. The nuclear membrane dissolves and the nucleolus becomes indistinct, allowing the spindle fibers access to the chromosomes. The spindle fibers attach themselves to each chromosome with a structure called the *kinetochore* that attaches in the middle of each chromosome at the centromere, setting the stage for metaphase.

Metaphase. In *metaphase*, the two centrioles are at opposite poles of the cell, with the spindle fully elongated between them, spanning the length of the cell, including the space previously occupied by the nucleus and now the location of the condensed chromosomes. The kinetochore fibers attached to the chromosomes at the centromere align the chromosomes at the metaphase plate, a plane halfway between the two ends of the cell. The alignment of the chromosomes prepares the cell to pull the chromosomes apart toward the two ends of the cell.

Anaphase. During S phase, each chromosome is replicated and the replicated copies stay bound together. From S phase up through metaphase, each chromosome contains two replicate copies, and each copy is called a sister chromatid. During *anaphase*, the two sister chromatids in each chromosome are pulled apart by the kinetochore and spindle fibers. The telomeres at the ends of the chromosomes are the last part of the chromatids to separate. Each chromatid then has its own centromere by which it is pulled towards opposite poles of the cell through the shortening of the kinetochore fibers. It is crucial that when the sister chromatids separate, each daughter cell receives one of each chromosome. If this does not happen, the cell lacking a chromosome or the other cell with an extra chromosome may not function normally if it is lacking a copy of many genes or has an extra copy of genes.

Telophase. At the beginning of *telophase*, the sister chromatids have been pulled apart so that one copy of each chromosome, a complete copy of the genome, is at one end of the cell and another copy is at the other end. At this point, the spindle apparatus disappears, a nuclear membrane reforms around each set of chromosomes, and the nucleoli reappear. The chromosomes uncoil, resuming their spread out interphase form. Each of the two new nuclei contains a complete copy of the genome identical to the original genome and to each other.

Cytokinesis. After the cell has divided its DNA in the processes described above, it enters the final stage of mitosis. In cytokinesis, the cytoplasm and all the organelles of the cell are divided as the plasma membrane pinches inward and seals off to complete the separation of the two newly formed daughter cells from each other.

Asexual Reproduction

Asexual reproduction is any method of producing new organisms in which fusion of nuclei from two individuals (fertilization) does not take place. The fusion of nuclei from two parent individuals to create a new individual is *sexual reproduction*. In asexual reproduction, only one parent organism is involved. The new organisms produced through asexual reproduction form daughter cells through mitotic cell division and are genetically identical, clones of their parents. Asexual reproduction serves primarily as a mechanism for perpetuating primitive organisms and plants, especially in times of low population density. Asexual reproduction can allow more rapid population growth than sexual reproduction, but does not create the great genetic diversity that sexual reproduction does.

Binary fission occurs in prokaryotes, algae, and bacteria. In this process, a single DNA molecule attaches to a plasma membrane during replication and duplication, while the cell continues to grow in size. Hence each daughter cell receives a complete copy of the original parent cell's chromosomes.

This type of reproduction occurs at a rapid pace. Undesirable, potentially harmful bacteria cells, for example, can reproductive every 20 minutes under optimal.

Don't Mix These Up on Test Day

Binary fission results in equal division of the cytoplasm and symmetrical daughter cells.

Budding involves an unequal division of the cytoplasm and asymmetrical daughter cells.

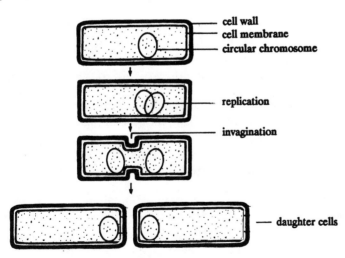

cell wall
cell membrane
circular chromosome

replication

invagination

daughter cells

Binary Fission

Budding is a form of mitotic asexual reproduction that involves an unequal division of cytoplasm (*cytokinesis*) between the daughter cells and equal division of the nucleus (*karyokinesis*). The parent cell forms a smaller daughter cell that sprouts off with less cytoplasm than the parent.

Eventually the daughter organism becomes independent and is released. Although budding is common in unicellular organisms like yeast, it also occurs in some multicellular organisms such as hydra, forming small identical copies of the parent.

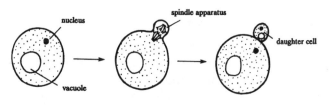

Budding

The Eggs of the Rotifer

The rotifer, a tiny aquatic animal, has two types of eggs that develop via parthenogenesis. These eggs mature to produce females and degenerate males that cannot feed themselves. These males produce sperm that will fertilize eggs to produce zygotes. These zygotes are so hardy that they can survive even if the pond in which they live dries up.

Asexual reproduction is not common in animals, although it does bring with it certain benefits. It is suitable for animal populations that are widely dispersed, as animals that practice asexual reproduction do not need to find another animal to fertilize them sexually. Asexual reproduction allows rapid growth of a population when conditions are suitable and has a much lower energetic cost than sexual reproduction. The two major types of asexual reproduction found in animals are *parthenogenesis* and *regeneration*.

When people think about asexual reproduction in animals, they usually have *parthenogenesis* in mind. During sexual reproduction, an ova does not develop without fertilization by a sperm, and the resulting zygote contains a diploid genome with one copy from each parent. In parthenogenesis, an egg develops in the absence of fertilization by sperm through mitotic cell division. This form of reproduction occurs naturally in bees: fertilized eggs develop into worker bees and queen bees, while unfertilized eggs become male drone bees. Artificial parthenogenesis can be performed in some animals. The eggs of rabbits and frogs, for example, can be stimulated to develop without fertilization, by giving them an electric shock or a pin prick.

Regeneration is the ability of certain animals to regrow a missing body part. Sometimes parts of an animal grow into a complete animal, resulting in reproduction. For example, the planaria (a flatworm), the earthworm, the lobster, and the sea star can all regenerate limbs or entire organisms. This process is similar in nature to vegetative propagation.

Both regeneration and vegetative propagation are poorly understood. One possible explanation is that both are enabled by the presence of undifferentiated stem cells in certain plants or animals; if stimulated properly, these stem cells can differentiate as they divide and grow to form new organs. Improved understanding of stem cells and the ability to use them to improve regeneration could lead to revolutionary advances in medicine, such as regeneration of damaged spinal nerves after a crippling accident, regeneration of heart tissue after a heart attack, healing of brain tissue after a stroke, or healing of limbs.

Sexual Reproduction

Most multicellular animals and plants reproduce sexually, as do as many protists and fungi. *Sexual reproduction* involves the union of a haploid cell from two different parents to produce diploid offspring. These haploid cells are the *gametes*, sex cells produced through mitosis in males and females. Gametes have a single copy of the genome (one of each chromosome), and diploid cells have two copies of the genome (two of each chromosome). In humans, all of the cells of the body are diploid, with the exception of the gametes. When the male gamete (the sperm) and the female gamete (the egg) join, a *zygote* is formed that develops into a new organism genetically distinct from both its parents. The zygote is the diploid single cell offspring formed from the union of gametes.

Sexual reproduction ensures genetic diversity and variability in offspring. Since sexual reproduction is more costly in energy than asexual reproduction, the reason for its overwhelming prevalence must be that genetic diversity is worth the effort. Sexual reproduction does not create new alleles, though. Only mutation can do that. Sexual reproduction increases diversity in a population by creating new combinations of alleles in offspring and therefore new combinations of traits. Genetic diversity is not an advantage to an individual, but allows a population of organisms to adapt and to survive in the face of a dynamic and unpredictable environment.

The diversity created by sexual reproduction occurs in part during meiotic gamete production and in part through the random matching of gametes to make unique individuals. The range of mechanisms involved in sexual reproduction in animals, including humans, are detailed below.

Gamete Formation

Specialized organs called *gonads* produce gametes through meiotic cell division. Male gonads, *testes*, produce male gametes, *spermatozoa*, while female gonads, *ovaries*, produce *ova*. A cell that is committed to the production of gametes, although it is not itself a gamete, is called a *germ cell*. The rest of the cells of the body are called *somatic cells*. Only the genome of germ cells contributes to gametes and offspring. A mutation in a somatic cell, for example, may be harmful to that cell or the organism if it leads to cancer, but a mutation in a somatic cell will not affect offspring since the mutation will not be found in germ cell genomes. Germ cells are themselves diploid and divide to create more germ cells by mitosis, but create the haploid gametes through meiosis.

The production of both male and female gametes involves meiotic cell division. Meiosis in both spermatogenesis and oogenesis involves two rounds of cell division, in which a single diploid cell first replicates its genome, then divides once into two cells each with two copies of the genome. Without replicating their DNA, these two cells divide again to produce four haploid gametes. Meiosis in both cases also involves recombination between the homologous copies of chromosomes during the first round of meiotic cell division. This recombination is one of the key sources of genet-

Hermaphrodites

Organisms like the hydra and the earthworm are *hermaphrodites*, with both functional male and female gonads.

ic diversity provided during sexual reproduction and is discussed in more detail in the genetics chapter.

There are also many differences in meiosis as conducted during sperm production and oogenesis, as outlined below.

Human Male Reproductive System

The human male produces sperm in the *testes*, gonads located in an outpocketing of the abdominal wall called the *scrotum*. The sperm develop in a series of small, coiled tubes within the testes called the *seminiferous tubules*. *Sertoli cells* in the seminiferous tubules support the sperm and *Leydig cells* make the *testosterone* that supports male secondary sex characteristics. The *vas deferens* carry sperm to the urethra that passes through the penis. During ejaculation, the *prostate gland* and *seminal vesicles* along the path add secretions to the sperm that carries and provides nutrients for the sperm as part of *semen*.

As gonads, the testes have a dual function; they produce both sperm and male hormones (such as testosterone). Leydig cells in the testis secrete testosterone beginning in puberty. *Testosterone* and other steroid hormones collectively called *androgens* induce secondary sexual characteristics of the male, such as facial and pubic hair, changes in body shape, and deepening voice changes.

Spermatogenesis is the meiotic development of sperm in males. Sperm production occurs throughout adult life in males, and meiosis in sperm production is continuous, proceeding forward without a significant pause. In the testes, diploid germ cells divide mitotically to create primary *spermatocytes*, which continuously undergo meiosis to form four haploid *spermatids* from each primary spermatocyte. The four spermatids are equivalent in size and function and all four result in viable gametes. Spermatids must mature further to develop the head with DNA and the tail for motility that are found in mature sperm. A specialized sac at the tip of the sperm called the *acrosome* is full of enzymes that allow the sperm to break through the protective layers around the egg. One strategy to develop birth control has been to inhibit these enzymes so that the sperm cannot reach the egg. The testes are located outside the abdominal cavity because they must remain 2–4 degrees cooler than the rest of the body to ensure proper development of sperm.

Don't Mix These Up on Test Day

Spermatogenesis:

- Produces 4 mature sperm; each sperm has an X or Y chromosomes and does not donate mitochondria to the embryo
- Is a continuous process
- Produces fresh sperm daily

Oogenesis:

- Produces one egg and 2–3 polar bodies
- Produces ova with only X chromosomes
- Is a discontinuous process
- Produces ova that donate mitochondria to the embryo
- Produces a limited supply of ova early in life that are arrested in development and finish meiosis later in life

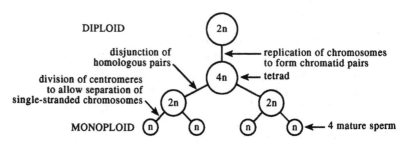

Spermatogenesis

Human Female Reproductive System

The *ovum* develops in a discontinuous process called *oogenesis*. Oogenesis is called discontinuous because it is not completed in a single continuous process as occurs during spermatogenesis. During development of female children ova progress to meiotic prophase I in the first round of meiotic cell division and then become arrested, stuck at this stage. These ova remain arrested in meiosis throughout the life of a woman, except for the ova that mature during each menstrual cycle and progress through this meiotic block. Women are born with all the eggs they will ever have, while males produce fresh sperm daily. This is the reason that genetic anomalies are more common in the eggs of older women; these anomalies have had years to accumulate in ova while sperm have a short life span.

The Ticking Clock

Because eggs are produced only once in a female's lifetime, they age and have a higher incidence of mutations than sperm. This is why older women tend to have more babies with birth defects than younger women.

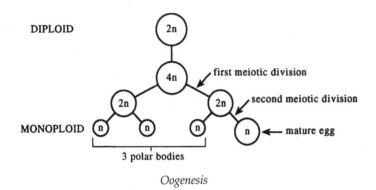

Oogenesis

The completion of the first meiotic cell division by maturing ova preparing for ovulation creates one cell with most of the cytoplasm and another smaller cell that has little cytoplasm. This smaller cell may itself divide later to create two smaller cells, but does not create viable ova and is called a *polar body*. The developing ovum becomes a secondary oocyte that pauses again in the second meiotic cell division even as it is released during ovulation. The ova in humans do not actually complete oogenesis until after fertilization, at which time it releases the last polar body and joins the nuclei of the male and female cells to create the diploid zygote. The unequal distribution of cytoplasm during oogenesis is another feature that is distinct from spermatogenesis.

Ovaries are paired structures in the lower portion of the abdominal cavity. As part of the menstrual cycle, one ova develops each month within a follicle in an ovary. The follicle is a collection of cells around the ova that support its development and secrete hormones. Each ovary is accompanied by a *fallopian tube*, also called an *oviduct*, one on each side of the abdomen. During ovulation, an ovum leaves the ovary from the follicle and is ejected into the upper end of the oviduct. At birth, all the eggs that a female will ovulate during her lifetime are already present in the ovaries, but these eggs develop and ovulate at a rate of one every 28 days (approximately), starting in puberty.

The ovaries also produce female sex hormones such as *estrogen*. Like male sex hormones, the female sex hormones regulate the secondary sexual characteristics of the female, including the development of the *mammary* (milk) *glands* and wider hip bones (pelvis). They also play an important role in the menstrual cycle, which involves the interaction of the pituitary gland, ovaries, and uterus.

The Menstrual Cycle

The *menstrual cycle* is a repeating sequence of events in the tissues and hormones of the female body. We will describe the process in humans. The key hormones in the menstrual cycle are GnRH from the hypothalamus, FSH and LH from the pituitary, and *estrogen* and *progesterone* from the ovary. These hormones each regulate the secretion of the other hormones as part of the menstrual cycle. GnRH stimulates FSH and LH secretion, which in turn stimulate the production of estrogen and progesterone. Estrogen and progesterone inhibit the production of FSH and LH as well as GnRH usually, with a key exception that is required for ovulation. Estrogen and progesterone also regulate the tissues in the uterus involved in the menstrual cycle.

There are four stages in the menstrual cycle:

- The follicular stage
- Ovulation
- The corpus luteum (luteal) stage
- Menstruation

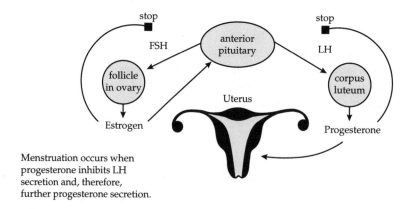

Menstruation occurs when progesterone inhibits LH secretion and, therefore, further progesterone secretion.

Menstrual Cycle

In the *follicular stage* of menstruation, FSH (follicle stimulating hormone) from the anterior pituitary gland stimulates a follicle to mature and produce estrogen. Estrogen promotes thickening of the uterine lining to support an embryo if fertilization occurs. This stage lasts approximately nine to ten days.

Menstruation Timeline

Number of Days	Menstrual Stage
1	Menstruation
4	Follicular Stage
13	Ovulation
14	
15	Corpus Luteum Stage
28	

When the follicle is mature, a surge in LH secretion from the pituitary causes *ovulation*, the release of the ovum from an ovary. The LH surge is a key factor in ovulation and ovulation will not occur without it. Constant high levels of estrogen block the LH surge and block ovulation. This is the mechanism by which the birth control pill acts.

After ovulation, the remains of the follicle in the ovary create the *corpus luteum*. Lutenizing hormone from the pituitary stimulates the corpus luteum to produce progesterone and estrogen, which stimulates vascularization (growth of blood vessels) and lining formation of the uterus in preparation for implantation of the fertilized egg. This stage lasts 12 to 15 days. Then, if no fertilization or implantation has occurred, the increased estrogen and progesterone block LH production. Without LH, the corpus luteum atrophies and progesterone levels fall. Without progesterone, the thickened, spongy uterine wall that had been prepared for implantation breaks down. The degenerating tissue, blood, and unfertilized egg are passed out as *menstrual flow*. This stage lasts approximately four days, bringing the total to 28 days for the entire cycle.

If fertilization occurs, the developing placenta produces HCG (*human chorionic gonadotrophic hormone*), which maintains the corpus luteum. The corpus luteum then continues to make progesterone and estrogen. Progesterone prevents menstruation and ensures that the uterine wall is thickened so that embryonic development can occur and pregnancy can continue. With time, the placenta develops and takes over the production of estrogen and progesterone for the duration of pregnancy.

Embryonic Development

The first step in development is *fertilization*. If sperm are present in the oviduct during ovulation, and a sperm succeeds in encountering the ovum, then fertilization can occur. *Zygotes*, single *diploid* cells, form as a result of fertilization. In fertilization, the egg nucleus (containing the *haploid number*, or n chromosomes) unites with the sperm nucleus (containing *n* chromosomes). This union produces a zygote of the original diploid or 2*n* chromosome number. In this way, the normal (2*n*) somatic number of chromosomes in a diploid cell is restored, and the cell has two homologous copies of each chromosome. Everything else in development up to adulthood consists of mitotic divisions.

If there are two or more eggs released by the ovaries, more than one can be fertilized. The result of multiple fertilizations will be fraternal (*dizygotic*) twins, which are produced when two separate sperm fertilize two eggs. Fraternal twins are related genetically in the same way that any two siblings are. Drugs to treat infertility often induce multiple ovulation and can lead to multiple birth pregnancies.

If there is only one fertilized egg, twins may still result through separation of identical cells during the early stages of cleavage (for example, the two-, four-, or eight-cell stage) into two or more independent embryos.

Pregnancy Giveaway

A pregnancy test looks for the presence of HCG, or human chorionic gonadotrophic hormone, the substance that maintains the uterine wall. If a woman is not pregnant, she will not produce HCG.

How Many Eggs?

Animals that practice external fertilization and external development must produce many eggs and sperm in order to ensure survival of the species. On the other hand, animals that practice internal fertilization and development invest their energy in taking very good care of a smaller number of offspring.

Ectopic Pregnancy

Sometimes the blastula implants itself outside the uterus and develops there, a situation referred to as an ectopic pregnancy.

These develop into identical (monozygotic) twins, triplets, and so forth, since they all came from the same fertilized egg and have identical genomes. Identical twins are often used in human genetic studies to determine what traits are genetically inherited, since differences between twins must be caused by their environment.

When the egg and the sperm join, they trigger a cascade of events that occur as the zygote begins to divide rapidly. These events, which are part of the process of fertilization, may occur either externally or internally.

External fertilization occurs in vertebrates that reproduce in water including most fish and amphibians. Eggs are laid in the water, and sperm are deposited near them in the water. The sperm have flagella, enabling them to swim through the water to the eggs. Since there is no direct passage of sperm from the male to the female, the sperm are likely to be diluted and the chances of fertilization for each ovum are reduced considerably. External fertilization also decreases the probability of survival of the young after fertilization since the developing animals are easy targets for predators. Internal fertilization is found in vertebrate land animals (like reptiles, birds, and mammals). The moist passageway of the female reproductive tract from the vagina through the oviducts provides a direct route to the egg for mobile sperm and increases the chance of fertilization.

The number of eggs produced depends upon a number of factors. One of these factors is the type of fertilization employed. Because very few sperm actually reach the egg during external fertilization, this process requires large quantities of eggs to ensure success. The type of development practiced by the organism is also significant. If development occurs outside the mother's body from the very beginning, many eggs are required to ensure survival of at least some of the offspring. Finally, the less care the parents provide, the more eggs are required to guarantee survival of enough offspring to continue the species.

Development of the Embryo. Cleavage of the embryo starts in the oviduct immediately after fertilization. The developing embryo travels down the oviduct, and, within five to ten days, implants itself in the uterine wall. Initially, the fertilized embryo divides into many undifferentiated cells. In the earliest stages, mitotic divisions result in one cell producing two cells, which produce four cells, which produce eight cells, and so on. This ultimately creates what is known as a *morula*, a solid ball of cells. Cells in the morula continue to rapidly divide mitotically to form the *blastula*, a hollow ball of cells (a single layer thick). The central cavity of the blastula is filled with fluid secreted by the cells, and is referred to as the *blastocoel*. More rapid division of cells at one end of the blastula causes an inpocketing or involution known as the two-layer *gastrula*. Two germ layers, the *ectoderm* and *endoderm*, are initially present, endoderm on the inside and ectoderm on the exterior. In a three-layer gastrula, *mesoderm* cells develop between the ectoderm and endoderm. This formation and rearrangement of the three germ layers is known as *gastrulation*.

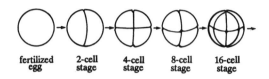

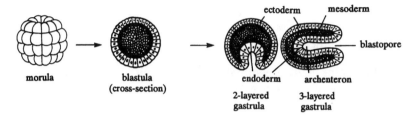

Cleavage of the Egg

Differentiation of Embryonic or Germ Layers. In the next stage of embryonic development, the cells of each germ layer begin to differentiate and specialize to form tissues, organs, and organ systems. Differentiation of cells occurs when the form and function of cells changes to reflect a distinct function or developmental fate. A cell is determined if its developmental fate is set in place even if it does not yet look differentiated. The ectoderm develops into the epidermis of skin, nervous system, and sweat glands. The endoderm becomes the lining of digestive and respiratory tracts, parts of the liver and the pancreas, and the bladder lining. Finally, the mesoderm develops into the muscles, skeleton, circulatory system, excretory system (except bladder lining), gonads, and the inner layer of skin (dermis).

The embryo may develop either outside or inside the animal. These two forms of development, external and internal development, are described below.

External Development of Embryo. External development occurs outside the female's body, in water or on land. The eggs of fish and amphibia, for example, are fertilized externally in water. The embryo then develops in water inside the egg, feeding on the yolk in egg. Such embryos are given very little parental care so large numbers of animals must be born to ensure survival of at least replacement numbers. External development on land occurs in reptiles, birds, and a few mammals, such as the duck-billed platypus. Fertilization in most animals must be internal, though, since sperm must have a watery environment to survive and move.

There are many adaptations for embryonic development within eggs and on land. One of these is a hard shell for protection, which is brittle in birds and leathery in reptiles. Embryonic membranes also help to provide a favorable environment for the developing embryo. Evolution of the egg was one of the adaptations that permitted terrestrial vertebrates to become more independent of water.

Types of embryonic membranes include the *chorion*, which lines the inside of the egg shell. This moist membrane permits gas exchange through the

Don't Mix These Up on Test Day

The *ectoderm* develops into the skin, the nervous system, and the eyes, hair, and teeth.

The *mesoderm* develops into the muscles, the skeleton, the circulatory system, the kidney, and the gonads.

The *endoderm* develops into the lining of the digestive and respiratory tracts, and the lining of the bladder, pancreas, and liver.

Egg-Laying Mammal

The duck-billed platypus, native to Australia, does not develop its young internally as do most other mammals. Instead, it lays a leathery egg. It is classified as a mammal because of its fur and milk glands.

In a Nutshell

There are four types of embryonic membranes:

- Yolk sac
- Allantois
- Chorion
- Amnion

shell. The *allantois*, a saclike structure developed from the digestive tract, is another embryonic membrane. It carries out functions like respiration and excretion, particularly the exchange of gases with the external environment. The allantois layer has many blood vessels to take in O_2 and give off CO_2, water, salt, and nitrogenous wastes. A third embryonic membrane, the *amnion*, encloses the amniotic fluid. Amniotic fluid provides a watery environment for the embryo to develop in, and provides protection against shock. Finally, the *yolk sac* encloses the yolk. Blood vessels in the yolk sac transfer food to the developing embryo.

Internal Development. In animals that develop internally, fertilization and embryo development occur within the mother. This internal development can take a number of different forms, depending on whether or not a placenta is utilized in sustaining the embryo. The *placenta* includes tissues of both the embryo and the mother. It is the site at which exchange of food, oxygen, waste, and water can take place.

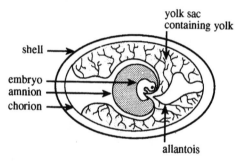

Egg

In some nonplacental animals, development occurs inside the mother, but the embryo lacks a placenta. Thus, there is no region of exchange of materials between the blood of the mother and the embryo. Eggs must therefore be relatively large, as their yolk must supply the developing embryo's needs. Tropical fish and opossums are examples of nonplacental animals. They develop inside the oviduct, obtaining food from the yolk of the egg, and are born alive after a relatively brief period of internal embryonic development.

A Tough First Challenge

Marsupials, such as the kangaroo, are nonplacental animals. The newborn kangaroo is very immature. After birth, the immature infant must climb up its mother's stomach to the pouch, where it is protected and fed milk.

In placental animals, there is no direct contact between the bloodstreams of the mother and the embryo. Transport is accomplished by diffusion and active transport between juxtaposed blood vessels of the mother and embryo in the placenta. The eggs of placental animals are very small, since the embryo requires only a small amount of yolk to be maintained until a placental connection is completed. Humans, for example, have no yolk, but they do have a yolk sac. The *umbilical cord* that attaches the embryo to the placenta is composed completely of tissues of embryonic, not maternal, origin. This cord contains the umbilical artery and vein. As in birds and reptiles, the amnion of placental mammals provides a watery environment to protect the embryo from shock.

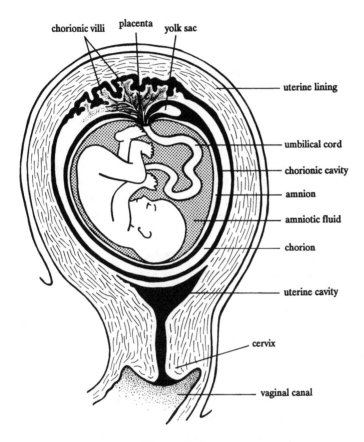

Human Embryo

Postembryonic Development. The development of the embryo to the adult is termed *maturation*. Maturation involves cell multiplication, differentiation, increase in size, and development of a distinctive adult shape. Maturation can be interrupted, such as in the metamorphosis of arthropods, or uninterrupted, as in mammals. Differentiation of cells is complete when all organs reach adult form. Further cell division is needed only for repair and replacement of tissues. In humans, growth occurs rapidly in children, followed by sexual maturation during puberty.

Physiology

All organisms must perform a variety of vital functions to survive. We have seen the various ways in which organisms perform one of these functions, reproduction. It is now time to turn to some of the ways in which organisms perform other essential physiological functions.

Nutrition

Animals are multicellular heterotrophic organisms that must get their energy and raw material through the consumption of food. Food comes in large chunks of insoluble material mostly bound up in biological polymers that cells cannot access to use for anything. Food must be digested to be absorbed and used by cells. Digestion involves *mechanical* breaking of food into small pieces, *chemical* breakdown of food into its molecular building blocks, followed by *absorption* of digested nutrients. Digestion can be intracellular, occurring through the action of intracellular enzymes. It can also be extracellular, using enzymatic secretions in a gut cavity to break down nutrients into simpler compounds that are absorbed by cells lining the gut.

In many organisms, including humans, the mechanical breakdown of large fragments of food into small particles occurs through cutting and grinding in the mouth and churning in the digestive tract. The molecular composition of these food particles is unchanged by breaking food into smaller pieces, but making the pieces smaller gives enzymes greater access to the molecules in the food.

Chemical breakdown of molecules in digestion is accomplished through enzymatic hydrolysis. Food in large part consists of large biological polymers, including proteins, nucleic acids and carbohydrates like starch. Organisms first hydrolyze these with enzymes into their building block components. The smaller digested nutrients (glucose, amino acids, fatty acids, and glycerol) are absorbed by cells lining the gut to be metabolized or transported to other parts of the body.

In addition to carbohydrates, fats and proteins, there are many small molecules that are essential nutrients although they are not used in the formation of energy. These include minerals and vitamins. Many of these act as coenzymes for enzymatic activity and are found associated with essential enzymes. Niacin, for example, is used to make NADH. These vitamins and minerals are essential in the diet since the body cannot produce these substances through biosynthesis.

Ingestion and Digestion in Protozoa and Cnidarians

Protozoans utilize *intracellular digestion*. In amoebae, pseudopods surround and engulf food through *phagocytosis* and enclose it in food vacuoles. *Lysosomes* (containing digestive enzymes) fuse with the food vacuole and release their digestive enzymes, which break down macromolecules like proteins, nucleic acids, and carbohydrates into their building blocks. The resulting simpler molecules diffuse into the cytoplasm and unusable end products are eliminated from the vacuoles.

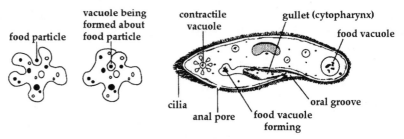

Protozoan Digestive System

In the *paramecium*, cilia sweep microscopic food such as yeast cells into the oral groove and cytopharynx. A food vacuole forms around food at the lower end of the cytopharynx. Eventually, the vacuole breaks off into the cytoplasm and progresses toward the anterior end of the cell. Enzymes are secreted into the vacuole and the products diffuse into the cytoplasm. Solid wastes are expelled at the anal pore.

Hydra (*phylum Coelenterata*, also called *Cnidarians*) employ both intracellular and *extracellular digestion*. Tentacles bring food to the mouth (ingestion) and release the particles into a cuplike sac. The endodermal cells lining this gastrovascular cavity secrete enzymes into the cavity. Thus, digestion principally occurs outside the cells (extracellularly). However, once the food is reduced to small fragments, the gastrodermal cells engulf the nutrients and digestion is completed intracellularly. Undigested food is expelled through the mouth. Every cell is exposed to the external environment, thereby facilitating intracellular digestion.

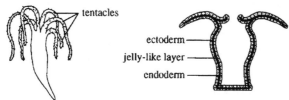

Hydran Digestive System (external view and in cross-section)

They Grab Their Food, Too

Many simple organisms, such as paramecia or hydra, sweep food into their oral grooves or mouths with their cilia or tentacles.

Ingestion and Digestion in Annelida

Since the earthworm's body is many cells thick, only the outside skin layer contacts the external environment. For this reason, this species requires a more advanced digestive system and circulatory system. Like higher animals, earthworms have a complete one-way, two-opening digestive tract. Their digestive tract is a tube that moves food through in one direction instead of a sac like in cnidarians. Having a tube is more efficient than a sac since food moves in one direction through the tube and digestion can become a stepwise process with specialization of parts of the tube for specific digestive. These parts of the digestive tube in annelids include the mouth, pharynx, esophagus, crop (to store the food), gizzard (to grind the food), intestine (which contains a large dorsal fold that provides increased surface area for digestion and absorption), and anus (where undigested food is released).

Crops, Stomachs, and Gizzards

Crops and stomachs store food, while **g**izzards **g**rind it.

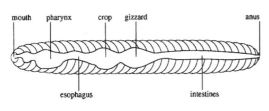

Digestive System of Annelida

Ephemeral Arthropoda

Mayflies belong to the order Ephemerata, which means short-lived. The adult flies emerge from the pupae, mate, and die within 24 hours. These insects do not eat during their adult lifetimes; in fact, they do not even possess functioning mouths.

Don't Mix These Up on Test Day

Mechanical digestion involves chewing in the mouth, grinding in the gizzard, and churning in the digestive tract.

Chemical digestion involves enzymes that break down carbohydrates (e.g., amylase), proteins (e.g., pepsin), or lipids (e.g., lipase).

Ingestion and Digestion in Arthropoda

Insects have a similar digestive system as annelids, except that they utilize jaws for chewing and salivary glands for better digestion.

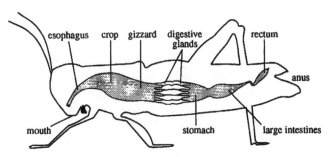

Digestive System of Arthropoda

Ingestion and Digestion in Humans

The human digestive system consists of the *alimentary canal* and the associated glands that contribute secretions into this canal. The alimentary canal is the entire path of food through the body: the oral cavity, *pharynx, esophagus, stomach, small intestine, large intestine,* and *rectum.* Many glands line this canal, such as the gastric glands in the wall of the stomach and intestinal glands in the small intestine. Other glands, like the pancreas and liver, are outside the canal proper, and deliver their secretions into the canal via ducts.

Mechanical Digestion. Food is crushed and liquefied by the teeth, tongue, and peristaltic contractions of the stomach and small intestine, increasing the surface area for the digestive enzymes to work upon. *Peristalsis* is a muscular action conducted by smooth muscle that lines the gut in the esophagus, stomach, small intestine and large intestine. Rings of muscle circling the gut contract, and move a ring of contraction down the gut, moving the food within the gut as well.

Chemical Digestion. Several exocrine glands associated with the digestive system produce secretions involved in breaking food molecules into simple molecules that can be absorbed. Polysaccharides are broken down into glucose, triglycerides are hydrolyzed into fatty acids and glycerol, and proteins are broken down into amino acids.

Chemical digestion begins in the mouth. In the mouth, the *salivary glands* produce saliva that lubricates food and begins starch digestion. *Saliva* contains *salivary amylase* (ptyalin), an enzyme that breaks the complex starch polysaccharide into maltose (a disaccharide). As food leaves the mouth, the *esophagus* conducts it to the stomach via the *cardiac sphincter* by means of peristaltic waves of smooth muscle contraction.

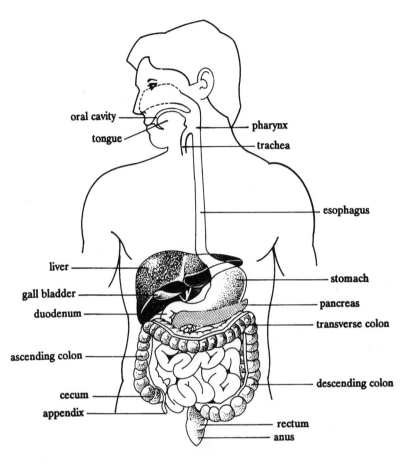

Human Digestive System

In the *stomach*, *gastric glands* produce *hydrochloric acid* and the enzyme *pepsin*. The acidity of the stomach provides the low pH environment necessary for the optimum enzymatic activity of pepsin. In addition, the acidity destroys ingested microorganisms, sterilizing the stomach. *Chyme* (partially digested food in the stomach) enters the *duodenum* through the *pyloric sphincter*.

The *liver* is also involved in digestion by producing *bile*, an important factor in fat digestion. Bile is stored in the *gall bladder* prior to its release into the small intestine. Bile salts are detergents that emulsify fats, breaking large fat globules into smaller droplets to expose a greater surface area of the fats to the action of pancreatic lipase.

The liver also helps regulate blood glucose levels and produces urea. Glucose and other monosaccharides absorbed in the *small intestine* during digestion are delivered to the liver directly from the intestine via the *hepatic portal vein* without passing first through the rest of the tissues. Glucose-rich blood is processed by the liver, converting excess glucose to glycogen for storage in the liver. If the blood has a low glucose concentration, the liver converts glycogen into glucose and releases it into the blood, restoring blood glucose levels to normal. The liver also synthesizes glucose from

Busy Liver

The liver is also responsible for:

- Detoxification of toxins
- Storage of iron and vitamin B$_{12}$
- Destruction of old erythrocytes
- Synthesis of bile
- Synthesis of various blood proteins
- Defense against various antigens
- Beta-oxidation of fatty acids to ketones
- Interconversion of carbohydrates, fats, and amino acids

Quick Quiz

Where does digestion of the following foods begin?

1) carbohydrates
2) proteins
3) lipids

Answers:

1) in the mouth (through the action of salivary amylase)
2) in the stomach (through the action of pepsin)
3) in the small intestine (through the action of bile and lipase)

noncarbohydrate precursors via the process of gluconeogenesis when blood glucose levels are low. Glycogen metabolism is under both hormonal and nervous control.

Pancreatic lipase is produced and secreted by the *pancreas*, which is also responsible for manufacturing amylase (for starch digestion), *trypsin,* and *chymotrypsin* (for protein digestion). Unlike pepsin, these enzymes have a pH optimum in the alkaline range. The necessary alkaline environment is created by the release of large quantities of bicarbonate ion (HCO_{3-}) by the pancreas along with the digestive enzymes. This bicarbonate neutralizes the acidity of the chyme released into the duodenum from the stomach.

Pancreatic proteases like trypsin are produced and stored in inactive forms called *zymogens.* Zymogens are activated after secretion when they are cleaved by another protease. The production of these digestive enzymes as inactive zymogens prevents damage to the pancreatic tissues that could occur if proteases were synthesized in their active state. This damage does occur in some people if pancreas secretions are blocked.

As food enters the small intestine, digestion continues. Glands in the wall of the intestine produce aminopeptidases (for polypeptide digestion) and disaccharidases (for digestion of maltose, lactose, and sucrose). The pancreatic and intestinal enzymes in the small intestine are responsible for the bulk of digestion in the gastrointestinal tract. In addition, most of the absorption of the digested nutrients occurs here. The large intestine is devoted mainly to water and Vitamin K absorption, and the rectum acts as a transient storage place for feces prior to their elimination through the anus.

Adaptations for Absorption

It is not enough for organisms to simply digest food. They must also be able to transport this food to their cells to be used for energy. To be transported to cells, food molecules must be absorbed out of the interior of the gut. As we ascend the evolutionary scale of organisms and as they become larger, we can observe that adaptations for absorption become increasingly complex.

Protozoa and Hydra. The single-cell protozoa is one of the simpler organisms under discussion in this book. In the protozoa, digested food in the food vacuole passes by simple diffusion into cell cytoplasm. The hydra absorbs extracellularly digested nutrients into its body cells; some nondigested material is also taken up into the cells and digested intracellularly.

Annelida and Arthropoda. The earthworm (annelida) and grasshopper (arthropoda) absorb nutrients in similar ways. Soluble food passes, by diffusion, through the walls of these creatures' small intestines into blood. The infolding (typholosole) of the digestive canal increases the absorptive surface.

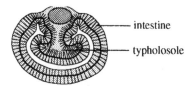

Absorption in Annelida

Humans. As for humans, most absorption of nutrients occurs in the small intestine, which is adapted for absorption. *Villi* lining the wall of the intestine are microscopic outpocketings of the lining of the intestine. The villi increase the surface area of the intestine, significantly increasing the area that is involved in absorption.

Villi contain capillaries and lacteals (projections of the lymphatic system) and are also covered with microvilli, "hairs" that further increase surface area and aid in absorption. These capillaries are in close proximity to the contents of the intestine, helping the movement of food molecules into the blood to be transported through the body. Amino acids, small fatty acids, and glucose pass through the villi walls into the capillary system. Some nutrients, such as glucose and amino acids, are actively transported into cells using energy, against a concentration gradient, while others are passively absorbed, flowing down a gradient into cells and the circulation.

Fats are not generally transported through capillaries. Triglycerides are broken down in the intestinal interior into fatty acids and glycerol, are absorbed by cells lining the intestine, then reformulated as triglycerides in fat containing particles called chylomicrons. Chylomicrons pass into the lymphatic system in the villi to be transported eventually into the blood.

Sites of Absorption

Amino acids and monosaccharides are absorbed into the bloodstream by the microvilli of the small intestine.

Lipid molecules, however, are picked up by the lymphatic system via lacteals.

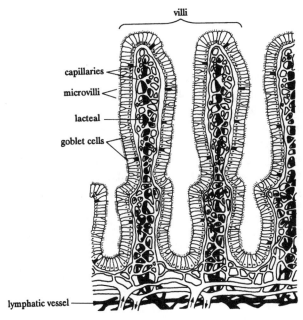

Absorption in Humans

A Simple Life

Note that in simple organisms, circulation occurs mainly through the process of simple diffusion. This implies that the processes of digestion and respiration are practically identical in such organisms.

Circulation

As we saw earlier, organisms make nutrients available to cells through absorption. But these nutrients, along with gases and wastes, must also be transported throughout the body to be used. The system involved in transport of these materials to different parts of the body is called the *circulatory system*. Small animals have their cells either directly in contact with the environment or in close enough proximity that diffusion provides for the movement of gases, wastes and nutrients making a specialized system for circulation unnecessary. Larger, more complex organisms require circulatory systems to move material within the body.

Circulation in Protozoans and Cnidarians

Protozoans are single-celled organisms and cnidarians often have only two cell layers, with both layers in contact with the environment. Circulation in these organisms occurs through simple diffusion across the plasma membrane between the cytoplasm and external environment, and streaming of cytoplasm within the cell (cyclosis). In hydra, for example, water circulates into and out of the body cavity, and all cells are in direct contact with the external environment.

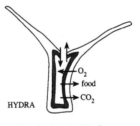

Circulation in Hydra

Circulation in Annelida

Annelids are larger and more complex animals than cnidarians and most cells in annelids (earthworms) are not in direct contact with the external environment. An internal closed circulatory system indirectly moves food, water, and oxygen within the body. In a closed circulatory system, blood travels always within defined blood vessels. In annelids blood travels forward to the head (anterior) through dorsal blood vessels. Five aortic arches or "hearts" force blood down to the ventral vessel, which carries blood to the posterior and up to complete the circuit (see figure).

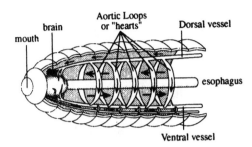

Circulation in Annelida

Annelids have no red blood cells, unlike vertebrates, but a hemoglobin-like pigment is dissolved in their blood. Nourishment in the form of food and gases is diffused into the cells from the capillaries.

Circulation in Arthropoda

Arthopods (grasshoppers) utilize an open circulatory system. In an open circulatory system, blood flows within vessels some of the time but in some areas of the body it is not contained, flowing instead through open spaces called sinuses. Arthropods have a simple beating tube for a heart, which moves blood through a dorsal vessel and then out into sinuses. In these sinuses, blood is not enclosed in blood vessels but directly bathes cells, and exchange of food takes place (air exchange, meanwhile, is accomplished through a tracheal system of air tubes). Blood then reenters blood vessels.

Circulation in Humans

Vertebrates have closed circulatory systems, with a chambered heart that pumps blood through arteries into tiny capillaries in the tissues. Blood from capillaries passes into veins that return to the heart. The chambers of vertebrate hearts include atria and ventricles. *Atria* are chambers where blood from veins collects and is pumped into ventricles. *Ventricles* are larger, more muscular chambers that pump blood out to the body.

Fish have a two chambered heart with one atria and one ventricle, in which the sole ventricle pumps blood into capillaries in the gills to collect oxygen. From the capillaries in the gills the blood is collected in arteries that pass move toward a second set of capillaries in the rest of the tissues of the body to deliver oxygen and nutrients. From there the blood passes back to the heart.

The problem with this set-up is that it takes a lot of pressure to pump blood through the first set of capillaries in the gills, and there is little pressure left afterward to pump the blood through the rest of the body. This set-up is not sufficient for the greater metabolic needs of terrestrial vertebrates. Amphibians have a three-chambered heart and birds and mammals have four-chambered hearts, with two atria and two ventricles. The right ventricle pumps deoxygenated blood to the lungs through the pulmonary artery. Oxygenated blood returns through the pulmonary vein to the left atrium.

From there it passes to the left ventricle and is pumped through the aorta and arteries to the rest of the body. Valves in the chambers of the heart keep blood from moving backward. There are in effect two separate circulations, one for the lungs called the pulmonary circulation, and the systemic circulation for the rest of the body. Using a four-chambered heart to split the pumping of blood through the lungs from the pumping of the blood to the rest of the body allows much greater pressure in the systemic circulation than would be possible with a two-chambered heart.

The heartbeat a doctor hears through a stethoscope is the sound of the chambers of the heart contracting in a regular pattern called the *cardiac cycle.* The heart is composed of specialized muscle tissue called *cardiac muscle.* Cardiac muscle cells are connected together in an electrical network that transmits nervous impulses throughout the muscle to stimulate contraction. The transmission and spreading of the signal is highly controlled to coordinate the beating of the chambers. During each cardiac cycle, the signal to contract initiates on its own in a special part of the heart called the *sinoatrial node,* or the pacemaker region. Cells from this region fire impulses in regular intervals all on their own, without stimulation from the nervous system. Once the signals start, they spread through both atria, which then contract, forcing blood into the ventricles. The signal then passes into the ventricles and spreads throughout their walls, causing contraction of the ventricles and the movement of blood into the major arteries. The ventricular contraction occurs during the *systole* part of the cardiac cycle, and the atria contract during the *diastole* part of the cardiac cycle.

The signal that causes the beating of the heart originates spontaneously within the heart without nervous stimulation, but the heart rate can be altered by nervous stimulation. The most important nervous stimulation of the heart is the vagus nerve of the parasympathetic system, which acts to slow the heart rate. The vagus nerve is more or less always stimulating the heart, and can increase the heart rate simply by stimulating the heart less than usual.

Arteries. The *arteries* carry blood from the heart to the tissue of the body. They repeatedly branch into smaller arteries (arterioles) until they reach the capillaries, where exchange with tissues occurs. Arteries are thick-walled, muscular, and elastic. Arteries conduct blood at high pressure and have a pulse caused by periodic surges of blood from the heart. Arterial blood is oxygenated except for blood in the pulmonary artery, which carries deoxygenated blood from the heart to the lungs to renew the oxygen supply.

Veins. *Veins* carry blood back to the heart from the capillaries. Veins are relatively thin-walled, conduct at low pressure because they are at some distance from the pumping heart, and contain many valves to prevent backflow. Veins have no pulse; they usually carry dark red, deoxygenated blood (except for the pulmonary vein, which carries recently oxygenated blood from the lungs). The movement of blood through veins is assisted by the

Don't Mix These Up on Test Day

Arteries:

- Are thick-walled
- Are usually oxygenated
- Conduct blood at high pressure
- Have a pulse
- Have no valves

Veins:

- Are thin-walled
- Are usually deoxygenated
- Conduct blood at low pressure
- Have no pulse
- Have valves to prevent backflow

contraction of skeletal muscle around the veins, squeezing blood forward. Once it moves forward in this way, valves keep the blood from going back.

Capillaries. *Capillaries* are thin-walled vessels that are very small in diameter. In fact, their walls are only one endothelial cell thick and red blood cells must pass through capillaries in single file. Capillaries, not arteries or veins, permit exchange of materials between the blood and the body's cells. Their small size and thin walls assist in the diffusion of material through their walls. Also, some of the liquid component of blood seeps from capillaries to directly bath cells with nutrients. Proteins and cells are too large to pass into the tissues and stay in the blood within the capillary walls. Some of the fluid that enters tissues passes directly back into the blood at the other end of the capillary, and the rest can circulate back in the lymphatic system. If the capillaries are too permeable or too much liquid stays in the tissues, swelling results.

At different times, different tissues require differing blood flow. The body regulates much of blood flow in tissues locally. Arterioles that feed capillaries in a tissue have smooth muscle in their walls that can relax or constrict to allow more or less blood into a specific area of tissue. Factors like the level of oxygen and carbon dioxide in blood that are affected by metabolic activity also act on the arteriole smooth muscle to match the blood flow to the metabolic need of the tissue.

Lymphatic system. *Lymph vessels* are a separate system independent of the blood system. This system carries extracellular fluid (at this stage known as lymph) at very low pressure, without cells. The *lymph nodes* are responsible for filtering lymph to rid it of foreign particles, maintain the proper balance of fluids in the tissues of the body, and are involved in the transport of chylomicrons as part of fat metabolism. The system ultimately returns lymph to the blood system via the largest lymph vessel, the thoracic duct, which empties lymph back into circulation shortly before it enters the heart.

The blood. The fluid moved through the body by the circulatory system is the blood. The blood is composed of a liquid component, the plasma, and cells. The cells include red blood cells called *erythrocytes*, *platelets*, and white blood cells, *lymphocytes*, each with specific functions.

The *plasma* is composed of water, salts, proteins, glucose, hormones, lipids and other soluble factors. The main salts in plasma are NaCl and KCl, in a composition that has been noted as similar to the composition of salt in sea water, our evolutionary origin. Calcium is another important salt in the extracellular fluid, including blood. The body regulates the blood volume and salt content through water intake and through excretion of urine. Oxygen is dissolved as a gas to a small extent in blood, although most oxygen is transported bound to hemoglobin in red blood cells. Carbon dioxide is converted to carbonic acid in the blood. Not only does this increase the solubility of carbon dioxide in the blood, but it creates a pH buffer that protects the body against large changes in the pH of blood. The glucose in blood is transported as a dissolved sugar for cells to uptake as needed. Hormones, both steroid hormones and peptide hormones, are transported in blood from one tissue where they are secreted to other tissues where they exert their actions. The protein component of plasma consists of antibodies for immune responses, fibrinogen for clotting, and serum albumin. The protein component of blood helps to draw water into the blood in the capillaries and prevent loss of fluid from the blood into the tissues to cause swelling.

Red blood cells are the most abundant cells in blood, and their primary function is to transport oxygen. After they are formed in the bone marrow, mature red blood cells (*erythrocytes*) lose their nuclei and become biconcave discs. They live for about four months in circulation before they are worn out and are destroyed in the spleen. Without a nucleus, mature red blood cells cannot make new proteins to repair themselves. Red blood cells also lose mitochondria, which renders them incapable of performing aerobic respiration. If they were able to carry on this form of respiration, they would themselves use up the oxygen that they carry to the tissues of the body. Instead, they produce energy in the form of ATP without using oxygen, through glycolysis.

The oxygen-carrying component of red blood cells is the protein *hemoglobin*. Hemoglobin is a tetrameric protein, in which each of the four polypeptide subunits has its own heme group with an iron that binds oxygen to form oxyhemoglobin. A given tetramer hemoglobin protein can have from 0-4 oxygen molecules bound to it. The binding of oxygen to hemoglobin is reversible and is cooperative. Cooperativity in binding oxygen means that when one of the subunits in hemoglobin binds an oxygen molecule, this increases the affinity of the remaining subunits to bind oxygen, resulting in a sigmoidal curve for the binding of oxygen by hemoglobin. In the lungs, where the partial pressure of oxygen is high, hemoglobin readily picks up oxygen. In the tissues, where the partial pressure of oxygen is low, oxygen leaves hemoglobin to diffuse into the tissues. The hemoglobin molecule has evolved to deliver oxygen more efficiently in response to changes in the tissues. In periods of great metabolic activity in muscle, the pH of the blood can decrease and carbon dioxide increase, both of which tend to reduce the affinity of hemoglobin for oxygen and cause it to leave more oxygen in the tissue.

Blood Types. Red blood cells manufacture two prominent types of antigens, antigen A (associated with blood type A) and antigen B (blood type B). In any given individual, one, both, or neither antigen may be present. The same pattern appears in every red blood cell.

The plasma of every individual also contains antibodies for the antigens that are not present in the individual's red blood cells (if an individual were to produce antibodies against his or her own red cells, they would agglutinate and the blood would clump). Type A individuals have anti-B antibody, and type B individuals have anti-A antibody. Type O individuals, who have neither A nor B antigens, have both anti-A and anti-B antibodies. Type AB individuals have neither type of antibody. These relationships are depicted in the graph below.

Blood Type	Antigen on RBC	Antibodies Found in Plasma
A	A	anti-B
B	B	anti-A
Universal Recipient AB	A, B	none
Universal Donor O	none	anti-A, anti-B

Clotting of blood involves soluble proteins, and small fragments of cells called platelets. Platelets are fragments of cells released into the circulation from cells called megakaryocytes. Platelets in an open wound release the enzyme thromboplastin, which initiates a series of reactions that ultimately lead to the formation of a fibrin clot. Thromboplastin, with the aid of calcium and vitamin K as cofactors, leads, in several steps, to the conversion of the inactive plasma prothrombin to its active form, thrombin. Thrombin in its turn converts fibrinogen (dissolved in plasma) into the fibrinous protein called fibrin. Threads of fibrin trap red blood cells to form clots. As the blood clots, serum is the liquid left over. Thus serum is essentially plasma, minus fibrinogen and other clotting factors.

Immune System

The interior of the body is an ideal growth medium for some pathogenic organisms like disease-causing bacteria and viruses. To prevent this, the body has defenses that either prevent organisms from getting into the interior of the body or stop them from proliferating if they are within the body. The system that plays this protective role is called the *immune system*. The trick for the immune system is to be able to mount aggressive defenses, and, at the same time, to distinguish foreign bodies to avoid attacking one's own tissues and causing disease. This is exactly what happens in autoimmune disorders; the immune system attacks one's own tissues as if they were foreign invaders.

Passive immune defenses are barriers to entry. These include the skin, the lining of the lungs, the mouth and stomach. The skin is a very effective barrier to most potential pathogens, but if wounded the barrier function of skin is lost. This is why burn patients are very susceptible to infection. The lungs are a potential route of entry, but are patrolled by immune cells, and have mucus to trap invaders and cilia lining the respiratory tract to remove the trapped invaders. The spleen plays a role in the immune system in adults, and in embryos plays a role in blood cell development.

Active immunity is conferred by the cellular part of the immune system. White blood cells are actually several different cell types that are involved in the defense of the body against foreign organisms in different ways. White blood cells include *phagocytes* that engulf bacteria with amoeboid motion, and various types of *lymphocytes*, B and T cells that are involved in

Vaccine Protection

When you are vaccinated, you are injected with weakened or dead pathogens. Your body has a protective immune response to these pathogens stimulating B cells and T cells, so that when the real pathogen comes along, you're already protected!

the immune response. B cells produce *antibodies,* or *immunoglobins,* which are secreted proteins specific to foreign molecules such as viral or bacterial proteins. Helper T cells coordinate the immune response and killer T cells directly kill cells that are infected with intracellular pathogens like viruses or cells that are aberrant like malignant cells. A given B or T cell responds to a specific antigen. Since the body does not know what antigens or pathogens may attack it, the immune system creates a varied population of B and T cells in which each cell recognizes only one antigen, but the population of cells contains a huge range of specificities. If a B cell or T cell encounters an antigen that matches its specificity, then it is stimulated to proliferate and create more cells with the same specificity. This amplification of a clone of cells that respond to the invading antigen helps the body to respond and to remain immune to infection in the future by the same pathogen. When a B cells encounters antigen that it recognizes, it proliferates to make more B cells that produce antibody. The stimulated B cells also produce memory cells that do not make antibody, but have the same specificity and will lie dormant for many years, ready to respond if the body is challenged again with the same antigen.

Phagocytes consist of neutrophils, which are the first cells to arrive at a site of inflammation to eat bacteria and other foreign particles. They are the primary component of pus. Macrophages and monocytes are also phagocytic cells and present foreign components, such as bacteria and viruses, to B cells and T cells to stimulate these parts of the immune system to respond.

Respiration

Cells performing aerobic respiration need oxygen and need to eliminate carbon dioxide. To do this, organisms must exchange gases with the environment. The respiratory system provides oxygen and removes CO_2. The oxygen is used to drive electron transport and ATP production and CO_2 is produced from burning glucose in the Krebs cycle. Even the most expert pearl divers cannot live without breathing for more than a few minutes. Gas exchange is accomplished via a variety of efficient ways, which range from simple diffusion to complex systems of respiration. We describe the adaptations employed by a range of organisms below.

Respiration in Protozoa and Cnidarians

Since every cell of these types of primitive organism is in contact with the external environment (in this case, water), respiratory gases can be easily exchanged between the cell and the outside by direct diffusion of these gases through the cell membrane. Lipid bilayer membranes are fully permeable to oxygen and carbon dioxide.

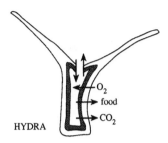

Respiration in Hydra

Respiration in Annelida

Mucus secreted by cells at the external surface of the annelid's body provides a moist surface for gaseous exchange from the air to the blood through diffusion. The annelid's circulatory system then brings O_2 to the cells and waste products such as CO_2 back to the skin, excreting them into the outside environment.

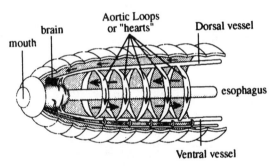

Respiration in Annelida

Respiration in Arthropoda

The arthropod respiratory system consists of a series of respiratory tubules called *tracheae*. These tubules open to the outside in the form of pairs of openings called *spiracles*. Inside the body, the tracheae subdivide into smaller and smaller branches, enabling them to achieve close contact with most cells. In this way, this system permits the direct intake, distribution, and removal of respiratory gases between the air and the body cells. No oxygen carrier is needed and specialized cells for this purpose are not found. Since a blood system does not intervene in the transport of gases to the body's tissues, this system is very efficient and rapid, enabling most arthropods to produce large amounts of energy relative to their weights. The direct diffusion of air through trachea is one factor that limits body size in arthropods.

Respiration in Arthropoda

Countercurrent Exchange

Respiration in fish is very efficient, due to a process called countercurrent exchange. Blood in the gills flows in the opposite direction as water flowing around the gills. This system makes the exchange of O_2 and CO_2 function smoothly.

Respiration in Fish

Water entering a fish's mouth travels over numerous thin-walled, thread-like *gill* filaments that are well fed by capillaries. As water passes over these gill filaments, O_2 diffuses into the blood, while CO_2 leaves the blood to enter the water. Arteries then transport the oxygenated blood through the body. The blood in the gills and the water moving over the gills move in opposite direction, creating what is known as a countercurrent exchange mechanism for the exchange of gases between the blood and the environment. Countercurrent exchange is a very efficient mechanism for exchange. After passing over the gills, water passes out of the body through openings on either side of the head, taking the discarded carbon dioxide with it. Marine invertebrates also have gills to exchange gases with their water environment.

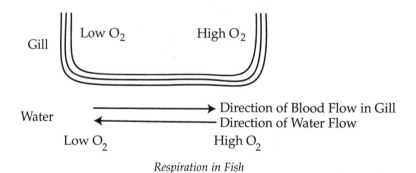

Respiration in Fish

Respiration in Humans

Humans have developed a complex system of respiration to transport oxygen to their cells and to rid their bodies of waste products like carbon dioxide. First, the *lungs* are designed to move air between the exterior atmosphere and an interior air space that is in close contact with capillaries. Here, oxygen and carbon dioxide diffuse between the blood and air. The blood then circulates through the body to exchange gases with the tissues, then returns to the lungs.

The lungs are found in a sealed cavity in the chest, bound by the ribs and chest wall and by the muscular *diaphragm* on the bottom. A membrane called a *pleura* surrounds the lungs and is held tightly against another membrane in the chest by a thin layer of liquid. The diaphragm is curved upward when released, and flattens when contracted, expanding the chest cavity. During inspiration, or inhalation, chest muscles move the ribs up and out as the diaphragm moves down; this creates both a larger chest cavity and a vacuum that forces air into the respiratory passages. The reverse process decreases the size of the chest cavity and forces air out of the lungs (exhalation). Exhalation is largely a passive process that does not require muscle contraction. During exhalation the elasticity of the lungs draws the chest and diaphragm inward when the muscles relax, decreasing the volume of the lungs and causing air to be forced out.

The breathing rate is controlled by a part of the brain, the *medulla oblongata,* that monitors carbon dioxide content in the blood. Excess CO_2 in the blood stimulates the medulla to send messages to the rib muscles and the diaphragm to increase the frequency of respiration.

The air passages involved in respiration consist of the *nose, pharynx, larynx, trachea, bronchi, bronchioles,* and the *alveoli.* The *nose* adds moisture and warmth to inhaled air, and helps to filter it, removing particulates and organisms. The *pharynx* is involved in diverting ingested material into the esophagus and away from the lungs to prevent choking. The *larynx* contains a membrane that vibrates in a controlled manner with the passage of air to create the voice. The *trachea* carries air through the vulnerable throat protected by flexible but strong rings of cartilage. At the end of the trachea the respiratory passage splits into the two lungs and into smaller and smaller passages that terminate in the tiny *alveoli,* tiny air sacs that are the site of gas exchange in the lungs.

The alveoli have thin, moist walls and are surrounded by thin-walled capillaries. Oxygen passes from the alveolar air into the blood by diffusion through the alveolar and capillary walls. CO_2 and H_2O pass out in the same manner. Note that all exchanges at the alveoli involve passive diffusion.

Since passive diffusion drives gas exchange, both in the lungs as well as the tissues, gases always diffuse from higher to lower concentration. In the tissues, O_2 diffuses into tissues and CO_2 leaves, while in the lungs this is reversed due to high oxygen pressure and low CO_2. CO_2 is carried in blood mainly as dissolved carbonate ions.

Thermoregulation and the Skin

In humans, the *skin* protects the body from microbial invasion and from environmental stresses like dry weather and wind. Specialized epidermal cells called *melanocytes* synthesize the pigment *melanin,* which protects the body from ultraviolet light. The skin is a receptor of stimuli, such as pressure and temperature. The skin is also an excretory organ (removing excess water and salts from the body) and a thermoregulatory organ (helping control both the conservation and release of heat).

Sweat glands secrete a mixture of water, dissolved salts, and urea via sweat pores. As sweat evaporates, the skin is cooled. Thus, sweating has both an

Don't Smoke

Emphysema is a disease characterized by the destruction of the alveolar walls. This results in reduced elasticity of the lungs, making exhalation difficult. Most cases can be traced to cigarette smoking.

Extra Sensitive

People who are albinos cannot synthesize the pigment melanin. This autosomal recessive disease results in an exquisite sensitivity to the sun.

excretory and a thermoregulatory function. Sweating is under autonomic (involuntary) nervous control.

Subcutaneous fat in the *hypodermis* insulates the body. Hair entraps and retains warm air at the skin's surface. Hormones such as epinephrine can increase the metabolic rate, thereby increasing heat production. In addition, muscles can generate heat by contracting rapidly (shivering). Heat loss can be inhibited through the constriction of blood vessels (*vasoconstriction*) in the *dermis*, moving blood away from the cooling atmosphere. Likewise, dilation of these same blood vessels (*vasodilation*) dissipates heat.

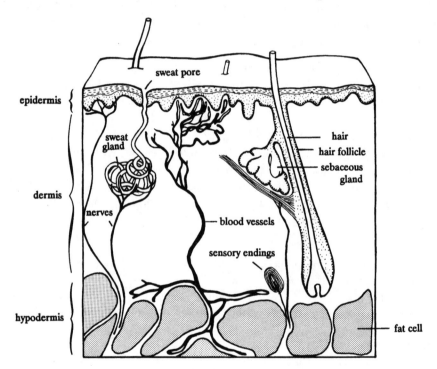

Human Skin

Don't Mix These Up on Test Day

Ectoderms warm their bodies by absorbing heat from their surroundings.

Endoderms derive their body heat from their own metabolism.

Alternate mechanisms are used by some mammals to regulate their body temperature. For example, *panting* is a cooling mechanism that evaporates water from the respiratory passages. Most mammals have a layer of fur that traps and conserves heat. Some mammals exhibit varying states of torpor in the winter months to conserve energy; their metabolism, heart rate, and respiration rate greatly decrease during these months. *Hibernation* is a type of intense or extreme torpor during which the animal remains dormant over a period of weeks or months with body temperature maintained below normal.

Excretion

Excretion is the term given to the removal of metabolic wastes produced in the body. (Note that it is to be distinguished from elimination, which is the removal of indigestible materials.) Sources of metabolic waste include:

Waste	Metabolic Activity Producing the Waste
Carbon dioxide	Aerobic respiration
Water	Aerobic respiration, dehydration synthesis
Nitrogenous wastes (urea, ammonia, uric acid)	Deamination of amino acids
Mineral salts	All metabolic processes

Our select group of organisms has developed various types of adaptations for excretion:

Excretion in Protozoa and Cnidarians

Remember that in these simple organisms, all cells are in contact with the external, aqueous environment. Water-soluble wastes such as the highly toxic ammonia produced by protein metabolism can therefore exit via simple diffusion through the cell membrane. Some freshwater protozoa, such as the paramecium, possess a contractile vacuole, an organelle specialized for water excretion by active transport. Excess water, which continually diffuses into the hyperosmotic cell from the hypo-osmotic environment (in this case, fresh water), is collected and periodically pumped out of the cell to maintain the cell's volume and pressure.

Excretion in Annelida

In annelids, two pairs of *nephridia* tubules in each body segment excrete water, mineral salts, and nitrogenous wastes in the form of urea. Fluid from the circulatory system is filtered out of the blood into fluid that fills the central body cavity. This fluid enters the nephridia tubules where some material is removed and other material is secreted into the urine before it is excreted from a pore with the nitrogenous wastes. Urine formation in this simple organism resembles to some extent the filtration and processing of urine that occurs in the mammalian kidney.

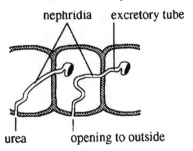

Excretion in Annelida

Excretion in Arthropods

Nitrogenous wastes are excreted in the form of solid uric acid crystals. The use of solid nitrogenous wastes is an adaptation that allows arthropods to conserve water. Mineral salts and uric acid accumulate in the *Malphigian tubules*; they are then transported to the intestine to be expelled along with solid wastes of digestion.

Human Excretory System

The principle organs of excretion in humans are the *kidneys*. The kidneys form urine to remove nitrogenous wastes in the form of urea as well as regulating the volume and salt content of the extracellular fluids. From the kidney the urine passes into an *ureter tube* that passes to the *urinary bladder* where urine is stored until urination occurs. During urination, the urine leaves the bladder through the *urethra*.

One of the main metabolic waste products that the kidneys remove is *urea*. When excess amino acids are present, or during a period of starvation when other energy sources are depleted, the body will break down amino acids from proteins and burn them for energy in the Krebs cycle. During this process, the nitrogen is enzymatically removed from the amino acid and released as ammonia, which is highly toxic. The liver converts the ammonia to urea, which is much less toxic than ammonia. The kidneys then remove the urea from the bloodstream.

Functions of the Kidney

The kidney is involved in the following activities:

- Filtering of blood
- Reabsorption of amino acids, glucose, and salts
- Secretion of urea, uric acid, and other wastes

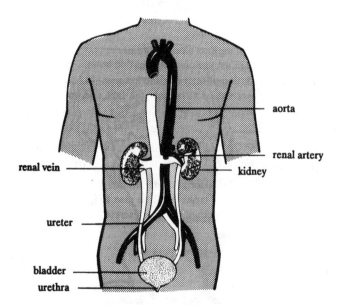

Human Excretory System

The basic functional unit of the kidney involved in urine formation is a tubelike small structure called the *nephron* (see diagram) that first filters blood to form a filtrate fluid and then selectively modifies the filtrate to produce urine. The blood that is to be filtered enters each nephron in a ball-shaped cluster of capillaries called the *glomerulus*. The pressure of blood in the glomerulus squeezes the liquid portion of the blood out of the glomerulus through a sievelike filtering structure. Blood cells are too large to pass through the sieve and most proteins are retained in blood by their size and charged nature. Other small molecules such as salts, amino acids, glucose, water, and urea pass easily into the filtrate. The filtrate that leaves the blood enters a cup-shaped structure around the glomerulus called *Bowman's capsule,* which forms the starting end of the tubelike nephron.

Visual Aid

Picture the *glomerulus* as a kind of sieve or colander. Small molecules, such as amino acids and glucose, can pass through (to be subsequently reabsorbed), while large molecules like proteins and blood cells cannot pass through.

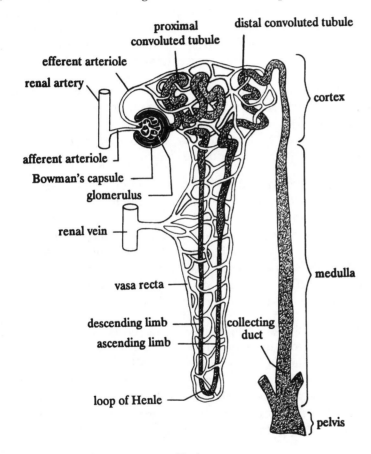

Nephron

From Bowman's capsule, the urinary filtrate will move down the nephron tubule, becoming increasingly modified as it progresses. Before urine is ready to be excreted, it passes through 1) the *proximal convoluted tubule,* 2) the *loop of Henle,* 3) the *distal tubule,* and 4) the *collecting ducts.* The first step in modifying the urinary filtrate occurs in the *proximal convoluted tubule.* In this region, active transport pumps glucose, amino acids, sodium, and proteins back out of the filtrate. Water follows these by osmosis, concentrating the urine and reducing the volume of filtrate. This step conserves necessary metabolites that would otherwise be wasted in urine at the same time that it concentrates the urinary filtrate. In diabetics with very high levels of glucose in blood, for example, the reabsorption mechanism for glucose can be overwhelmed, leading to the loss of glucose in urine.

From the proximal tubule, the filtrate passes to the *loop of Henle*. While the glomerulus and Bowman's capsule of each nephron are located in the outer region of the kidney (the cortex), the loop of Henle dips down into the inner kidney region called the *medulla*. The medulla has a very high concentration of extracellular sodium. As the filtrate passes down the loop of Henle, water is drawn out of the filtrate due to osmosis, passing from the low ion concentration in the filtrate to the high ionic strength of the extracellular fluid in the medulla. When the filtrate passes back up the loop of Henle, sodium is pumped out into the medulla. These two steps further reduce the volume of the urinary filtrate, drawing out the water along with the sodium, and help to preserve the high concentration of sodium in the medulla.

After passing through the *distal tubule*, the filtrate must pass through the *collecting duct* before passing out to the ureter and the urinary bladder. The collecting duct passes back down through the high ionic strength medulla. To make concentrated or dilute urine, the hormone *ADH* (antidiuretic hormone) regulates the permeability of the collecting duct walls.

When a person is dehydrated and their extracellular fluid volume is low, they will secrete ADH and excrete more concentrated urine, saving water. ADH acts on the walls of the collecting ducts to make them more permeable to water. Since the fluid of the medulla is very concentrated with ions, water will flow out of the collecting ducts if the walls of the collecting duct are water-permeable and allow osmosis, reabsorbing water from the urine as it forms. If no ADH is present, the walls of the collecting ducts do not permit osmosis, and the urine will remain dilute.

Another hormone that regulates urine formation is the steroid hormone *aldosterone*. Aldosterone is secreted in response to low extracellular sodium and acts on the distal tubule to increase the reabsorption of sodium from the urinary filtrate. The reabsorption of sodium causes water to be removed from the filtrate by osmosis as well, reducing the urine volume and increasing the volume of the extracellular fluids of the body. Thus, aldosterone is another means the body can use to conserve water as well as sodium.

Endocrine System

The body has two communication systems to coordinate the activities of different tissues and organs. One communication system is the nervous system and the other is the *endocrine system*. The endocrine system is the network of glands and tissues that secrete *hormones*, chemical messengers produced in one tissue and carried by the blood to act on other parts of the body. Compared to the nervous system, the signals conveyed by the endocrine system take much more time to take effect. A nervous impulse is produced in a millisecond and travels anywhere in the body in less than a second. Hormones require time to be synthesized, can travel no more quickly than the blood can carry them, and often cause actions through inducing protein synthesis or transcription, activities that require time. However, hormone signals will tend to be more long-lasting than nerve impulses. When the nerve impulse ends, a target such as skeletal muscle returns quickly to its starting state. When a hormone induces protein synthesis, the proteins remain long after the hormone is gone. Often the two systems work together. The *endocrine glands*, such as the pancreas or the adrenal cortex, can be the direct targets (effectors) of the autonomic nervous system. The hormone adrenaline acts in concert with the sympathetic nervous system to produce a set of results similar to those produced directly by sympathetic neurons.

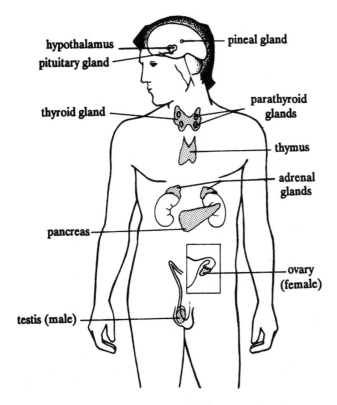

Human Endocrine System

Endocrine glands secrete hormones directly into the bloodstream. This is in contrast to *exocrine secretions* that do not contain hormones and are released through ducts into a body compartment. An example of exocrine secretion is the secretion by the pancreas of digestive enzymes into the small intestine through the pancreatic duct. Both endocrine and exocrine functions can be found in the same organ. The pancreas simultaneously produces exocrine secretions like digestive enzymes and endocrine secretions like insulin and glucagon that are released into the blood to exert their effects throughout the body.

Don't Mix These Up on Test Day

Peptide hormones, such as insulin and ADH, are proteins, have membrane receptors, act through intercellular second messengers, and have a quick onset time.

Steroid hormones, such as testosterone and estrogen, are cholesterol-derived, have intracellular receptors, act as first messengers, affect transcription through receptors they bind to, and have a longer onset time.

Two Types of Hormones: Steroid Hormones and Peptide Hormones

Based on their chemical nature, hormones can be classified into two groups: *steroid hormones* and *peptide hormones.* In both cases, the hormone must bind to a protein receptor on the target cell to affect that cell. The types of receptors they affect are different, however. Peptide hormones are large, hydrophilic, charged, and cannot diffuse across the plasma membrane. As a result, the receptors they bind to are located on the cell surface. When a peptide hormone binds to its receptor on the surface of a target cell, it activates the receptor and causes it to transmit a signal into the cellular interior. The nature of this signal can be to turn on a protein kinase that phosphorylates certain proteins and changes their activity, or to release second messengers in the cell, such as calcium or cyclic AMP, that amplify the signal and alter many different cellular activities. This form of indirect signaling by a hormone is called a *signal transduction cascade* because of the amplification by downstream signaling factors.

Peptide hormones can be small peptides such as ADH, with just a few amino acid residues, or large complex polypeptides like insulin. Peptide hormones are often produced as large inactive precursors, or pro-hormones, that are cleaved by proteases into smaller active peptides before they are released. Since peptide hormones are secreted proteins, they are synthesized on the rough ER in the cell, then packaged and processed in the Golgi before they are delivered to the plasma membrane for secretion. Since hormone signaling is usually regulated, the release of hormones is usually regulated. The hormones are stored in secretory vesicles in the cytoplasm, waiting for the signal that fuses the vesicle with the plasma membrane, dumping the hormones into the extracellular fluid and blood and transporting them by the circulatory system to distant target tissues in the body.

Steroid hormones are small and hydrophobic, and most hormones of this class are derived from cholesterol, including estrogen, progesterone, testosterone, and cortisol. Since they are small and hydrophobic, they can diffuse through the cell membrane. These hormones bind to steroid hormone receptors after they have diffused into the cell through the plasma membrane. The hormone-bound receptors enter the nucleus and bind to target regions in genes that regulate transcription, turning the genes on or off. Steroid hormone signals are changes in gene transcription and protein expression caused by the steroid hormone receptors.

Since steroid hormones freely diffuse through membranes, they are not stored after production. They are usually secreted at a rate equal to their production. It is the rate of production for these hormones that is highly regulated, by controlling the activity of the enzymes that produce the hormones.

Endocrine Glands

Hormones are secreted by a variety of endocrine glands, including the *hypothalamus, pituitary, thyroid, parathyroids, adrenals, pancreas, testes, ovaries, pineal, kidneys, gastrointestinal glands, heart,* and *thymus.* It is likely that additional tissues like skin and fat not traditionally considered glands also have endocrine functions. Some hormones regulate a single type of cell or organ, while others have more widespread actions. The specificity of hormonal action is determined by the presence of specific receptors on or in the target cells.

A common principle that regulates the production and secretion of many hormones is the *feedback loop*. Often several hormones regulate each other in a chain. For example, the hypothalamic hormone corticotropin acts on the anterior pituitary to release ACTH, which acts on the adrenal cortex to release cortisol. In a feedback loop, the level of the last hormone regulates the production of earlier hormones in the loop. When cortisol blood levels increase, cortisol acts on the pituitary to decrease further ACTH secretion, which leads to a decrease in cortisol production. Acting in this way, feedback loops act to maintain hormones and the body's internal state at a relatively constant level.

Hypothalamus and Pituitary Gland. The *hypothalamus,* a section of the posterior forebrain, is located above the pituitary gland and is intimately associated with it via a portal circulation that carries blood directly from the hypothalamus to the pituitary. In most parts of the circulatory system, blood flows directly back to the heart from capillaries, but in a portal system blood flows from capillaries in one organ to capillaries in another. When the hypothalamus is stimulated (by feedback from endocrine glands or by neurons innervating it), it releases hormonelike substances called *releasing factors* into the anterior-pituitary-hypothalamic portal circulatory system. These hormones are carried directly to the pituitary by the portal system. In their turn, these releasing factors stimulate cells of the anterior pituitary to secrete the hormone indicated by the releasing factor.

The *pituitary gland* is a small gland with two lobes lying at the base of the brain. The two lobes, anterior and posterior, function as independent glands. The anterior pituitary secretes the following hormones:

- *Growth hormone* fosters growth in a variety of body tissues.

- *Thyroid stimulating hormone* (TSH) stimulates the thyroid gland to secrete its own hormone, thyroxine.

- *Adrenocorticotrophic hormone* (ACTH) stimulates the adrenal cortex to secrete its corticoids.

- *Prolactin* is responsible for milk production by the female mammary glands.

- *Follicle-stimulating hormone* (FSH) spurs maturation of seminiferous tubules in males and causes maturation of ovaries in females. It also encourages maturation of follicles in the ovaries.

- Finally, *luteinizing hormone* (LH) induces interstitial cells of the testes to mature by beginning to secrete the male sex hormone

Mnemonic: Flat Pig

The anterior pituitary gland secretes the following hormones:

FSH
LH
ACTH
TSH

Prolactin
I(gnore)
Growth

testosterone. In females, a surge of LH stimulates ovulation of the primary oocyte from the follicle. LH then induces changes in the follicular cells and converts the old follicle into a yellowish mass of cells rich in blood vessels. This new structure is the corpus luteum, which subsequently secretes progesterone and estrogen.

The *posterior pituitary* is a direct extension of nervous tissue from the hypothalamus. Nerve signals cause direct hormone release. The two hormones secreted by the posterior pituitary are ADH and oxytocin.

- *ADH* acts on the kidney to reduce water loss.

- *Oxytocin* acts on the uterus during birth to cause uterine contraction.

Thyroid Gland. The thyroid hormone, *thyroxine*, is a modified amino acid that contains four atoms of iodine. It accelerates oxidative metabolism throughout the body. An abnormal deficiency of thyroxine causes goiter, decreased heart rate, lethargy, obesity, and decreased mental alertness. In contrast, hyperthyroidism (too much thyroxine) is characterized by profuse perspiration, high body temperature, increased basal metabolic rate, high blood pressure, loss of weight, and irritability.

Parathyroid Glands. The parathyroid glands are small pealike organs located on the posterior surface of the thyroid. They secrete parathyroid hormone, which regulates the calcium and phosphate balance between the blood and other tissues. Increased parathyroid hormone increases bone resorption and elevates plasma calcium. Plasma calcium must be maintained at a constant level for the function of muscles and neurons.

Pancreas. The pancreas is a multifunctional organ. It has both an *exocrine* and an *endocrine function*. The exocrine function of the pancreas secretes enzymes through ducts into the small intestine. The endocrine function, on the other hand, secretes hormones directly into the blood stream.

The endocrine function of the pancreas is centered in the *islets of Langerhans*, localized collections of endocrine alpha and beta cells that secrete glucagon and insulin respectively. *Insulin* stimulates the muscles to remove glucose from the blood when glucose concentrations are high, such as after a meal. Insulin is also responsible for spurring both muscles and liver to convert glucose to glycogen, the storage form of glucose. The islets of Langerhans also secrete *glucagon*, which responds to low concentrations of blood glucose by stimulating the breakdown of glycogen into glucose, keeping the level of glucose in blood high enough to supply tissues.

Adrenal Glands. The adrenal glands are situated on top of the kidneys and consist of the *adrenal cortex* and the *adrenal medulla*.

Adrenal cortex: In response to stress, ACTH stimulates the adrenal cortex to synthesize and secrete the steroid hormones, which are collectively known as *corticosteroids*. The corticosteroids, derived from cholesterol, include glucocorticoids, mineralocorticoids, and cortical sex hormones.

Glucocorticoids, such as cortisol and cortisone, are involved in glucose regulation and protein metabolism, and their presence is important to deal with stress. Glucocorticoids raise blood glucose levels by promoting gluconeogenesis and decrease protein synthesis. They also reduce the body's immunological and inflammatory responses. ACTH from the pituitary induces cortisol production. Elevated cortisol represses ACTH expression and lowers cortisol levels, acting as a feedback loop to maintain relatively constant cortisol levels. Although cortisol is an important hormone to deal with stress, prolonged high levels of corticosteroids probably repress the immune system. Corticosteroids are effective anti-inflammatory medicines, but their use is limited by their alterations of fat metabolism and their repression of the immune system.

Mineralocorticoids, particularly aldosterone, regulate plasma levels of sodium and potassium, and consequently, the total extracellular water volume. Aldosterone causes active reabsorption of sodium and passive reabsorption of water in the nephron, a topic we dealt with in more detail earlier in this book.

The adrenal cortex also secretes small quantities of androgens (male sex hormones) in both males and females. Since, in males, most of the androgens are produced by testes, the physiologic effect of the adrenal androgens is quite small. In females, however, overproduction of the adrenal androgens may have masculinizing effects, such as excessive facial hair.

Adrenal Medulla: The secretory cells of the adrenal medulla can be viewed as specialized sympathetic postganglionic nerve cells that secrete hormones into the circulatory system. This organ produces *epinephrine* (adrenaline) and *norepinephrine* (noradrenaline), both of which belong to a class of amino acid-derived compounds called *catecholamines*.

Epinephrine increases the conversion of glycogen to glucose in liver and muscle tissue, causing a rise in blood glucose levels and an increase in the basal metabolic rate. Both epinephrine and norepinephrine increase the rate and strength of the heartbeat, and dilate and constrict blood vessels. These in turn increase the blood supply to skeletal muscle, the heart, and the brain, while decreasing the blood supply to the kidneys, skin, and digestive tract. These effects are known as the *"fight or flight response,"* and are elicited by sympathetic nervous stimulation in response to stress. Both of these hormones are also neurotransmitters.

Ovaries and Testes. The gonads are important endocrine glands, with testes producing testosterone in males and ovaries producing estrogen in females. See the section on reproduction earlier in this chapter for more details.

Fight or Flight

Hormones like *epinephrine* and *norepinephrine* are responsible for the body's physical reactions to stress. They temporarily increase heart beat and blood supply to the active organs in what is known as "the fight or flight response."

Nervous System

The nervous system enables organisms to receive and respond to stimuli from their external and internal environments. Your brain and spinal column regulate your breathing, your movement, and provide perception of sight, sound, touch, smell and taste. They allow organisms to not only perceive their environment but to respond to their experience, and to alter their behavior over time through learning.

Functional Units of the Nervous System

To understand the nervous system, it is best to start with the basic functional unit of the nervous system—the *neuron*. The neuron is a specialized cell that is designed to transmit information in the form of electrochemical signals called *action potentials*. These signals are generated when the neuron alters the voltage found across its plasma membrane. The property that neurons have that allows them to carry an action potential is an excitable membrane.

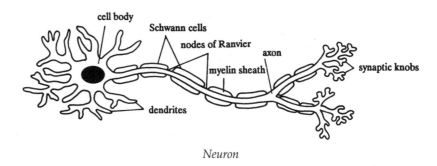

Neuron

The basic parts of the neuron's cell structure are the *cell body*, the *dendrites*, and the *axon* (see the figure below). The cell body contains the nucleus and most of the organelles and is the site of most protein synthesis and energy production in neurons. The dendrites receive chemical information from other neurons as changes in membrane potential and carry this information to the cell body. The axon is a very long, slender projection of the neuron that transmits the action potential from the cell body to the target the neuron is to communicate with. The axon can be as long as a meter when the cell body is located in the central nervous system and the axon must carry the action potential to a target in the extremities.

Resting Potential. The action potential involves manipulating the voltage across the plasma membrane to carry information. All cells have a voltage across their plasma membrane that is generated through the actions of a protein called the Na^+/K^+ ATPase. Using the hydrolysis of ATP for energy, this protein pumps sodium ions out of the cell and potassium into the cell. This activity is essential to maintain the osmotic balance of cells. Some of the potassium leaks back out of the cell through an ion channel called the potassium leak channel. With more positive ions on the outside of the cell, a net negative voltage of about –70 mVolts is found across the plasma membrane of most cells and is called the resting potential.

Action Potential. Action potentials are a wave of electrochemical information that moves through axons and muscle tissue, creating a response in those target tissues. All cells have a resting potential, but not all cells have an excitable membrane that creates action potentials. The difference is that neurons, as well as muscle cells, have a protein in their plasma membrane that lets sodium ions through the membrane in response to a decrease in the membrane potential. This protein is called a voltage-gated sodium channel and gives neurons an excitable membrane. If the membrane voltage becomes less negative than the resting potential, changing from –70 mVolts to perhaps –50 mVolts, then the voltage-gated sodium channels in the neuron's plasma membrane will open. The voltage at which the voltage-gated channels open is called the *threshold potential*. When these channels are open, sodium will diffuse freely through the channel to cross the plasma membrane, flowing down a gradient from the outside of the cell into the cytoplasm. The opening of these channels in one region of the membrane and the entry of the sodium through the channels causes membrane *depolarization* (the membrane is less polarized, moving toward 0 potential).

After the voltage-gated sodium channels have opened and depolarization is complete, the channels close again, allowing the membrane voltage to return to its normal negative voltage. The return of the voltage to its normal negative state is called *repolarization*. At this point, the axon is ready to begin the process again and transmit a new action potential if necessary.

This describes depolarization in only one section of membrane, not the length of a neuron however. As the depolarization occurs in one section of membrane, the depolarization triggers the opening of voltage-gated sodium channels in the neighboring section of membrane further down the axon. The sodium channels further along the neuron then open, creating a new region of depolarization. Very rapidly the region of depolarization moves along the neuron by triggering the opening of voltage-gated ion channels in one region after another, moving like a wave along the length of an axon in a neuron.

A substance called *myelin* allows action potentials to travel more quickly by surrounding the axons of most mammalian neurons. Myelin is an insulating agent that coats discrete patches of the plasma membrane. Small spaces between the myelin coating are called *nodes of Ranvier* (see the figure of the neuron). In a myelinated neuron, the action potential jumps from node to node, bypassing the insulated myelin regions where no ions cross the membrane. The action potential can travel much more rapidly this way in myelinated neurons, since it can jump forward instead of traversing the plasma membrane of the entire length of the axon. This method of jumping forward is called *saltatory conduction*. Another factor that affects the speed of an action potential is the size of the neuron: Larger neurons allow action potentials to travel more quickly.

Size and Frequency of Action Potentials. There are two important characteristics of action potentials affecting the way nerves carry information.

Malfunctioning Myelin

When the body mounts a reaction against its own myelin, it begins to impede the myelination of nerves. This results in weakness, lack of balance, vision problems, and/or incontinence. This condition is termed multiple sclerosis (MS).

One factor is that every action potential in a neuron is the same size. Once the neuronal membrane reaches the threshold for depolarization, it will fully depolarize. Neurons do not have half of a depolarization or half of an action potential - the action potential is an all-or-nothing response. Either a neuron fires an action potential or it does not. The strength of a stimulus does not change the size of the action potential depolarization or the duration of the depolarization.

Not every nervous signal is the same strength, though, we know from experience. A touch of a finger is different from a blow from a hammer, and the organism must recognize this. If the size of the action potential cannot carry this information, then what can? The answer is that each action potential may be the same size, but the neuron carries action potentials more frequently to indicate a stronger stimulus. A weak signal like the touch of a finger may trigger one action potential in a second, while a strong stimulus like a hammer may trigger many more in the same period.

When an action potential has just passed through a section of membrane, it cannot fire an action potential again immediately. First it must finish depolarization and repolarization. This limit to the frequency of action potential firing in a neuron is called the *refractory period*. The refractory period places an upper limit on the frequency of action potentials that can pass through a neuron. The refractory period and the directional nature of the neuron also mean that neurons only carry action potentials in one direction, from the cell body out to the end of the axon. Action potentials do not move back up the axon to the cell body.

Synapses. The nervous system is not simply a network of electrical wires with the brain as the switchboard. There is a strong chemical component to the signals that neurons convey. Neurons do not usually directly carry the action potential all the way to the membrane of the target cell. When a neuron reaches a target cell, the axon ends in a synaptic terminal, with a gap called the *synapse* between the neuron and the target cell. At this point, the membrane potential is converted to a chemical signal, or *neurotransmitter* that is released across a small gap between the neuron and the target cell. This gap between the neuron and the target cell is called the *synaptic cleft*. The target cell that the neuron is communicating with then receives the chemical signal by binding the neurotransmitter and starting a signal of its own.

When an action potential travels down an axon to reach the synaptic terminal, it causes the neuron to release neurotransmitter into the synaptic cleft between the neuron and the target cell. This neurotransmitter will diffuse across the gap between the cells and bind to receptors on the target cell plasma membrane. The nature of the response on the target cell depends on the neurotransmitter released and the receptor that receives the signal. Some receptors open ion channels in response to binding neurotransmitter, allowing specific ions through the membrane in response to neurotransmitter. An excitatory neurotransmitter is one that binds to a receptor that allows sodium to flow through a channel into the cell. When the sodium

Drugs and Nerves

Many drugs modify processes that occur in the synapse. Cocaine, for example, blocks neuronal uptake carriers, prolonging the action of neurotransmitters. Nerve gases, meanwhile, are potent acetylcholinesterase inhibitors that cause rapid death by exaggerating the action of skeletal muscles (most important, the diaphragm), which leads to respiratory arrest.

enters, it depolarizes the plasma membrane of the target cell. If the depolarization of the target reaches the threshold to open voltage-gated sodium channels, an action potential will be initiated in the target cell. A neurotransmitter can also allow chloride to enter the postsynaptic membrane, causing a hyperpolarization, with the membrane potential moving away from the threshold for triggering an action potential. Chloride is an inhibitory neurotransmitter, since it makes it more difficult for an action potential to start in the target cell.

Neurons do not usually exist with only one neuron forming a single synapse with another neuron. Neurons in the CNS form a dense network of interactions in which each neuron receives information in its dendrites from synapses with many other neurons and each neuron in turn sends out an axon that can terminate in synapses with many other neurons. The information from all of the synapses a neuron interacts with is combined in the cell body of a neuron in a process called *summation*. Summation is the means that a single neuron uses to process information from all of the neurons that form synapses with it and decide whether or not to initiate an action potential itself. If all of the combined changes in the potential of the neuron cause it to reach the threshold depolarization to open the voltage-gated channels, then it will fire an action potential. If not, the neuron will remain silent (no action potential). The result is called summation since it is determined by adding up the contributions to the membrane potential created by many different synapses that stimulate any given neuron.

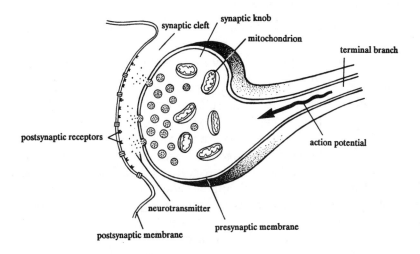

Synapse

There are many types of synapses of neurons with other neurons. For motor neurons in the somatic nervous system, a specialized synapse of motor neurons with skeletal muscle cells is called the *neuromuscular junction*. When an action potential reaches the neuromuscular junction, acetylcholine is released into the synaptic cleft and binds to postsynaptic receptors in the muscle cell. These receptors open sodium channels, depolarizing the muscle cell membrane and triggering an action potential in the muscle cell that is propagated throughout the membrane of this cell. This action potential triggers contraction of the muscle cell. The greater the number of action potentials that reach the muscle and the more muscle cells involved, the stronger the muscle contraction.

As important as turning on the signal caused by neurotransmitter is turning the signal back off again. At the neuromuscular junction, for example, if the acetylcholine is not removed, the muscle cell will continue to contract uncontrollably. Once the neurotransmitter is released into the synaptic cleft, it will continue binding to the postsynaptic receptors unless it is removed from the synapse in some way. There are several ways to remove the neurotransmitter from the synapse. One mechanism is for the neurotransmitter to diffuse away into the surrounding fluid. The synapse is small, making diffusion fairly rapid, but in most cases other mechanisms are involved. Another way to remove neurotransmitter from the synapse is with an enzyme that degrades the neurotransmitter. At the neuromuscular junction, acetylcholinesterase is an enzyme that acts on acetylcholine to degrade and inactivate it. Pesticides or nerve gas that inactivate this enzyme can be deadly. A third way to remove neurotransmitter is to take it up into cells at the synapse. This occurs for adrenaline or serotonin at synapses. The drug Prozac prevents depression by blocking the uptake of serotonin from synapses, elevating the level of serotonin found at synapses.

Organization of the Nervous System

As organisms evolved and became more complex, their nervous systems undergo corresponding increases in complexity. Simple organisms can only respond to simple stimuli, while complex organisms like humans can discern subtle variations of a stimulus, such as a particular shade of color.

Invertebrates. Protozoa are single-celled and have no organized nervous system, although they do have receptors that respond to stimuli like heat, light, and chemicals. Cnidarians have a nervous system consisting of a network of cells, the nerve net, located between the inner and outer layers of the cells of its body. With a limited network, the responses of cnidarians to their environment are limited to simple actions like retracting tentacles and stimulating swimming. Annelids possess a primitive central nervous system consisting of a solid ventral nerve cord and an anterior "brain" of fused ganglia. These clusters of cells allow a richer network of neurons than a simple nerve net, and more sophisticated information processing that results in more complex behavior. Arthropods also have ganglia and a more complex nervous system than annelids with more specialized sense organs, including simple or complex eyes and a tympanum for detecting sound. The sensory input and information processing capability of arthropods allows for amazing rich and complex behavior like social behavior in these small organisms.

Insensitive

Simple organisms have very primitive nervous systems and can only respond to strong stimuli like touch, heat, light, and chemicals. Higher organisms like vertebrates, on the other hand, may respond to a wide range of stimuli.

Vertebrate Nervous System

Vertebrates have a *brain* enclosed within the *cranium* and a *spinal cord* that together form the *central nervous system* (CNS) that processes, and stores information. Throughout the rest of the body is the *peripheral nervous system,* containing motor or efferent neurons that carry signals to effector organs like muscles or glands to take actions in response to nervous impulses. Sensory neurons in the peripheral nervous system convey information back to the CNS for processing and storage. Another division of the nervous system is into the autonomic and the voluntary components of the efferent pathways.

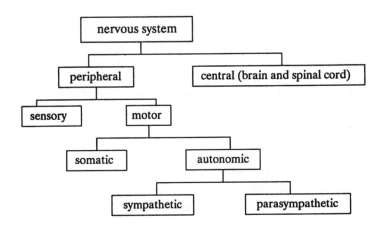

Organization of the Vertebrate Nervous System

The peripheral nervous system. The peripheral nervous system carries nerves from the CNS to target tissues of the body and includes all neurons that are not part of the CNS. The peripheral nervous system consists of 12 pairs of cranial nerves, which primarily innervate the head and shoulders, and 31 pairs of spinal nerves, which innervate the rest of the body. Cranial nerves exit from the brainstem and spinal nerves exit from the spinal cord. The peripheral nervous system has two primary divisions, the somatic and the autonomic nervous systems.

The somatic motor neuron system. This system innervates skeletal muscle and is responsible for voluntary movement, generally subject to conscious control. Motor neurons release the neurotransmitter acetylcholine (ACh) onto ACh receptors located on skeletal muscle. This causes depolarization of the skeletal muscle, leading to muscle contraction. In addition to voluntary movement, the somatic nervous system is also important for reflex action.

The autonomic nervous system. The autonomic nervous system is neither structurally nor functionally isolated from the CNS or the peripheral system. Its function is to regulate the involuntary functions of the body including the heart and blood vessels, the gastrointestinal tract, urogenital organs, structures involved in respiration, and the intrinsic muscles of the eye. In

Don't Mix These Up on Test Day

The *sympathetic nervous system*:

- Is associated with the fight-or-flight response
- Increases the heart rate
- Increases the breathing rate
- Lowers the digestive rate
- Causes pupil dilation

The *parasympathetic nervous system*:

- Is associated with the rest-and-digest response
- Lowers the heart rate
- Does not affect the breathing rate
- Increases the digestive rate
- Does not cause pupil dilation

general, the autonomic system innervates glands and smooth muscle, but not skeletal muscles. It is made up of the sympathetic nervous system and the parasympathetic nervous system.

- *The sympathetic nervous system.* This system utilizes norepinephrine as its primary neurotransmitter. It is responsible for activating the body for emergency situations and actions (the fight-or-flight response), including strengthening of heart contractions, increases in the heart rate, dilation of the pupils, bronchodilation, and vasoconstriction of vessels feeding the digestive tract. One tissue regulated by the sympathetic system is the adrenal gland, which produces adrenalin in response to stimulation. Adrenalin produces many of the same fight-or-flight responses as the sympathetic system alone.

- *The parasympathetic nervous system.* Here, acetylcholine serves as the primary neurotransmitter. One of this system's main functions is to deactivate or slow down the activities of muscles and glands (the rest-and-digest response). These activities include pupillary constriction, slowing down of the heart rate, bronchoconstriction, and vasodilation of vessels feeding the digestive tract. The principal nerve of the parasympathetic system is the vagus nerve. Most of the organs innervated by the autonomic system receive both sympathetic and parasympathetic fibers, the two systems being antagonistic to one another.

Human Brain

The human brain is divided into several anatomical regions with different functions including the following regions:

- **Cerebral cortex.** The cerebral cortex controls all voluntary motor activity by initiating the responses of motor neurons present within the spinal cord. It also controls higher functions, such as memory and creative thought. The cortex is divided into hemispheres, left and right, with some specialization of function between them ("left-brain, right-brain"). The cortex consists of an outer portion containing neuronal cell bodies (gray matter) and an inner portion containing axons (white matter).

- **Olfactory lobe.** This serves as the center for reception and integration of olfactory input.

- **Thalamus.** Nervous impulses and sensory information are relayed and integrated en route to and from the cerebral cortex by this region.

- **Hypothalamus.** Such visceral and homeostatic functions as hunger, thirst, pain, temperature regulation, and water balance are controlled by this center.

- **Cerebellum.** Muscle activity is coordinated and modulated here.

- **Pons.** This serves as the relay center for cerebral cortical fibers en route to the cerebellum.

- **Medulla oblongata.** This influential region controls vital functions like breathing, heart rate, and gastrointestinal activity. It has receptors for carbon dioxide; when carbon dioxide levels become too high, the medulla oblongata forces you to breathe. This is why when you hold your breath until carbon dioxide levels rise so high in your blood that you pass out, you will breathe involuntarily to bring an influx of oxygen into your body.

- **Reticular Activating System.** This network of neurons in the brain stem is involved in processing signals from sensory inputs and in transmitting these to the cortex and other regions. This system is also involved in regulating the activity of other brain regions like the cortex to alter levels of alertness and attention.

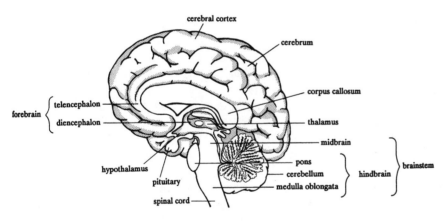

Human Brain

Don't Mix These Up on Test Day

In the spinal cord, the *dorsal horn* serves as the entrance point for sensory nerve fibers or afferent neurons.

The *ventral horn*, meanwhile, contains the cell bodies of motor or efferent neurons.

The *spinal cord* is also part of the CNS. The spinal cord acts as a route for axons to travel out of the brain. It also serves as a center for many reflex actions that do not involve the brain, such as the *knee-jerk reflex*. The spinal cord consists of two parts. The *dorsal horn* is the entrance point for sensory nerve fibers whose cell bodies are contained within the dorsal root ganglion. The *ventral horn*, on the other hand, contains the cell bodies of motor neurons. Fibers from the cerebral cortex synapse on the ventral horn motor neurons, thereby initiating muscular contractions.

Sensory Systems of the Nervous System

All complicated nervous systems are made more useful through input mechanisms that we know as our senses. Sight, hearing, balance, taste, smell, and touch provide an influx of data for the nervous system to assimilate. Other sensory information that we are not consciously aware of is also provided to the CNS, including internal conditions such as temperature and carbon dioxide content. All of these sensory detection systems use cells that are usually specialized, modified neurons to receive information and alter their membrane potential in response to this information. This altered membrane potential can then trigger an action potential to carry information back to the CNS. Some sensory cells detect chemical information (taste and smell), some detect electromagnetic energy (vision), and others detect mechanical information (sound, pressure). All sensation is caused by action potentials that are sent to the CNS by sensory cells. An action potential from the eye is the same as an action potential from the ear. The difference in perception, how we experience the information, is determined by how the information is received by the CNS, and how it is processed. An action potential from the eye is perceived as sight because it passes to the visual center of the brain for processing.

Sight. The eye detects light energy and transmits information about intensity, color, and shape to the brain. The transparent *cornea* at the front of the eye bends and focuses light rays. These rays then travel through an opening called the *pupil*, whose diameter is controlled by the pigmented, muscular *iris*. The iris responds to the intensity of light in the surroundings (light makes the pupil constrict). The light continues through the *lens*, which is suspended behind the pupil. This lens focuses the image onto the *retina*, which contains photoreceptors that transduce light into action potentials. The image on the retina is actually upside down but revision in the cerebral cortex and interpretation result in the perception of the image right-side up. The image from both eyes is also integrated in the cortex to produce the binocular vision with depth perception that allows us to throw, catch and drive with improved ability. The shape of the lens is changed to focus images from nearby or far objects. To see nearby objects, the muscles attached to the lens are relaxed and the lens rounds up, focusing light more sharply. If the shape of the eye is either too short or too long, or if the lens becomes stiff with age, then the eye is unable to focus the image and corrective lenses may be required to bring images into focus.

Cones and Rods. There are two types of specialized *photoreceptor cells* in the eye that respond to light: cones and rods. Cones respond to high-intensity illumination and are sensitive to color, while rods detect low-intensity illumination and are important in night vision. There are three different types of cones with different color sensitivity. The visual signal is received by the protein rhodopsin, which has a retinal group bound to it. This retinal group isomerizes when it is hit by a photon, starting the signal that leads to an action potential. The photoreceptor cells synapse onto bipolar cells, which in their turn synapse onto ganglion cells. Axons of these cells bundle to form the *optic nerves*, which conduct visual information to the brain.

Don't Mix These Up on Test Day

Cones are photoreceptors that respond to high intensity illumination and color.

Rods, on the other hand, respond to low intensity illumination (they are important in night vision), but do not detect color well.

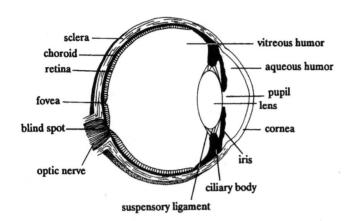

Human Eye

Hearing and Balance. The ear transduces sound energy into impulses that are perceived by the brain as sound. Sound waves pass through three regions as they enter the ear. First, they enter the outer ear, which consists of the *auricle* (pinna) and the *auditory canal.* Located at the end of the auditory canal is the *tympanic membrane* (eardrum) of the middle ear, which vibrates at the same frequency as the incoming sound. Next, three bones, or *ossicles* (malleus, incus, and stapes), amplify the stimulus, and transmit it through the oval window, which leads to the fluid-filled inner ear.

This inner ear consists of the cochlea and semicircular canals. The *cochlea* contains the *organ of Corti,* which has specialized sensory cells called hair cells. Vibration of the ossicles vibrates the cochlea, causing specific regions within the cochlea to vibrate depending on the frequency of the tone. Louder sounds increase the strength of the vibration and the response. The stimulation of a specific set of hair cells triggers the hair cells to transduce the mechanical pressure into action potentials, which travel via the auditory nerve to the brain for processing.

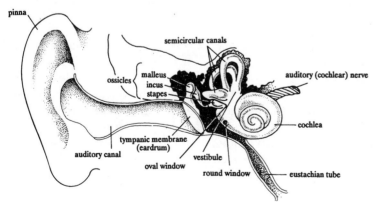

Human Ear

The *semicircular canals* are used for balance. Each of the three semicircular canals in the inner ear is perpendicular to the other two and filled with a fluid called endolymph. At the base of each canal is a chamber with senso-

Mnemonics

These two mnemonics will help you to get a handle on the following concepts:

With the exceptions of sexual activity and digestion,

The **s**ympathetic nervous system **s**peeds things up.

The parasympathetic nervous system slows things down.

and:

Cones respond to **c**olor.

Rods respond to black and white.

It's Too Spicy!

We lose taste buds as we age. This is why young children enjoy bland food such as macaroni and cheese, while adults tend to be more interested in spicier food.

ry hair cells; rotation of the head displaces endolymph in one of the canals, putting pressure on the hair cells in it. This changes the nature of the impulses sent by the vestibule nerve to the brain. The brain interprets this information to determine the position of the head.

Taste and Smell. *Taste buds* are chemical sensory cells located on the tongue, the soft palate, and the epiglottis. The outer surface of a taste bud contains a taste pore, from which microvilli, or taste hairs, protrude. Interwoven around the taste buds is a network of nerve fibers that are stimulated by the taste buds, and these neurons transmit impulses to the brainstem. There are four main kinds of taste sensations: sour, salty, sweet, and bitter.

Olfactory receptors are chemical sensors found in the olfactory membrane, which lies in the upper part of the nostrils over a total area of about 5 cm^2. The receptors are specialized neurons from which olfactory hairs project. When odorous substances enter the nasal cavity, they bind to receptors in the cilia, depolarizing the olfactory receptors. Axons from the olfactory receptors join to form the olfactory nerves, which project direction to the olfactory bulbs in the base of the brain.

Motor Systems

One of the key systems of the body is the system of muscles that are effectors for the CNS. To exert an effect muscles also require something to act against, which is the skeletal system. Read on for explanations of the characteristic motor systems of members of the select group of organisms we have been returning to throughout this book.

Cilia and Flagella

Ciliates and *flagellates* are unicellular organisms that do not have discrete skeletal-muscular systems. Protozoans and primitive algae move by the beating of cilia or flagella. Amoebae, meanwhile, use cell extensions called *pseudopodia* for locomotion; the advancing cell membrane extends, and the cytoplasm liquifies and flows into the pseudopods.

The cilia and flagella of all eukaryotic cells possess the same basic structure, which is different from prokaryotic structures. Each eukaryotic cilia and flagella contains a cylindrical stalk of eleven microtubules, nine pairs arranged in a peripheral circle with two single microtubules in the center of the stalk. Movement is affected by means of the power stroke, a thrusting movement in which the microtubule cylinders slide past each other with work provided by protein motors. Return of the cilium or flagellum to its original position is called the recovery stroke.

Effectors of the Nervous System

Motor systems allow organisms to respond to stimuli from their nervous systems.

Hydrostatic Skeletons

The muscles within the body wall of advanced flatworms such as planaria are arranged in two antagonistic layers, longitudinal and circular. As the muscles contract against the resistance of the incompressible fluid within the animal's tissues, this fluid functions as a hydrostatic skeleton.

Contraction of the circular layer of muscles causes the incompressible interstitial fluid to flow longitudinally, lengthening the animal. Conversely, contraction of the longitudinal layer of muscles shortens the animal. This movement of muscle is similar to the peristaltic motion involved in the digestive system.

The same type of hydrostatic skeleton assists in the locomotion of annelids. Each segment of this animal can expand or contract independently. Annelids advance principally through the action of muscles on a hydrostatic skeleton, as well as through bristles in the lower part of each segment. These bristles, called *setae,* anchor the earthworm temporarily in earth while muscles push the earthworm ahead.

Exoskeleton

An *exoskeleton* is a hard external skeleton that covers all the muscles and organs of some invertebrates. Exoskeletons in arthropods are composed of chitin. In all cases, the exoskeleton is composed of noncellular material secreted by the epidermis. Although it serves the additional function of protection, an exoskeleton imposes limitations on growth. Periodic molting and deposition of a new skeleton are necessary to permit body growth. To cause movement, muscles attach to the interior of the exoskeleton. Despite the giant insects in monster movies, this is not possible since the muscles in these monsters would not be strong enough to move the weight of the skeleton that would be required.

Endoskeleton.

Vertebrates have an *endoskeleton* as a framework for the attachment of skeletal muscles, permitting movement when a muscle contracts by bringing two bones together. All voluntary movement involves muscle contraction bringing bones together. The endoskeleton also provides protection, since bones surround delicate vital organs. For example, the rib cage protects the thoracic organs (heart and lungs), while the skull and vertebral column protect the brain and spinal cord. The vertebrate skeleton contains *cartilage* and *bone,* both formed from connective tissue.

Cartilage, although firm, is also flexible and is not as hard or as brittle as bone. It makes up the skeletons of lower vertebrates, such as sharks and rays. In higher animals, cartilage is the principle component of embryonic skeletons, and is replaced during development by the aptly termed replacement bone. Because cartilage has no vessels or nerves, it takes longer to heal than bone.

Bone makes up most of the skeleton of mature higher vertebrates, including humans. Bone is made of calcium and phosphate salts and strands of the protein collagen. Bones are produced as a balance between deposition of new bone and reabsorption of old bone by cells that live in bone. *Osteoclasts* are cells that reabsorb bone and *osteoblasts* create new bone. Imbalance

Quick Quiz

Match the numbered organism with the correct lettered motor system below.

1. earthworm
2. primitive algae
3. bear

    (A) endoskeleton
    (B) hydrostatic skeleton
    (C) cilia and flagella

Answers:
1. = (B)
2. = (C)
3. = (A)

Build and Destroy

Bone tissue is made up of *osteoblasts*, which build bone, and *osteoclasts*, which destroy bone. The reason that the body builds bone is obvious; the reason it destroys it is less obvious. Bone acts as a calcium reservoir for the entire body. When the body needs calcium, bone is destroyed in order to provide it. The two cell types work together constantly to remodel bone.

Mnemonic

Remember:

Osteo**b**lasts **b**uild **b**one!

between these processes can lead to weakening of the bones such as in osteoporosis. During growth bone arises through the replacement of cartilage or through direct ossification. Bone produced through the latter process is called *dermal bone;* the bones of the skull are examples of this. In *replacement bone,* such as the long bones of the legs and arms, osteoblasts replace the cartilage that has already formed. A hollow cavity within each long bones is subsequently filled with *bone marrow,* the site of formation of blood cells. The long bones originate as cartilage, with ossification beginning in the middle. The bones grow in the cartilaginous regions at the ends. As the bone grows, the region of ossification extends until growth ceases during adulthood and the bone become fully ossified.

While the division between dermal and replacement bone is based on embryologic origin, the division between spongy and compact bone is based on function and internal structure. *Spongy bone* is located in the central portions of bone. It consists of a network of hard spicules separated by marrow-filled spaces. The low density and the ability to withstand lateral stress are characteristics of bone that may be attributed to this type of spongy bone. *Compact bone,* located on the outer surfaces and articular surfaces, is responsible for the hardness of bone and its ability to withstand longitudinal stress. It consists of cylindrical units called *Haversian systems.* These cells radiate around a central capillary within a Haversian canal.

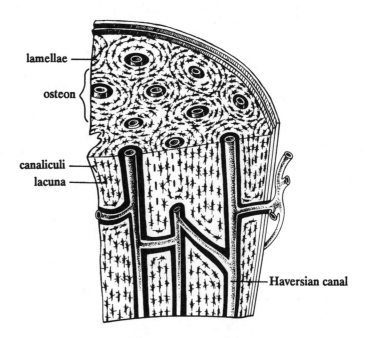

Microscopic Bone Structure

Bones are connected at joints, either immovable joints as in the skull, or movable joints like the hip joint. In the latter type, ligaments serve as bone-to-bone connectors, while tendons attach skeletal muscle to bones and bend the skeleton at the movable joints. In the vertebrate skeleton, the *axial skeleton* is the midline basic framework of the body, consisting of the skull, vertebral column, and the rib cage. The *appendicular skeleton,* on the other hand, includes the bones of the appendages and the pectoral and pelvic girdles.

Muscle System

The muscle system serves as an effector of the nervous system. Muscles contract to implement actions after they receive nervous stimuli. For example, your arm muscles will automatically contract if you touch a hot stove. A skeletal muscle originates at a point of attachment to the stationary bone. The insertion of a muscle is the portion attached to the bone that moves during contraction. An *extensor* extends or straightens the bones at a joint-as in, for example, straightening out a limb. A *flexor* bends a joint to an acute angle, as in bending the elbow to bring the forearm and upper arm together. Bones and muscles work together like a lever system.

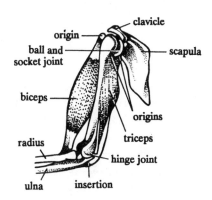

Muscle Movement

Types of Muscles. Vertebrates possess three different types of muscle tissues: *smooth*, *skeletal*, and *cardiac*. In all three types, muscles cause movement by contraction, and the contraction is caused by the sliding of actin and myosin filaments past each other within cells. The differences between the types of muscle include where they are located, what they do, and what the cells look like.

- *Smooth muscle,* or involuntary muscle, is generally found in visceral systems and is innervated by the autonomic nervous system. Each muscle fiber consists of a single cell with one centrally located nucleus. Smooth muscle is nonstriated, meaning it does not have clearly organized arrays of actin and myosin filaments. Smooth muscle is located in the walls of the arteries and veins, the walls of the digestive tract, the bladder, and the uterus. Smooth muscle contracts in response to action potentials, and the contraction is mediated by actin-myosin fibers like in other muscle, although the fibers do not have the clear organization they

BIOLOGY REVIEW

Don't Mix These Up on Test Day

Smooth muscle is involuntary muscle in the arteries, the gastrointestinal tract, and elsewhere.

Skeletal muscles are voluntary muscles that causes body movement.

Cardiac muscle is the tissue that makes up the heart.

display in other muscle types. Smooth muscle cells in a tissue are connected to each other through junctions that allow electrical impulses to pass directly from one cell to the next without passing through chemical synapses.

Smooth Muscle

- *Skeletal muscles,* or voluntary muscles, produce intentional physical movement. A skeletal muscle cell is a single large multinucleated fiber containing alternating light and dark bands called *striations.* The striations are caused by overlapping strands of thick myosin protein filaments that slide past thin actin protein filaments during muscle contraction. The actin and myosin filaments in skeletal muscle are organized into sections called *sarcomeres* that form contractile units within each muscle cell. The somatic nervous system innervates skeletal muscle. Each skeletal muscle fiber is stimulated by nerves through neuromuscular synapses. When a muscle cell is stimulated by a nerve, an action potential moves over the whole muscle fiber, releasing calcium in the cytoplasm of the cell. This calcium causes the actin and myosin to slide over each other, shortening the fibers and the cell. Many muscle cells are bundled together to create muscles.

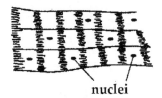

Skeletal Muscle

- *Cardiac muscle* is the tissue that makes up the heart. It has characteristics of both skeletal and smooth muscle. Cardiac muscle cells have a single nucleus, like smooth muscle, and are striated like skeletal muscle. Cardiac muscle cells are connected by gap junctions like smooth muscle, so that cells can pass action potentials directly between cells throughout the heart and do not require chemical synapses. Cardiac muscle contraction is regulated by the autonomic nervous system, which increases the rate and strength of contractions through sympathetic stimulation and decreases their rate through the parasympathetic system. Cardiac muscle has an internal pacemaker responsible for the heartbeat that is modified by the nervous system but does not require the nervous system to maintain a regular heartbeat.

KAPLAN

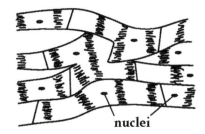

Cardiac Muscle

With that, we've covered the basics of the physiology you might be called upon to know on Test Day. The final topics associated with organismal biology that we need to discuss are those of animal behavior and plants.

Animal Behavior

Animals respond to their environment in a great variety of ways to increase their ability to survive and reproduce. As animals have evolved more complex nervous systems and motor systems, they have also evolved more complex behaviors. Some of these behaviors are inherited genetically while others are learned. Reflexes and fixed action patterns are examples of behaviors that are genetically determined, ingrained responses hardwired in the nervous system that do not require any learning. Many behaviors are a mixture of learned and instinctive actions. Through behavioral experiments and genetics scientists try to discern the role of "nature vs. nurture" in specific behaviors.

Patterns of Animal Behavior

Animal behavior is characterized by a number of different patterns. Simple organisms are capable only of simple, automatic responses to their environment, while more complex organisms are characterized by an increased reliance on mental processes like learning.

Simple Reflexes

Reflexes are simple, automatic responses to simple stimuli. A simple reflex is controlled at the spinal cord level of a vertebrate, involving a direct pathway from the receptor (afferent nerve) to the efferent or motor nerve. The efferent nerve innervates the effector, a muscle or gland. Thus a simple reflex involves a two-neuron pathway. Monosynaptic reflex pathways have only one synapse between the sensory neuron and the motor neuron. The classic example is the knee-jerk reflex. When the tendon covering the patella (kneecap) is hit, stretch receptors sense this and action potentials are sent up the sensory neuron and into the spinal cord. The sensory neuron synapses with a motor neuron in the spinal cord, which, in turn, stimulates the quadriceps muscle to contract, causing the lower leg to kick forward. In polysynaptic reflexes, sensory neurons synapse with more than one neu-

Don't Mix These Up on Test Day

An example of a *simple reflex* is the knee-jerk reflex.

An example of a *complex reflex* is the startle response.

Fixed-action patterns, meanwhile, are the characteristic movements of a herd or flock of animals.

ron. A classic example of this is the withdrawal reflex. When a person steps on a nail, the injured leg withdraws in pain, while the other leg extends to retain balance.

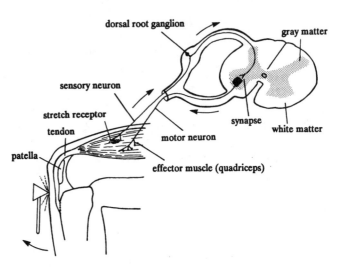

Reflex Arc for Knee Jerk

Simple reflexes can be changed over time to become more or less sensitive to stimulation of sensory neurons. This is a simple form of learning that even very simple organisms like snails display called *habituation*. The opposite of habituation, *sensitization*, occurs when a repeated stimulus creates a stronger reflex response over time.

Fixed-Action Patterns

Some behaviors such as avoiding predators are so important to survival that they cannot depend on learning. Learning is not possible in some circumstances where individuals grow without parents or adults and yet must immediately know how to eat, avoid predators and reproduce. *Fixed-action patterns* are used for behaviors that are important and cannot depend on learning. A fixed-action pattern is a complex, coordinated behavioral response triggered by specific stimulation from the environment. The stimulus that elicits the behavior is referred to as the releaser. Fixed-action patterns are not learned and are not usually modified by learning. The particular stimuli that trigger a fixed-action pattern are more readily modified, provided certain cues or elements of the stimuli are maintained.

An example of a fixed-action pattern is the retrieval-and-maintenance response of many female birds to an egg of their species. Certain kinds of stimuli are more effective than others in triggering a fixed-action pattern. Hence an egg with the characteristics of that species will be more effective in triggering the response than one that only crudely resembles the natural egg. The characteristic movements made by animals that herd or flock together, such as the swimming actions of fish and the flying actions of locusts are fixed-action patterns, as is the spinning of webs by spiders.

Characteristics of the stimulus for the fixed-action pattern can be altered to determine what part of the stimulus is the most important factor. It can either be the size of eggs, their shape, their color, or the pattern of speckling that elicits care of a bird for eggs. The characteristic of the stimulus that triggers the response can be artificially manipulated to create a larger than normal response.

Other physical factors like hormones can exert important influences on behavior, particularly the influence of estrogen and testosterone on sex-specific behavior. Even if a behavior is learned, genetic biological components such as hormone expression and sexual maturation might be required for the behavior to be manifested. For example, hormones and sexual maturation play a key role in the ability of songbirds to perform their characteristic songs.

Learning

Learned behavior involves a change in the way an animal behaves based on experience. Learning is a complex phenomenon that occurs, to some extent, in all animals. In lower animals, instinctual or innate behaviors are the predominant determinants of behavior patterns, and learning plays a relatively minor role in the modification of these predetermined behaviors. Higher animals, on the other hand, with their well-developed brains and neurological systems, learn the major share of their repertoire of responses to the environment.

Habituation

Habituation is a simple learning pattern in which repeated stimulus creates a decreased responsiveness to that stimulus. If the stimulus is no longer regularly applied, the response tends to recover over time. If you poke a snail once, it retracts into its shell, but if you poke it repeatedly, it will learn to ignore the stimulus. The receptors and sensory neurons involved become down-regulated over time and no longer signal, causing the habituation. However, if you leave the snail alone for a while and then touch it again, its retraction response will return; the response neurons have recovered.

Classical Conditioning

Classical or *Pavlovian* conditioning involves the association of a physical response with an environmental stimulus. Pavlov studied the salivation reflex in dogs. In 1927, he discovered that if a dog was presented with an arbitrary stimulus (e.g., a bell) and then presented with food, it would eventually salivate on hearing the bell alone. He came up with the following terminology:

- An established (innate) reflex consists of an *unconditioned stimulus* (e.g., food for salivation), and the response that it naturally elicits is the *unconditioned response* (e.g., salivation).

- A *neutral stimulus* is a stimulus that will not by itself elicit a

Learning and Nerves

The capacity for learning adaptive responses is closely correlated with the degree of neurologic development of the organism—that is, the capacity of its nervous system, particularly its cerebral cortex, for flexibility to form new synapses.

response, prior to conditioning (e.g., a bell). During *conditioning,* the establishment of a new reflex, a neutral stimulus is presented with an unconditioned stimulus. When the neutral stimulus elicits a response in the absence of the unconditioned stimulus, it becomes the *conditioned stimulus.*

- The product of the conditioning experience is termed the *conditioned reflex* (e.g., salivation at the sound of a ringing bell).

Operant or Instrumental Conditioning. Instrumental conditioning involves conditioning responses to stimuli with the use of reward or reinforcement. When the organism exhibits a behavioral pattern that the experimenter would like to see repeated, the animal is rewarded, with the result that it exhibits this behavior more often. B.F. Skinner used the well-known "Skinner Box" to show that animals in a cage could be conditioned to push down a lever to release food from a food dispenser. Instrumental conditioning can be performed through *positive reinforcement* (such as a food reward) or *negative reinforcement* (such as giving an animal a painful electric shock whenever it exhibits a certain behavior).

Modifications of Conditioned Behavior. Organisms eventually "unlearn" conditioned responses, if they are not reinforced. *Extinction* is the gradual elimination of conditioned responses in the absence of reinforcement. The recovery of such a conditioned response after extinction is termed spontaneous recovery.

Limits of Behavioral Change

There are limits, however, to what an organism is capable of learning. These limits can be either neurologic (the organism simply doesn't have the brain power) or chronologic (learning must occur during a narrow window during the organism's development in order to be successful).

These are specific time periods called *critical periods* during an animal's early development when it is physiologically capable of developing specific behavioral patterns. If the proper environmental pattern is not present during this critical period, the behavioral pattern will not develop correctly. In some animals there is also a visual critical period; if light is not present during this period, visual effectors will not develop properly later on, no matter how much stimulus is given.

Imprinting

Imprinting is a process in which environmental patterns or objects presented to a developing organism during a brief critical period in early life become accepted permanently as an element of its behavioral environment. To put it another way, these patterns are "stamped in" and included in an animal's behavioral response. A duckling, for example, passes through a critical period in which it learns that the first large moving object it sees is

Quick Quiz

A dog is rapped hard on the nose with a newspaper every time he steals cinnamon buns from the kitchen table. Eventually, he stops stealing the buns. This is an example of

(A) imprinting

(B) instrumental conditioning through negative reinforcement

(C) a circadian rhythm

(Answer = (B))

its mother. In the natural environment, its mother usually is the first thing a duckling sees. However, if a large, moving object other than its mother is the first thing it sees, the duckling will follow it, as Konrad Lorenz discovered when he was pursued by newborn ducklings that assumed that he was their mother.

Intraspecific Interactions

Just as an organism communicates within itself via nervous and endocrine systems, it also requires methods to communicate with other members of its species. These methods include behavioral displays, pecking order, territoriality, and responses to chemicals. Pheromones can mark a territory or attract a mate, for example, creating a long-lasting message that can be detected over long distances in some cases. Visual communication between animals is quite common, since most animals have vision of some sort, and visual signals are a rapid way to communicate.

The olfactory sense is immensely important as a means of communication. Many animals secrete substances called pheromones that influence the behavior of other members of the same species. One type of pheromone, the releaser pheromone, triggers a reversible behavioral change in the recipient. For example, female silkworms secrete a very powerful attracting pheromone that will attract a male from a distance of two miles or more. In addition to their sex-attracting purposes, releaser pheromones are secreted as alarm and toxic defense substances.

Behavioral Displays

A *display* may be defined as an innate behavior that has evolved as a signal for communication between members of the same species. According to this definition, a song, a call, or an intentional change in an animal's physical characteristics is considered a display. Many animals have evolved a variety of complex actions that function as signals in preparation for mating. Agonistic displays are intended to appease the observer, such as a dog's display of appeasement when it wags its tail. Meanwhile, antagonistic displays are intended to imply hostility and to instill fear, as when a dog directs its face straight at its opponent and raises its body.

Other displays include various dancing procedures exhibited by honeybees, especially the scout honeybee, which is able to convey information to workers in the hive concerning the quality, direction, and distance from the hive of food sources. Displays utilizing auditory, visual, chemical, and tactile elements are often used as a means of communication.

Pecking Order

Frequently, the relationships among members of the same species living as a contained social group become stable for a period of time. When food, mates, or territory are disputed, a dominant member of the species will pre-

Follow the Leader

The trailing behavior of ants is based on scent. Scout ants release scent trails to guide the other ants to food sources.

Round and Waggle Dancing

Honey bees have developed a complex system of communication that involves a form of dancing. A "round dance" (rapid sideways movement in tight circles) indicates that food is near. A "waggle dance" (a half-circle swing in one direction, followed by a straight run and another half circle swing in the other direction) indicates that food is further away in a specified direction.

vail over a subordinate one. This social hierarchy is often referred to as the *pecking order.* This established hierarchy minimizes violent intraspecific aggressions by defining the stable relationships among members of the group; subordinate members only rarely challenge dominant individuals.

Territoriality

Members of most land-dwelling species defend a limited area or territory from intrusion by other members of the same species. These territories are typically occupied by a male or a male-female pair. The territory is frequently used for mating, nesting, and feeding.

Territoriality serves the adaptive function of distributing members of the species so that the environmental resources are not depleted in a small region. Furthermore, intraspecific competition is reduced. Although there is frequently a minimum size for a species' territory, that size varies with population size and density. The larger the population, and the scarcer the resources the available to it, the smaller the territories are likely to be.

Behavioral Cycles

Daily cycles of behavior are called *circadian rhythms.* Animals who exemplify these behavior cycles lose their exact 24-hour periodicity if they are isolated from the natural phases of light and dark. Cyclical behavior, however, will continue with approximate day-to-day phasing. The cycle is thus initiated intrinsically, but modified by external factors. Daily cycles of eating, maintained by most animals, provide a good example of cycles characterized by both internal and external control. The internal controls are the natural bodily rhythms of eating and satiation. External modulators include the elements of the environment that occur in familiar cyclic patterns, such as dinner bells and clocks, but most particularly the light of the sun. Sleep and wakefulness are the most obvious examples of cyclical behavior. These behavior patterns have been associated with particular patterns of brain waves.

Even Plants Need to Sleep Sometimes

The first evidence noted for the existence of 24-hour circadian rhythms was the movements of bean plants.

Plants

Plants are so distinct in their body form and so important to life on earth that we present their physiology separately. Plants are multicellular autotrophs that use the energy of the sun, carbon dioxide, water, and minerals to manufacture carbohydrates through photosynthesis. The chemical energy plants produce is used for respiration by the plants themselves and is the source of all chemical energy in most ecosystems. The plant life cycle is distinct from animals, in that animals are diploid as adults with only gametes as haploid while plants alternate between diploid and haploid forms in each generation.

Plant Organs

Although we may not usually think of plants having organs in the same manner as animals, roots, stems, and leaves each have a defined function and are composed of tissues united around that function in the same manner as animal organs. The stems provide support against gravity and allow for the transport of fluid through vascular tissue. Water travels upward from the roots to the leaves and nutrients travel from the leaves down through the rest of the plant. The roots provide anchoring support, and remove water and essential minerals from the soil. In some plants the roots have a symbiotic relationship with bacteria that fix nitrogen from the atmosphere into a biologically available form that plants can use. The leaves are the primary photosynthetic tissue, generating glucose that can be used to drive all of the plants biochemical energy needs.

One of the key plant tissues is the *xylem*, which contains cells that carry water upward from the roots to the rest of the plant. The xylem is structured differently in flowering and nonflowering plants. In nonflowering plants, or *gymnosperms*, cells called *tracheids* in the stem form a connected network. It is not the cells themselves that conduct water, however. When the cells die, they leave behind their cell walls connected together in one long channel for water transport. Flowering plants, or *angiosperms*, also conduct water through their xylem using the cell walls of dead cells, but in angiosperms the cell walls are more tubelike, making water transport more efficient. In trees, older xylem cells at the innermost layer die, forming the heartwood used for lumber. The outer layer of xylem is alive and is called the sapwood.

Transport in plants, as in animals, encompasses both absorption and circulation. In plants, circulation is called *translocation*, and mainly involves transporting water and carbohydrates. The rise of water up the xylem is caused by transpiration pull (as water evaporates from the leaves of plants, a vacuum is created), capillary action (the rise of any liquid in a thin tube because of the surface tension of the liquid), and root pressure exerted by water entering the root hairs.

Another important plant tissue is the *phloem*. The phloem transports nutrients from the leaves to the rest of the plant. This nutrient liquid is commonly called sap. In the phloem, cells are alive when they perform their transport function. The phloem cells are tube-shaped, moving the sap through the tube. Like terrestrial animals, plants need a protective coating provided by an external layer of epidermis cells. Another plant tissue is the *ground tissue*, involved in storage and support.

Mnemonic

Remember:

Xylem cells transport water, and **ph**loem cells transport **f**ood!

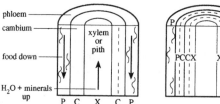
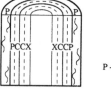

Parts of the Plant

Plant Cells

Plant cells have all of the same essential organelles as other eukaryotic cells, including the mitochondria, ER, Golgi and nucleus. A major distinction of plant cells is the presence of the photosynthetic organelle, the *chloroplast* (see chapter 3, Cellular and Molecular Biology). Some plant cells contain large storage vacuoles not found in animal cells. Another distinct feature of plant cells is their cell wall. Each plant cell is surrounded outside of its plasma membrane by a stiff cell wall made of *cellulose.* The cellulose cell wall helps to provide structure and support for the plant. From grasses to trees, plants rely on cellulose from cell walls to help support the plant against gravity.

Plant Phyla

Within the plant kingdom there are several major phyla. One of the major distinctions for these plant groups is whether or not a plant has vascular tissue for the transport of fluids. Plants without vascular tissue are small simple plants called *nontracheophytes* and include mosses. The *tracheophytes* are the rest of the plants, including pines, ferns, and flowering plants. The evolution of vascular tissue was an important step in the colonization of land by plants, since it increases the support of plants against gravity, and increases their ability to survive dry conditions. Ferns are a phylum of tracheophytes that do not produce seeds, using spores instead for reproduction.

Gymnosperms

Gymnosperms represent the development of the seed, although they have no flowers, and the angiosperms are the flowering plants, the dominant plants in many ecosystems today. Each of these represented major evolutionary steps. *Gymosperms* were the first plants to evolve the use of the seed in reproduction. Gymnosperms do not have flowers, and thus their name, which means "naked seed." Gingko trees are one example of gymnosperms, but by far the most common gymnosperms are the conifer trees such as pine, fir, and redwood. These trees have cones that are involved in reproduction, with one cone that produces male spores and separate cones that produce female spores. The female spores are enclosed in eggs that later develop into seeds after fertilization. Male spores are released as pollen that reaches female cones by wind dispersal usually. The seeds of gymnosperms, without a flower, also are not enclosed in a fruit and are quite small. When released by the female cone, the seeds may fall to the ground.

Angiosperms

Ferns were once dominant land plants and gymnosperms were dominant after that. Today the flowering plants, the angiosperms, are dominant. Most of the discussion of plant physiology will focus on the flowering plants due to their importance in ecosystems, and the great number of species they represent. The two main groups of angiosperms are the *monocots* and the *dicots*. The monocots usually have narrow leaves and include the grasses while dicots have broad leaves. Monocots and dicots differ in many other ways as well such as seed structure.

Angiosperm Structure and Tissues

The stems of angiosperms are arranged with bundles of vascular tissues, separating the xylem and the phloem into organized layers. The *vascular bundles* in monocots are scattered through the stem cross section, while the vascular bundles in dicot stems are organized into a ring. *Phloem* cells are thin-walled and are found on the outside of the vascular bundle while *xylem* cells are found on the inside. Additional stem tissues are the *pith* involved in storage and the cortex to provide strength and structure. The *epidermis* on the outside of the stem protects tissues from the environment. Stems also have a layer of *cambium*, a tissue involved in growth. As cambium cells grow and divide, some of the cells differentiate to form xylem and others form phloem. This cambium contributes to growth of the plant, allowing stems to grow in thickness over time. Another layer of cambium lies beneath the bark of trees.

Like the stem, the root has an epidermis, cortex, phloem cells, xylem cells, and cambium cells. The epidermis contains the *root hair* cells. Root hairs are specialized cells of the root epidermis with thin-walled projections. They provide increased surface area for absorption of water and minerals from the soil through diffusion and active transport. The main functions of the root are absorption, which is accomplished through these root hairs, and anchorage of the plant in the ground. Some roots additionally function in the storage of food (such as the roots of turnips and carrots).

Don't Mix These Up on Test Day

Xylem cells are found in the center of the vascular bundle of the plant, and are responsible for transporting water and minerals **up** the stem.

Phloem cells are found outside the vascular bundle of the plant, and transport nutrients **down** the stem.

Cambium is undifferentiated tissue that can develop into either xylem or phloem.

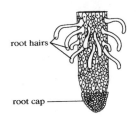

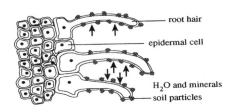

Root

Leaves are the other plant organ. To perform photosynthesis, leaves have adapted in various ways. First of all, the leaf has a waxy cuticle on top to conserve water. Its upper epidermis, the top layer of cells, has no openings, an adaptation which is also intended to inhibit water from being released. Another photosynthetic adaptation, the *palisade layer*, is the term given to elongated cells that are spread over a large surface area and contain chloroplasts. They are directly under the upper epidermis and are well exposed to light.

The leaf also possesses a *spongy layer*, where stomates open into air spaces that contact an internal moist surface of loosely packed spongy layer cells. Spongy cells contain chloroplasts. As in animals, this moist surface is necessary for diffusion of gases into and out of cells in both photosynthesis and respiration. Air spaces in leaves increase the surface area available for gas diffusion by the cells, allowing gaseous interchange of CO_2, H_2O, and O_2. The lower epidermis of the leaf is punctuated by *stomate openings*, which further regulate the loss of water through transpiration and permit diffusion of carbon dioxide, water vapor, and oxygen between the leaf and the atmosphere. The size of these stomate openings is controlled by guard cells. These cells open during the day to admit CO_2 for photosynthesis and close at night to limit loss of water vapor through transpiration.

One explanation for the mechanism by which the guard cells open and close is as follows: During the day, the guard cells, which contain chloroplasts, produce glucose. High glucose content in the cells causes them to swell up via osmosis. This condition is known as *turgor*. Because the inner wall of the guard cell is thickened, the swelling produces a curvature of the opening between the guard cells, and the stomate opening increases. At night, photosynthesis ceases, cell turgor decreases, and the stomate opening closes. During a drought, the stomates will also close during the day to prevent loss of water by transpiration. In this case, photosynthesis ceases because of a lack of CO_2.

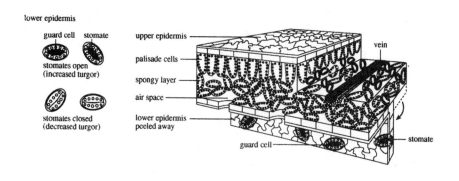

Leaf

Growth of Tissues in the Developing Plant

Growth in higher plants is restricted to embryonic or undifferentiated cells called *meristem*. These tissues undergo active cell reproduction. Gradually, the cells elongate and differentiate into cell types characteristic of the species. Different types of meristem include the *apical meristem,* which is found in the tips of roots and stems. These cells provide for growth in length, which occurs only at the root and stem tips. The *lateral meristem* or cambium is located between the xylem and phloem. This tissue permits growth in diameter. It is not an active tissue in monocots (grasses) or herbaceous dicots (alfalfa). Instead, it is predominant in woody dicots (like oak).

Control of Growth in Plants

The regulation of growth patterns is largely accomplished by plant hormones, which are almost exclusively devoted to this function. These hormones are produced by actively growing parts of the plant, such as the meristematic tissues in the apical region (apical meristem) of shoots and roots. They are also produced in young, growing leaves and developing seeds. Some of these hormones and their specific functions are discussed below.

Auxins

Auxins are an important class of plant hormone associated with several growth patterns, including *phototropism* (growth toward light) and *geotropism* (growth directed by gravity).

Auxins are responsible for phototropism, the tendency of the shoots of plants to bend toward light sources (particularly the sun). When light strikes the tip of a plant from one side, the auxin supply on that side is reduced. Thus the illuminated side of the plant grows more slowly than the shaded side. This asymmetrical growth in the cells of the stem causes the plant to bend toward the slower growing light side; thus the plant turns toward the light. Indoleacetic acid is one of the auxins associated with phototropism.

Geotropism is the term given to the growth of portions of plants towards or away from gravity. With negative geotropism, shoots tend to grow upward, away from the force of gravity. If the plant is turned on its side (horizontally), the shoots will eventually turn upward again. Gravity itself increases the concentration of auxin on the lower side of the horizontally placed plant, while the concentration on the upper side decreases. This unequal distribution of auxins stimulates cells on the lower side to elongate faster than cells on the upper side. Thus the shoots turn upward until they grow vertically once again.

With positive geotropism, roots, unlike shoots, grow toward the pull of gravity. In a horizontally placed stem, however, the effect on the root cells is the opposite. Those exposed to a higher concentration of auxin (the lower side) are inhibited from growing, while the cells on the upper side continue to grow. In consequence, the root turns downward.

Sun Worshippers

When sunflowers bend to follow the sun, they are exhibiting phototropism.

Treating the Fruit You Eat

Ethylene and one of its most important inhibitors, abscissic acid, are commonly used to treat fruits for human consumption. Bananas, for example, are picked unripe, then treated with ethylene so that the ripening process will continue after the fruit is off the tree. As bananas are transported to markets and sold, they are treated with abscissic acid to prevent rotting and extend their shelf life.

Quick Quiz

What kinds of vegetative propagation do the following plants utilize?

(1) daffodil
(2) strawberry
(3) fern
(4) yam

Answers:

(1) bulb
(2) runner
(3) rhizome
(4) tuber

Auxins produced in the terminal bud of a plant's growing tip move downward in the shoot and inhibit development of lateral buds. Auxins also initiate the formation of lateral roots, while they inhibit root elongation.

Gibberellins

This second class of hormones stimulates rapid stem elongation, particularly in plants that normally do not grow tall (for example, dwarf plants). *Gibberellins* also inhibit the formation of new roots, and stimulate the production of new phloem cells by the cambium (where the auxins stimulate the production of new xylem cells). Finally, these hormones terminate the dormancy of seeds and buds, and induce some biennial plants to flower during their first year of growth.

Kinins

Kinins make up another general class of hormone compounds that promote cell division. Kinetin is an important type of cytokinin, and is involved in general plant growth, breaking seed dormancy, and expanding leaves. The action of kinetin is enhanced when auxin is present. The ratio of kinetin to auxin is of particular importance in the determination of the timing of the differentiation of new cells.

Ethylene

Ethylene stimulates the ripening of fruit and induces aging. Its functioning is slowed by inhibitors, which block cell division as part of a number of important control mechanisms. Inhibitors are particularly important to the maintenance of dormancy in the lateral buds and seeds of plants during autumn and winter. They break down gradually with time (and, in some cases, are destroyed by the cold), so that buds and seeds can become active in the next growing season.

Antiauxins

Antiauxins regulate the activity of auxins. For example, indoleacetic acid oxidase regulates the concentration of indoleacetic acid. An increase in the concentration of indoleacetic acid increases the amount of indoleacetic acid oxidase produced.

Asexual Reproduction in Plants

Many plants utilize asexual reproduction, such as *vegetative propagation*, to increase their numbers.

Undifferentiated tissue (*meristem*) in plants provides a source of cells from which new plants can develop. Vegetative propagation offers a number of advantages to plants, including speed of reproduction, lack of genetic variation, and the ability to produce seedless fruit. This process can occur either naturally or artificially.

Natural forms of vegetative propagation include:

- **Bulbs.** These are parts of the root that split to form several new bulbs (an example is the tulip).

- **Tubers.** These modified underground stems have buds, such as the eye of a potato, which develop into new plants.

- **Runners.** Runners are plant stems that run above and along the ground, extending from the main stem. Near the main plant, new plants develop which produce new roots, as well as upright stems at intervals (as in lawn grasses).

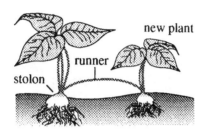

Runner

- **Rhizomes.** *Stolons* is another term used for these woody, underground stems. They reproduce through new upright stems that appear at intervals, eventually growing into independent plants. The iris is a rhizome.

Meanwhile, artificial forms of vegetative propagation include:

- **Cutting.** When cut, a piece of stem of some plants will develop new roots in water or moist ground. Examples include the geranium and the willow. Plant growth hormones like auxins accelerate root formation in cuttings.

- **Layering.** The stems of certain plants, when bent into the ground and covered by soil, will take root. The connection between the main plant and this offshoot can then be cut, resulting in the establishment of a new plant. Blackberry and raspberry bushes reproduce in this manner.

- **Grafting.** Desirable types of plants can be developed and propagated using this method, in which the stem of one plant (*scion*) is attached to the rooted stem of another closely related plant (stock). One prerequisite for successful grafting is that the cambium (the tissue in stem that is not differentiated and allows stems to grow thicker) of the scion must be in contact with the cambium of the stock, since these two masses of undifferentiated cells must grow together to make one. Grafting does not allow for any mixing of hereditary characteristics, since the two parts of the grafted plant remain genetically the same.

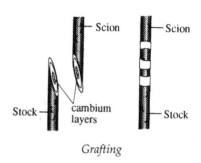

Grafting

The Birds, the Bees, and the Flowers

We enjoy sexual reproduction in flowering plants when we admire blooming trees or eat fruits and vegetables. Flowering is the elaborate mechanism that plants have evolved to attract pollinators such as birds and bees, guaranteeing for themselves the genetic diversity that sexual reproduction provides.

Sexual Reproduction in Plants

Most plants are able to reproduce both sexually and asexually; some do both in the course of their life cycles, while others do one or the other. In the life cycles of mosses, ferns, and other vascular plants, there are two kinds of individuals associated with different stages of the life cycles: the *diploid* and the *haploid*.

Diploid and Haploid Generations

In the diploid or *sporophyte* generation, the asexual stage of a plant's life cycle, diploid nuclei divide meiotically to form haploid spores (not gametes) and the spores germinate to produce the haploid or gametophyte generation.

The haploid or *gametophyte* generation is a separate haploid form of the plant concerned with the production of male and female gametes. Union of the gametes at fertilization restores the diploid sporophyte generation. Since there are two distinct generations, one haploid and the other diploid, this cycle is sometimes referred to as the *alternation of generations*. The relative lengths of the two stages vary with the plant type. In general, the evolutionary trend has been toward a reduction of the gametophyte generation, and increasing importance of the sporophyte generation.

How do these generations express themselves in common plants? In moss, the gametophyte is the green plant that you see growing on the north side of trees. The sporophyte variety is smaller, nongreen (nonphotosynthetic), and short-lived. It is attached to the top of the gametophyte, and is dependent upon it for its food supply. Spores from the sporophyte germinate directly into gametophytes.

In ferns, on the other hand, the reverse pattern may be observed, with the sporophyte of the species dominant. The gametophyte is a heart-shaped leaf the size of a dime. Fertilization produces a zygote from which the commonly seen green fern sporophyte develops. The sporophyte fern's leaves (the fronds) develop spores beneath the surface of the leaf. These spores germinate to form the next generation of gametophyte. In gymnosperms and angiosperms the haploid gametophyte is small and is not independent and is orders of magnitude smaller and more transient than the diploid plants, continuing the evolutionary trend over time for the sporophyte to increase in dominance of the cycle.

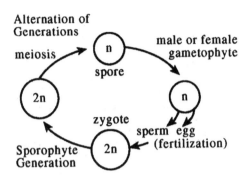

Alternation of Generations

Sexual Reproduction in Flowering Plants

In flowering plants or angiosperms, the evolutionary trend mentioned above continues; the gametophyte consists of only a few cells and survives for a very short time. The woody plant that is seen (for example, a rose) is the sporophyte stage of the species.

The Flower. The flower is the organ for sexual reproduction of angiosperms and consists of male and female organs. The flower's male organ is known as the *stamen*. It consists of a thin, stalklike filament with a sac at the top. This structure is called the *anther*, and produces haploid spores. The haploid spores develop into pollen grains. The haploid nuclei within the spores will become the sperm nuclei, which fertilize the ovum.

Meanwhile, the flower's female organ is termed a *pistil*. It consists of three parts: the *stigma*, the *style*, and the *ovary*. The stigma is the sticky top part of the flower, protruding beyond the flower, which catches the pollen. The tubelike structure connecting the stigma to the ovary at the base of the pistil is known as the style; this organ permits the sperm to reach the ovules. And the ovary, the enlarged base of the pistil, contains one or more ovules. Each ovule contains the monoploid egg nucleus.

Petals are specialized leaves that surround and protect the pistil. They attract insects with their characteristic colors and odors. This attraction is essential for cross-pollination—that is, the transfer of pollen from the anther of one flower to the stigma of another (introducing genetic variability).

Note that some species of plants have flowers that contain only stamens ("male plants") and other flowers that contain only pistils ("female plants").

Don't Mix These Up on Test Day

You will be expected to know that in sexual reproduction in animals, the haploid cell (gamete) is always unicellular, while in the alternation of generations in plants, the haploid gametophyte is multicellular.

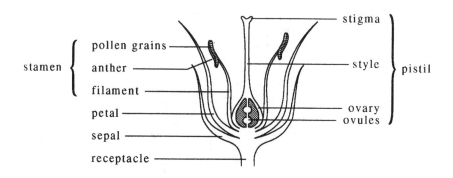

Parts of the Flower

The Male Gametophyte (Pollen Grain). The pollen grain develops from the spores made by the sporophyte (for example, a rose bush). Pollen grains are transferred from the anther to the stigma. Agents of cross-pollination include insects, wind, and water. The flower's reproductive organ is brightly colored and fragrant in order to attract insects and birds, which help to spread these male gametophytes. Carrying pollen directly from plant to plant is more efficient than relying on wind-born pollen and helps to prevent self-pollination, which does not create diversity. When the pollen grain reaches the stigma (pollination), it releases enzymes that enable it to absorb and utilize food and water from the stigma and to germinate a pollen tube. The pollen tube is the remains of the evolutionary gametophyte. The pollen's enzymes proceed to digest a path down the pistil to the ovary. Contained within the pollen tube are the tube nucleus and two sperm nuclei; all are haploid.

Female Gametophyte. The female gametophyte develops in the ovule from one of four spores. This embryo sac contains nuclei, including the two polar (*endosperm*) nuclei and an egg nucleus.

Fertilization. The gametes involved in this cycle of reproduction are nuclei, not complete cells. The sperm nuclei of the male gametophyte (pollen tube) enters the female gametophyte (embryo sac), and a double fertilization occurs. One sperm nucleus fuses with the egg nucleus to form the diploid zygote, which develops into the embryo. The other sperm nucleus fuses with the two polar bodies to form the endosperm (triploid or $3n$). The endosperm provides food for the embryonic plant. In dicotyledonous plants, the endosperm is absorbed by the seed leaves (*cotyledons*).

1 sperm nuclei + 1 egg nuclei = zygote = embryo

1 sperm nucleus + 2 polar nuclei = $3n$ endosperm

Don't Mix These Up on Test Day

The *endosperm* is a nutrient-rich structure formed when a sperm nucleus fuses with two polar nuclei. This fusion provides nourishment for the developing *embryo*.

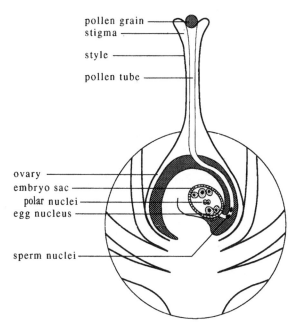

Fertilization in Angiosperms

Development of the Plant Embryo-Seed Formation. The zygote produced in the sequence above divides mitotically to form the cells of the embryo. This embryo consists of the following parts, each with its own function:

- The *epicotyl* develops into leaves and the upper part of the stem.

- The *cotyledons* or seed leaves store food for the developing embryo.

- The *hypocotyl* develops into the lower stem and root.

- The *endosperm* grows and feeds the embryo. In dicots, the cotyledon absorbs the endosperm.

- The *seed coat* develops from the outer covering of the ovule. The embryo and its seed coat together make up the seed. Thus, the seed is a ripened ovule.

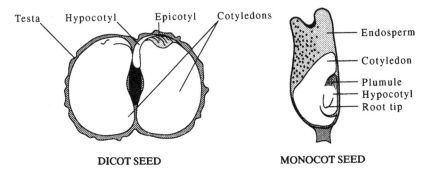

DICOT SEED   **MONOCOT SEED**

Types of Plant Embryos

Fruit Ovaries

Fruits may be thought of as ripened ovaries. They nourish their seed embryos until these can germinate under the right conditions.

Seed Dispersal and Development of Fruit. The fruit, in which most seeds develop, is formed from the ovary walls, the base of the flower, and other consolidated flower pistil components. Thus, the fruit represents the ripened ovary. The fruit may be fleshy (as in the tomato) or dry (as in a nut). It serves as a means of seed dispersal; it enables the seed to be carried more frequently or effectively by air, water, or animals (through ingestion and subsequent elimination). Eventually, the seed is released from the ovary, and will germinate under proper conditions of temperature, moisture, and oxygen.

This chapter aimed to provide you with a broad coverage of the different systems that keep organisms going. You should now feel comfortable with the basics of organismal biology, including concepts like the mechanics of sexual reproduction, nutrition, circulation, respiration, and the endocrine system. Find out just how comfortable you have become with these concepts by seeing how you do on the following quiz questions. Good luck!

Organismal Biology Quiz

1. Which enzyme breaks down starch to disaccharides?

   (A) amylase
   (B) gastrin
   (C) secretin
   (D) pepsin
   (E) maltase

2. Of the choices below, which statement about the blastula stage of embryonic development is false?

   (A) It consists of a solid ball of cells.
   (B) It contains a fluid-filled center called the blastocoel.
   (C) It is the stage of development that precedes the gastrula.
   (D) It is a more advanced stage than a morula.
   (E) It is a less advanced stage than a neurula.

3. Which of the following associations of germinal tissues and developed tissue is INCORRECT?

   (A) mesoderm: heart
   (B) ectoderm: nervous sytem
   (C) endoderm: lining of intestinal tract
   (D) mesoderm: pancreas
   (E) ectoderm: epidermis of skin

4. To ensure survival of their species, animals which do not spend a large amount of time caring for their offspring must

   (A) have the ability to live in water and on land
   (B) lay eggs
   (C) produce many offspring
   (D) have protective coloring
   (E) have internal fertilization

5. Basal metabolism disorders are most likely caused most directly by

   (A) impairment of the pituitary
   (B) impairment of the gonads
   (C) impairment of the thyroid
   (D) impairment of the thymus
   (E) impairment of the parathyroid

6. In humans, the site of successful fertilization is most commonly the

   (A) ovary
   (B) fallopian tube
   (C) uterus
   (D) cervix
   (E) vagina

GO ON TO THE NEXT PAGE

7. Which of the following statements about skeletal muscle tissue is true?

    (A) In the muscle fiber, actin is the thick filament.
    (B) The sarcoplasmic reticulum regulates the level of Ca^{2+} within a muscle cell.
    (C) Unlike a nerve cell, a muscle cell does not possess an excitable membrane.
    (D) In a muscle fiber, myosin is the thin filament.
    (E) Contraction of a muscle fiber can occur in the absence of Ca^{2+}.

8. Which of the following statements is NOT true of the digestive system?

    (A) Digestive enzymes from the pancreas are released via a duct into the duodenum.
    (B) Peristalsis is a wave of smooth muscle contraction that proceeds along the digestive tract.
    (C) In the small intestine, villi absorb nutrients into both the lymphatic and circulatory systems.
    (D) The low pH of the stomach is essential in order for carbohydrate digestive enzymes to function.
    (E) The release of bile from the gall bladder is triggered by the hormone cholecystokinin.

9. Which of the following statements about blood is false?

    (A) Mature red blood cells are not nucleated.
    (B) Blood platelets are involved in the clotting process.
    (C) The adult spleen is a site of red blood cell development.
    (D) White blood cells are capable of phagocytosing foreign matter.
    (E) New red blood cells are constantly developing in the bone marrow.

10. Which of the following is a normal pathway of blood flow?

    (A) right ventricle to aorta
    (B) pulmonary veins to left atrium
    (C) inferior vena cava to left atrium
    (D) pulmonary veins to left ventricle
    (E) left ventricle to pulmonary artery

11. Which of the following associations of brain structure and function is false?

    (A) hypothalamus: appetite
    (B) cerebellum: motor coordination
    (C) cerebral cortex: higher intellectual function
    (D) reticular activating system: sensory processing
    (E) medulla: basic emotional drives

GO ON TO THE NEXT PAGE

12. Which statement about the respiratory system is NOT true?

   (A) Ciliated nasal membranes warm, moisten, and filter inspired air.
   (B) Contraction of the diaphragm enlarges the thoracic cavity.
   (C) When the thoracic cavity enlarges, the pressure of air within the lungs falls.
   (D) When the pressure of air within the lungs is less than the atmospheric pressure, air will flow out of the lungs.
   (E) The respiratory process consists of inspiratory and expiratory acts following one another.

13. Which of the following statements about hormones is NOT true?

   (A) They are transported by the circulatory system.
   (B) They bind to receptors on target cells.
   (C) They must be present in large quantities to have an effect.
   (D) They are secreted by endocrine glands.
   (E) They can affect organs of the body that are far removed from their site of synthesis.

14. Which of the following is an INCORRECT pairing of an endocrine gland and hormone secretion?

   (A) posterior pituitary: LH
   (B) adrenal cortex: aldosterone
   (C) anterior pituitary: TSH
   (D) adrenal medulla: epinephrine
   (E) hypothalamus: FSH releasing factor

15. Which of the following statements about acetylcholine is NOT true?

   (A) ACH is released at the neuromuscular junction.
   (B) ACH binds to specific receptors on the postsynaptic membrane.
   (C) In a synaptic cleft, there are enzymes that degrade ACH.
   (D) ACH diffuses through the presynaptic membrane after its synthesis.
   (E) A synapse that is subjected to many action potentials may be depleted of ACH granules.

16. The sympathetic nervous system causes which of the following?

   (A) constriction of the pupil
   (B) decreased heart rate
   (C) increased gastric section
   (D) reduction of adrenaline secretion
   (E) increased respiration

GO ON TO THE NEXT PAGE

17. In the simple reflex arc

    (A) the sensory neuron synapses directly with the motor neuron
    (B) sensory and motor neurons can synapse outside of the spinal cord
    (C) sensory neurons synapse in the brain
    (D) the motor response occurs without synaptic delay
    (E) a minimum of three neurons must participate

18. Hyperthyroidism is always associated with

    (A) low blood pressure
    (B) severely diminished mental activity
    (C) high metabolic rate
    (D) low body temperature
    (E) decreased heart rate

19. Geotropism is

    (A) the growth of parts of plants towards or away from gravity
    (B) the tendency of plants to bend towards light
    (C) the tendency of plants to have branched roots
    (D) the formation of new phloem cells
    (E) none of the above

20. Gibberellins

    (A) inhibit lateral buds
    (B) stimulate fruit ripening
    (C) determine timing and differentiation of new cells
    (D) stimulate rapid stem elongation
    (E) all of the above

21. Rhizomes are

    (A) underground stems with buds
    (B) stems running above and along the ground
    (C) spores
    (D) the result of sexual reproduction
    (E) woody, underground stems

22. After Konrad Lorenz swam in a pond with newly hatched ducklings separated from their mother, they followed him around as if he were their mother. This is an example of

    (A) discrimination
    (B) response to pheromones
    (C) imprinting
    (D) instrumental conditioning
    (E) none of the above

STOP

Answers and Explanations to the Organismal Biology Quiz

1. (A) Amylase is the enzyme that breaks down complex carbohydrates into disaccharides. Amylase is found in both salivary and pancreatic form. Gastrin (B) is a hormone secreted by the stomach wall of mammals when food makes contact with the wall; it stimulates other parts of the wall to secrete gastric juice. Secretin (C), secreted by the small intestine, stimulates the pancreas to secrete pancreatic juice, which contains bicarbonate ions, which buffer the chyme. Pepsin (D) breaks down proteins into amino acids, while (E) maltase breaks down maltose into glucose.

2. (A) The blastula is not a solid ball of cells, as this answer asserts; it is the hollow ball stage of embryonic development. It develops from the morula, a solid, "mulberrylike" cluster of cells. The center of the blastula is termed the blastocoel. This inpockets during gastrulation to form the gastrula. At this stage, the mesoderm, the archenteron, and the blastopore are formed. During the next stage, the neurula, the ectoderm forms the nerve cord. Therefore, all of the remaining answers accurately represent the various stages of embryonic development.

3. (D) The ectoderm develops into the skin, the nervous system, and the eyes, while the mesoderm develops into the musculoskeletal system and the internal organs. The endoderm develops into the digestive tract and its associated organs such as the pancreas and liver, the respiratory tract, and the bladder lining. The pancreas would therefore develop from the endoderm, and not from the mesoderm as (D) states.

4. (C) There are many types of fertilization, development, and care of offspring in nature. At one end of the spectrum would be internal fertilization, internal development, and lots of care. Organisms that practice internal fertilization, such as elephants, produce few offspring, but a large percentage of the offspring produced reach adulthood. At the other end of the spectrum would be external fertilization, external development, and no care. Organisms at this end of the spectrum, such as many species of fish, must produce large numbers of both eggs and sperm, as relatively few eggs and sperm will interact to produce a zygote. Lacking protection, these zygotes are also susceptible to predation. Therefore, millions of eggs and sperm must be released in order to perpetuate the species. Whether or not a species lives in water or on land (A) does not affect its chances of survival at all, and while protective coloring (D) might be useful, it has no relation to the type of care a species' young receive.

5. (C) The thyroid gland controls the basal metabolic rate through release of thyroxin. It is stimulated by TSH released by the anterior pituitary gland. (A) is too general an answer; impairment of the pituitary would result in a myriad of disorders. The anterior and posterior pituitary glands secrete a large number of hormones, such as ACTH, LH, FSH, ADH, and GH, which control many bodily functions. Meanwhile, (B) impairment of the gonads would result in a loss of FSH and LH from the anterior pituitary. (D) The thymus is involved in the development, maturation, and education of T cells, and the parathyroid hormone (E) stimulates calcium resorption from bone, increasing plasma Ca^{2+} levels as a result.

6. (B) The eggs are released from the ovaries and picked up by the fallopian tubes. In the case of fertilization, a haploid sperm will swim to the egg and fertilize it through the use of a special granule, termed the acrosome, that allows the sperm to penetrate the cell membrane of the egg. The fertilized egg will then continue down the fallopian tube until it reaches the uterus, where it will implant. If it implants in the fallopian tube, an ectopic pregnancy results. (C) Fertilization in the uterus can occur, but it is rare and will sometimes be associated with complications like cervical implantation. (D) and (E) would both disallow implantation of the fertilized egg into the uterus. The vagina is the birth canal; the cervix is the connection between the uterus and the birth canal that expands during delivery.

7. (B) Muscle cells are composed of myofibrils of thick (myosin) and thin (actin) filaments. Contraction occurs through the sliding of these filaments over each other through reversible bridges. Acetylcholine released from the motor neuron excites the membrane of the muscle cell, causing a muscle action potential, and stimulates the sarcoplasmic reticulum to release Ca^{2+} into the cytoplasm. This increase in calcium causes troponin and tropomyosin to expose their myosin binding sites. As this occurs, muscle filaments slide over each other and the muscle contracts, which makes (B) the only correct answer.

8. (D) A low stomach pH is essential in order for the stomach enzyme pepsin that breaks down proteins into amino acids to function, not enzymes that break down carbohydrates, as this answer suggests. The other choices are all true. In (A), amylases, lipases, and bicarbonate are released through the pancreatic duct, while (B) peristalsis does propel food and waste through the system. (C) Glucose and amino acids are picked up by the blood, while fats are picked up by lacteals (special vessels that connect with the lymphatic system). Finally, cholecystokinin (E) is a hormone released by the wall of the small intestine. It travels through the blood to the gall bladder to release bile into the duodenum, emulsifying fats and decreasing the pH of the chyme.

9. (C) In the adult, all hematopoiesis (blood cell development) occurs in the bone marrow. In the fetus, hematopoiesis occurs in the fetal liver and spleen. The spleen acts as a reservoir for red blood cells in adults and filters blood. (A) Mature red blood cells, meanwhile, are not nucleated, in order to create more space for hemoglobin. Blood platelets (B) are crucial to the clotting of blood, and certain white blood cells (D), such as macrophages and neutrophils, engulf foreign matter in a process known as phagocytosis. (E) New blood cells are continually being produced in the bone marrow to replace the cells after their life span of 120 days has ended.

10. (B) Blood traveling from the left ventricle flows into the aorta, then goes to all areas of the body except the lungs. For example, blood in the brachiocephalic artery travels to the head and the shoulders, while blood in the renal artery travels to the kidney to be filtered. Blood flows from these arteries into arterioles, then into capillaries, where food, waste, and energy will be exchanged. Next, the blood continues into venules and collects in veins, which then transport this blood to the superior and inferior vena cavas. This process is known as systemic circulation. In pulmonary circulation, the blood enters the right atrium and flows into the right ventricle. It is then transported to the lungs via the pulmonary artery, where capillary beds surround the alveoli so that gas exchange can occur. At this point, the pulmonary veins bring the blood back to the left atrium to start the process all over again.

11. (E) The medulla monitors blood carbon dioxide levels and pH and adjusts breathing, temperature, and heart rate. It is also the center for reflex activities such as coughing, sneezing, and swallowing, and is not associated with emotional drives. The other answer choices are true: The hypothalamus (A) is the center that controls thirst, hunger, sleep, blood pressure, and water balance; (B) the cerebellum controls muscle coordination and tone and maintains posture; (C) the cerebral cortex is the center for vision, hearing, smell, voluntary movement, and memory; and (D) the reticular activating system receives and sorts sensory input.

12. (D) There is low pressure inside the thoracic cavity due to expansion of the thoracic volume when the diaphragm contracts (as in (B) and (C)). When this pressure drops, air rushes in, and (A) the ciliated membrane warms, moistens, and filters the inspired air. Air then travels through the bronchi, into the bronchioles, and finally into the alveoli, where diffusion occurs to oxygenate the blood and release CO_2 carried back from the tissue. (D) is incorrect because when the pressure of air within the lungs is less than the atmospheric pressure, air actually rushes into the lungs rather than flowing out of them.

13. (C) Hormones are capable of being effective at picomolar, or very small, concentrations. (A) They are also transported by the circulatory system after being secreted by the endocrine (ductless) glands (D), and they may travel far from their site of synthesis to have an affect on their target organ (E). As for (B), cells will respond to hormones only if they have the appropriate receptor on their cell surfaces (in the case of peptide hormones) or inside their cytoplasm (in the case of steroid hormones).

14. (A) The posterior pituitary secretes ADH and oxytocin, not LH. (B) The adrenal cortex does secrete aldosterone, a mineral cortisone that causes increased absorption of Na^+, while (C) TSH, secreted by the anterior pituitary, stimulates thyroid production of thyroxin, which raises the basal metabolic rate. (D) Epinephrine increases heart rate, blood pressure, and the flow of blood to skeletal muscle. Finally, (E) the hypothalamus-secreted FSH releasing factor stimulates the anterior pituitary.

15. (D) ACH is released at the neuromuscular junction when vesicles containing this neurotransmitter merge with the membrane. ACH diffuses across the synapse and binds to specific receptors on the post-synaptic membrane. An enzyme known as acetylcholinesterase degrades ACH very quickly after release in order to prevent constant stimulation of the synapse.

16. (E) The sympathetic nervous system is known as the "fight or flight" response. Stimulation of this branch of the autonomic nervous system is characterized by an increase in heart rate, dilation of the pupils, an increase in respiration and bronchial dilation, and more blood flow to the skeletal muscles and away from the digestive organs. An increase in the amount of adrenaline secretion also occurs. (A)–(D) are all characteristics of the anatagonistic branch of the autonomic nervous system, the parasympathetic nervous system, which is known as the "rest and digest" response.

17. (A) A reflex arc is a quick sensation response coupled to a motor response for quickness or protection. Examples of this are sneezing, coughing, blinking, or moving your hand away from a hot stove. The simplest version of this type of arc would be one sensory neuron linked to one motor neuron. As for (B) and (C), the synapse between the sensory and motor neuron is located in the spinal cord. There is a brief synaptic delay between the motor neuron and the corresponding muscle (D), while this reaction involves a minimum of two neurons (E) (although other neurons might be present, such as neurons that feel pain, they are not considered part of the arc).

18. (C) Hyperthyroidism results in an excess of thyroxin and other thyroid hormones. These hormones increase the basal metabolic rate and blood pressure. Patients with hyperthyroidism are often characterized by sensitivity to heat and nervousness. The other choices are all characteristics of hypothyroidism.

19. (A) Geotropism is the growth of portions of plants towards or away from gravity. Negative geotropism occurs as the shoot grows away from the force of gravity, and positive geotropism describes the growth of roots toward the pull of gravity. Both occur as a result of differential concentrations of auxins (plant hormones associated with growth patterns) on upper and lower sides of the plant. The tendency of plants to bend toward light is termed phototropism (B); the formation of new phloem cells is stimulated by gibberellins, another class of plant hormones (D).

20. (D) Gibberellins are a class of plant hormones that stimulate rapid stem elongation, inhibit the formation of new roots, stimulate the production of new phloem cells by the cambium, and terminate the dormancy of seeds and buds. Auxins, not gibberellins, inhibit the growth of lateral buds (A), while ethylene stimulates fruit ripening (B).

21. (E) Rhizomes are a type of natural form of vegetative propagation and are characterized by woody, underground stems. At intervals, new, upright stems appear. Examples of plants that utilize rhizomes are ferns and irises. Other forms of natural vegetative propagation include bulbs (such as tulips), tubers (underground stems with buds such as potatoes), and runners (stems running above and along the ground, such as strawberries). All of the above forms exemplify asexual reproduction in plants.

22. (C) Imprinting is a process in which environmental patterns or objects presented to a developing organism during a brief "critical period" in early life become accepted as permanent elements of their behavioral environment. A duckling, for example, passes through a critical period in which it learns that the first large moving object it sees is its mother. Discrimination (A) involves the ability of the learning organism to differentially respond to slightly different stimuli, while (B) pheromones are substances secreted by many animals that influence the behavior of other members of the same species. Finally, instrumental conditioning (D) involves conditioning responses to stimuli through reward or reinforcement.

CHAPTER FIVE

CLASSICAL GENETICS

Classical genetics is the study of the patterns and mechanisms of the transmission of inherited traits from one generation to another. The foundations for this field were laid by the monk Gregor Mendel, who in the mid-19th century performed a series of experiments to determine the rules of inheritance in garden pea plants.

The study of classical genetics requires an understanding of *meiosis*, the mechanism of gamete formation. Mendel knew that alleles were inherited from each parent, and that these alleles were somehow linked to the various characteristics he studied in his peas, but it was not until meiosis was truly elucidated that the mechanisms behind heredity were understood.

Meiosis

In asexual reproduction, a single diploid cell (or cells) is used to create new identical copies of an organism. In sexual reproduction, two parents contribute to the genome of the offspring and the end result of union is genetically unique offspring. To do this requires that each parent contribute a cell with one copy of the genome. *Meiosis* is the process whereby these sex cells are produced.

As in mitosis, the gametocyte's chromosomes are replicated during the S phase of the cell cycle, and the centrioles replicate at some point during interphase. The first round of division (*meiosis I*) produces two intermediate daughter cells. The second round of division (*meiosis II*) involves the separation of the sister chromatids, resulting in four genetically distinct haploid gametes. In this way, a diploid cell produces haploid daughter cells. Since meiosis reduces the number of chromosomes in each cell from $2n$ and $1n$, it is sometimes called "reductive division."

Each meiotic division has the same four stages as mitosis, although it goes through each of them twice (except for DNA replication). The stages of meiosis are detailed in the following paragraphs.

Don't Mix These Up on Test Day

Mitosis:
- Produces $2n$ cells from $2n$ cells
- Occurs in all dividing cells
- Does not involve the pairing up of homologous chromosomes
- Does not involve crossing over (recombination)

Meiosis:
- Produces n cells from $2n$ cells
- Occurs only in sex cells (gametocytes)
- Involves the pairing up of homologous chromosomes at the metaphase plate, forming tetrads
- Involves crossing over

Interphase I

Gametocyte chromosomes are replicated during the S phase of the cell cycle, while the centrioles replicate at some point during interphase.

Prophase I

During this stage, chromatin condenses into chromosomes, the spindle apparatus forms, and the nucleoli and nuclear membrane disappear. Homologous chromosomes (matching chromosomes that code for the same traits, one inherited from each parent), come together and intertwine in a process called *synapsis*. Since at this stage each chromosome consists of two sister chromatids, each synaptic pair of homologous chromosomes contains four chromatids, and is therefore often called a *tetrad*.

Sometimes chromatids of homologous chromosomes break at corresponding points and exchange equivalent pieces of DNA; this process is called *crossing over* or recombination. Note that crossing over occurs between homologous chromosomes and not between sister chromatids of the same chromosomes. The chromatids involved are left with an altered but structurally complete set of genes.

The chromosomes remain joined at points called chiasmata where the crossing over occurred. Such genetic recombination can "unlink" linked genes, thereby increasing the variety of genetic combinations that can be produced via gametogenesis. Recombination among chromosomes results in increased genetic diversity within a species. Note that sister chromatids are no longer identical after recombination has occurred.

Metaphase I

Homologous pairs (tetrads) align at the equatorial plane of the dividing cells, and each pair attaches to a separate spindle fiber by its kinetochore.

Anaphase I

Homologous pairs separate and are pulled to opposite poles of the cell. This process is called *disjunction,* and it accounts for a fundamental Mendelian law. During disjunction, each chromosome of paternal origin separates (or disjoins) from its homologue of maternal origin, and either chromosome can end up in either daughter cell. Thus, the distribution of homologous chromosomes to the two intermediate daughter cells is random with respect to parental origin. Each daughter cell will have a unique pool of alleles provided by a random mixture of maternal and paternal origin. These genes may code for alternative forms of a given trait.

Telophase I and Cytokinesis

A nuclear membrane forms around each new nucleus. At this point, each chromosome still consists of sister chromatids joined at the centromere. Next, the cell divides through cytokinesis into two daughter cells, each of which receives a nucleus containing the haploid number of chromosomes. Between cell divisions there may be a short rest period, or interkinesis, during which the chromosomes partially uncoil.

Prophase II

The centrioles migrate to opposite poles and the spindle apparatus forms.

Metaphase II

The chromosomes line up along the equatorial plane once again. The centromeres divide, separating the chromosomes into pairs of sister chromatids.

Anaphase II

The sister chromatids are pulled to opposite poles by the spindle fibers.

Telophase II

Finally, a nuclear membrane forms around each new haploid nucleus. Cytokinesis follows and two daughter cells are formed. Thus, by the time meiosis is completed, four haploid daughter cells are produced per gametocyte. In females, only one of these three becomes a functional gamete.

The diagram on the following page summarizes the various stages of meiosis I and meiosis II. Notice that the random distribution of homologous chromosomes in meiosis, coupled with crossing over in prophase I, enables an individual to produce gametes with many different genetic combinations. Every gamete gets one copy of each chromosome, but the copy of each chromosome found in a gamete is random. For example, each gamete has a chromosome #9, but this chromosome can be either of the two copies of this chromosome. With 22 autosomal chromosomes, this factor alone allows for 2^{22} possible gametes, not including the additional genetic diversity created by recombination. This is why sexual reproduction produces genetic variability in offspring, as opposed to asexual reproduction, which produces identical offspring. The possibility of so many different genetic combinations is believed to increase the capability of a species to evolve and adapt to a changing environment.

The Problem with Nondisjunction

If, during anaphase I or II of meiosis, homologous chromosomes or sister chromatids fail to separate (in what is termed *nondisjunction*), one of the resulting gametes will have two copies of a particular chromosome and the other gamete will have none. Subsequently, during fertilization, the resulting gamete may have one too many (47) or one too few (45) copies of the chromosome in question. Few of these mutated gametes survive. Those that do survive encounter difficulties associated with conditions like Down's syndrome (Trisomy 21).

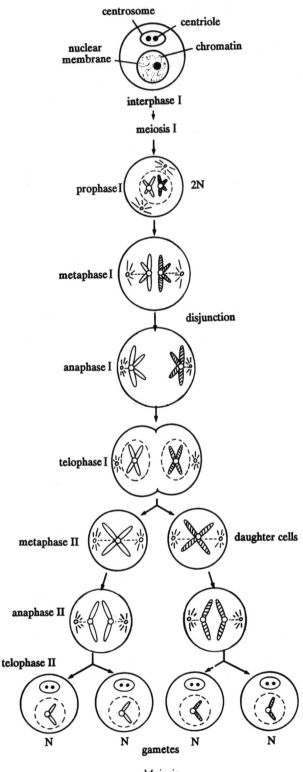

Meiosis

Mendelian Genetics

Around 1865, based on his observations of seven characteristics of the garden pea, Gregor Mendel developed the basic principles of genetics—*dominance, segregation*, and *independent assortment*. Although Mendel formulated these principles, he was unable to propose any mechanism for hereditary patterns, since he knew nothing about chromosomes or genes. Hence his work was largely ignored until the early 1900s.

After Mendel's work was rediscovered, Thomas H. Morgan tied the principles of genetics to the chromosome theory. He linked specific traits to regions of specific chromosomes visible in the salivary glands of *Drosophila melanogaster*, the fruit fly. Morgan brought to light the giant chromosomes, at least 100 times the size of normal chromosomes, that are found in the fruit fly's salivary glands. These chromosomes are banded, and the bands coincide with gene locations, allowing geneticists to visibly follow major changes in the fly genome. Morgan also described sex-linked genes.

The fruit fly is a highly suitable organism for genetic research. With its short life cycle, it reproduces often and in large numbers, providing large sample sizes. It is easy to grow in the laboratory, but has a fairly complex body structure. Its chromosomes are large and easily recognizable in size and shape. They are also few in number (eight chromosomes/four pairs of chromosomes). Finally, mutations occur relatively frequently in this organism, allowing genes for the affected traits to be studied.

Some of the basic rules of gene transmission and expression are:

- *Genes* are elements of DNA that are responsible for observed traits.

- In eukaryotes, genes are found in large linear chromosomes, and each chromosome is a very long continuous DNA double-helix. Humans have twenty-three different chromosomes, with two copies of each chromosome in most cells.

- Each chromosome contains a specific sequence of genes arranged along its length.

- Each gene has a specific location on a chromosome.

- Diploid organisms have two copies of each chromosome and therefore two copies of each gene (except for the *X* and *Y* chromosomes in males).

- The two copies of each gene can have a different sequence in an organism and a gene can have several different sequences in a population. These different versions of a gene are called *alleles*.

- The type of alleles of an organism has, its genetic composition, is called the *genotype*.

Perfect Peas

It is suspected that Mendel might have doctored the results of his experiments, because the data from his pea crosses is almost too perfect.

Flashback

As we discussed in our chapter on cellular and molecular biology, DNA sequence is responsible for genotype, while the protein produced during the translation of mRNA transcribed from the DNA sequence is responsible for phenotype.

- The appearance and physical expression of genes in an organism is called the *phenotype*.

- Types of alleles include dominant and recessive alleles. A dominant allele is expressed in an organism regardless of the second allele in the organism. A recessive allele will not be expressed if the other allele for the gene is a dominant one.

- A homozygous individual has two copies (two alleles) of a gene that are identical and a heterozygous individual has two different alleles for a gene.

- The phenotype of an individual is determined by the genotype.

Dominance of Phenotypic Traits

If two members of a pure-breeding strain are mated, their offspring will always have the same phenotype as the parents since they are all homozygous for the same allele. What happens if two different pure-breeding strains that are homozygous for two different alleles are crossed? In an example such as two different alleles for flower color, what often occurs is that all of the offspring of the cross match the phenotype of one parent and not the other. For example, if a pure-breeding red strain is crossed with a pure-breeding white one, perhaps all of the offspring are red. Where did the allele coding for the white trait go? Did it disappear from the offspring?

If it is true that both parents contribute one copy of a gene to each of their offspring, then the allele cannot disappear. The offspring must all contain both a white allele and a red allele. Despite having both alleles, however, they only express one—the red allele. Red is then a dominant allele and white a recessive allele, since it is not expressed in heterozygotes such as the offspring in this cross of two pure-breeding strains.

Every human has two copies of each of their 23 chromosomes, with the exception of the X and Y chromosome in men. Thus, each gene is present in two copies that can either be the same, or different. For example, a gene for eye color could have two alleles: B or b. B is a dominant allele for brown eye color and b is a recessive allele for blue eye color. There are three potential genotypes: BB, Bb, or bb. BB individuals and Bb individuals have brown eyes, and only bb homozygous people have blue eyes. Bb people have brown eyes since the B allele is dominant and the recessive b allele is not expressed in the heterozygote.

Test Crosses

Often, a geneticist will study the transmission of a trait in a species such as flies or pea plants by performing crosses (matings) between organisms with defined traits. For example, an investigator may identify two possible phenotypes for flower color in pea plants: pink and white. Pink plants bred together always produce pink offspring and white plants bred together always produce offspring with white flowers. It is likely that the differences in flower color are caused by different alleles in a gene that controls flower color. Which of these traits is determined by a recessive or dominant allele however? You cannot tell based on the color alone which trait will be dominant or recessive. Either pink or white could be dominant, or neither.

The way to determine the dominant or recessive nature of each allele is by performing a test cross. Since the pink plants always produce pink plants and the white plants always produce white plants, these are both termed "pure-breeding" plants and are each homozygous for either the P allele (PP genotype has a pink phenotype) or for the p allele (pp genotype has a white phenotype). What will be the phenotype of a plant with the Pp genotype?

When performing a test cross, a useful tool is called a *Punnett Square*. To perform a Punnett Square, first determine the possible gametes each parent in the cross can produce. In the example above, a PP parent can make gametes with either of the two P alleles and the pp parent can only make gametes with the p allele:

PP parent: Gametes have either one P allele or the other P

pp parent: Gametes have either one p allele or the other p allele.

The next step is to examine all of the ways that these gametes could combine if these two parents were mated together in a test cross. This is where the Punnett Square comes in. On one side of the square, align the gametes from one parent, and on the other side of the square align the gametes from the other parent. At the intersection of each potential gamete pairing, fill in the square with the diploid zygote produced by matching the alleles. In this example, all of the offspring of this cross are going to be heterozygous.

If all of the offspring are pink, what does this reveal about the nature of these alleles? If the heterozygous Pp plant has the same phenotype as the homozygous PP plant, then the P allele is dominant over the p allele. If the p allele is not expressed in the heterozygote, the p allele is recessive and the P allele is dominant. The offspring of this cross (shown within the box) can be called the F_1 generation.

A Cross Between Two Pure-Breeding Strains (F_1 generation):

	P	P
p	Pp	Pp
p	Pp	Pp

The F_1 offspring all have the Pp genotype and the pink phenotype. What will occur if two of these F_1 plants are crossed? A Punnett Square can be used again to predict the genotypes in the F_2 generation.

Punnett Square

A *Punnett square*, as shown in the text at left, is a useful tool. It provides a quick way to determine the probable traits of offspring produced from particular crosses.

Parent 1: P and p gametes are produced

Parent 2: P and p gametes are produced

F_2 Generation Punnett Square:

	P	P
P	PP	Pp
p	Pp	pp

Since we know that the P allele for pink is dominant, we can use the genotypes to predict phenotypes of the F_2 generation. PP homozygotes will be pink, and Pp heterozygotes will also be pink since P is dominant. pp plants will be white like the original pure-breeding white plants. Filling in the square above with these phenotypes:

	P	p
P	PP (pink)	Pp
p	Pp (pink)	pp (white)

The ratios of the different genotypes and phenotypes in the Punnett Square should match the statistical probability of producing these in real life by a cross of this type. For example, if two heterozygous Pp plants are crossed, 75 percent of the offspring will be pink and 25 percent white. This is predicted from the Punnett Square based on the ratio of 3:1 for phenotypes that will produce pink (3) to white (1).

The behavior of different pea plant traits helped Mendel to formulate two fundamental rules of Mendelian genetics, the Law of Segregation and the Law of Independent Assortment. Mendel derived these rules based purely on his knowledge of the transmission of traits, without knowing anything about the molecular basis for his observations in the mechanisms of meiosis.

Law of Segregation

The Law of Segregation states that if there are two alleles in an individual that determine a trait, these two alleles will separate during gamete formation and can act independently. For example, when a heterozygous Pp plant is forming gametes, the P and the p alleles can separate into different gametes and act independently during a cross. If this was not the case, and the P and p alleles could not separate, then all of the offspring would remain Pp and all of the F_2 would be pink still. The fact that white offspring are produced indicates that alleles do indeed segregate into gametes independently. The molecular basis for this observation is that during meiosis, each homologous chromosome carrying the two different alleles will end up in a different haploid gamete.

Mendel's Laws

Mendel's laws are:

- Law of Segregation
- Law of Independent Assortment

The dominance of phenotypic traits is also sometimes referred to as one of Mendel's laws, the "Law of Dominance."

Law of Independent Assortment

The Law of Independent Assortment describes the relation between different genes. If the gene that determines plant height is on a diff chromosome than the gene for flower color, then these traits will act independently during test crosses. The two alleles for tallness are the dominant allele T for tall plants and the recessive t allele for short plants. The two alleles for color are the dominant Y allele for yellow and the recessive y allele for white. When plants are crossed, the alleles for the tall gene act independently of the alleles for the color gene.

Example of a dihybrid cross in which tall and yellow are both hybrids:

	TY	Ty	tY	ty
TY	TTYY	TTYy	TtYY	TtYy
Ty	TTYy	Ttyy	TtYy	Ttyy
tY	TtYY	TtYy	ttYY	ttYy
ty	TtYy	Ttyy	ttYy	ttyy

Results of the cross:

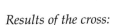

Phenotype ratio:

9 tall yellow $\left(\dfrac{9}{16}\right)$: 3 tall green $\left(\dfrac{3}{16}\right)$: 3 short yellow $\left(\dfrac{3}{16}\right)$: 1 $\left(\dfrac{1}{16}\right)$ short green

The simplest approach to an independent assortment problem is to consider each of the genes separately, determine the predicted Mendelian ratios for each of the traits alone, and then use the laws of probability to combine these. For example, in the cross above, the predicted Mendelian phenotype ratios are $\dfrac{3}{4}$ for tall and $\dfrac{1}{4}$ for green. The probability of observing these phenotypes together is the product of their independent probabilities—that is, $\dfrac{3}{4} \times \dfrac{1}{4}$, or $\dfrac{3}{16}$. A significant variation from this ratio indicates linkage and a failure to assort independently.

Three to One

Note that in the dihybrid cross under discussion here, each trait assorts individually in a 3:1 ratio, as is generally the case in a monohybrid cross. There are 9 tall yellow and 3 tall green for a total of 12 tall. There are also 3 short yellow and 1 short green, which amounts to a total of 4 short. Hence both the tall:short ratio (12:4) and the yellow:green ratio are 3:1.

Linkage

There is a significant exception to the law of independent assortment. For genes to assort independently into gametes during meiosis, they must be on different chromosomes. If two genes are located near each other on the same chromosome, then the alleles for these genes will stay together during meiosis. This phenomenon in which alleles fail to assort independently because they are on the same chromosome is called *linkage*.

Another factor that affects linkage is the recombination between homologous chromosomes that occurs during meiosis. Even if two genes are on the same chromosome, they will not necessarily be linked 100 percent through meiosis. If recombination occurs in the DNA between the two genes in the chromosome, then this will tend to reduce the linkage between genes. The further apart the genes, the more recombination that will occur between them, and the less linkage that will be observed. If the genes are far enough apart on the chromosome, then recombination between the genes may be so frequent that they will display almost no linkage and will assort independently despite being located on the same chromosome.

The fact that the linkage between genes is related to their distance from each other can be used to map the position of genes relative to each other on a chromosome. By performing a test cross and counting the number of offspring, a geneticist can determine how many recombination events occurred between the genes and from this can estimate the distance of two genes from each other.

Inheritance Patterns

Ethical restraints forbid geneticists to perform test crosses in human populations. Instead, they must rely on examining matings that have already occurred, using tools such as pedigrees. A *pedigree* is a family tree depicting the inheritance of a particular genetic trait over several generations. By convention, males are indicated by squares, and females by circles. Matings are indicated by horizontal lines, and descendants are listed below matings, connected by a vertical line. Individuals affected by the trait are generally shaded, while unaffected individuals are unshaded. When carriers of sex-linked traits have been identified (typically, female heterozygotes), they are usually half shaded in family traits.

The following pedigrees illustrate two types of heritable traits: recessive disorders and sex-linked disorders. When analyzing a pedigree, look for individuals with the recessive phenotype. Such individuals have only one possible genotype—homozygous recessive. Matings between them and the dominant phenotype behave as test crosses; the ratio of phenotypes among the offspring allows deduction of the dominant genotype. In any case in which only males are affected, sex-linkage should be suspected.

Study Tip

When faced with a difficult pedigree on your Biology E/M exam, remember that recessive phenotypes can only have one possible genotype—homozygous recessive.

Recessive Disorders

Note how the trait skips a generation in the autonomal recessive disorder depicted in the figure on the following page. Albinism is an example of this form of disorder.

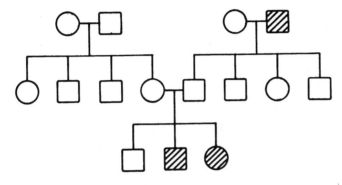

Recessive Disorder

Sex-Linked Disorders

Gender skewing is evident in this type of disorder, which includes traits such as hemophilia. Sex-linked recessive alleles are almost always expressed only by males and transmitted from one generation to another by female carriers.

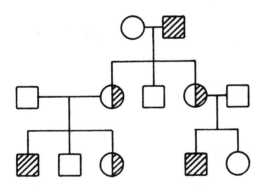

Sex-linked Disorder

Quick Quiz

Match each of the following numbered traits with the correct lettered definition.

1. recessive trait

2. sex-linked recessive trait

3. dominant trait

(A) appears in every generation

(B) skips generations

(C) is almost always found in males

Answers:

1. = (B)
2. = (C)
3. = (A)

Non-Mendelian Inheritance Patterns

While Mendel's laws hold true in many cases, these laws cannot explain the results of certain crosses. Sometimes an allele is only incompletely dominant or, perhaps, codominant. The genetics that enable the human species to have two genders would also not be possible under Mendel's laws.

Incomplete Dominance

Incomplete dominance is a blending of the effects of contrasting alleles. Both alleles are expressed partially, neither dominating the other.

An example of incomplete dominance is found in the four-o'clock plant and in the snapdragon flower. When a red flower (RR) is crossed with a white flower (WW), a pink blend (RW) is created. When two pink flowers are crossed, the yield is 25 percent red, 50 percent pink, and 25 percent white (phenotypic and genotypic ratio 1:2:1).

Codominance

In codominance, both alleles are fully expressed without one allele dominant over the other. An example is blood types. Blood type is determined by the expression of antigen proteins on the surface of red blood cells. The A allele and the B allele are codominant if both are present and combine to produce AB blood.

The allele for blood type A, I_A, and the allele for blood type B, I_B, are both dominant to the third allele, i. I_A and I_B may appear together to form blood type AB; however, when both are absent, blood type O results.
To summarize:

- I_A = gene for producing antigen A on the red blood cell

- I_B = gene for producing antigen B on the red blood cell

- i = recessive gene; does not produce either antigen

And these genes combine in various ways to form the following possible genotypes and blood types (phenotypes):

- $I_A I_A$ or $I_A i$ = Type A blood

- $I_B I_B$ or $I_B i$ = Type B blood

- $I_A I_B$ = Type AB blood, with A and B alleles codominant

- ii = Type O blood

Don't Mix These Up on Test Day

In *incomplete dominance*, two traits are blended together; both are partially expressed, and neither dominates.

In *codominance*, both traits are fully expressed; neither dominates.

Sex Determination

Most organisms have two types of chromosomes: *autosomes*, which determine most of the organism's body characteristics, and *sex chromosomes*, which determine the sex of the organism. Humans have 22 pairs of autosomes and one pair of sex chromosomes. The sex chromosomes are known as X or Y. In humans, XX is present in females and XY in males. The Y chromosome carries very few genes. Sex is determined at the time of fertilization by the type of sperm fertilizing the egg, since all normal eggs contain X chromosomes only. If the sperm carries an X chromosome, the offspring will be female (XX); if the sperm carries a Y chromosome, the offspring will be male (XY).

This process is illustrated in the Punnett square below:

	X	Y
X	XX	XY
Y	XX	XY

(From the mother) ... (From the father)

The ratio of the sex of the offspring is 1:1.

Sex Linkage

Genes for certain traits, such as color blindness or hemophilia, are located on the X chromosomes. Hence these genes are linked with the genes controlling sex determination. These genes seem to have no corresponding allele on the Y chromosome, with the result that the X chromosome contributed by the mother is the sole determinant of these traits in males. Genes determining two afflictions—hemophilia and red-green color blindness—are sex-linked (on the X chromosome). They are recessive, implying that they can be hidden by a dominant normal allele on the other X-chromosome in a female. For this reason, the female with two X chromosomes may carry, but will rarely exhibit, these afflictions. The male, on the other hand, with his Y chromosome, has no dominant allele to mask the recessive gene on his X chromosome. As a consequence of having a single copy of X-linked genes, males exhibit sex-linked traits much more frequently than females do.

Cross 1: Let's see what happens when we cross a hemophilia-carrying female and a normal male:

$XX_h \times XY$:

	X	X_h
X	XX	XX_h
Y	XY	X_hY

Results of the cross:

XX = healthy female
XX_h = carrier but healthy female
XY = healthy male
X_hY = hemophiliac male

There are no male carriers of this trait since all males that have the hemophilia allele express it.

Cross 2: Here's a cross between a carrier female and a male hemophiliac:

$XX_h \times X_hY$:

X	XX_h	XhX_h
X	XY	X_hY

Results of the cross:

X_hX_h = hemophiliac female (very rare)
XX_h = carrier but healthy female
XY = healthy male
X_hY = hemophiliac male

Mutations

Mutations can create new alleles, the raw material that drive evolution via natural selection. Mutations are changes in the genes that are inherited. To be transmitted to the succeeding generation, mutations must occur in sex cells—eggs and sperm—rather than somatic cells (body cells). Mutations in nonsex cells are called somatic cell mutations and affect only the individual involved, not subsequent generations. A somatic mutation can cause cancer, but will have no affect on offspring since it is not present in gametes. Most mutations are recessive and deleterious (harmful). Because they are recessive, these mutations can be masked or hidden by the dominant normal genes.

Chromosomal Mutations

These mutations result in changes in chromosome structure or abnormal chromosome duplication. In crossing over, segments of chromosomes switch positions during meiotic synapsis. This process breaks linkage patterns normally observed when the genes are on the same chromosome. A translocation is an event in which a piece of a chromosome breaks off and rejoins a different chromosome.

Nondisjunction is the failure of some homologous pairs of chromosomes to separate following meiotic synapsis. The result is an extra chromosome or a missing chromosome for a given pair. For example, Down's syndrome is due to an extra chromosome #21 (Trisomy 21). The number of chromo-

Flashback

DNA mutations can occur in two ways:

- Point mutations involve changes in single nucleotide bases in the DNA sequence

- Frameshift mutations involve the insertion or deletion of nucleotides, changing the reading frame of the protein

somes in a case of single nondisjunction is $2n + 1$ or $2n − 1$. In Trisomy 21, the individual has 47 chromosomes instead of the usual 46.

Polyploidy ($3n$ or $4n$) involves a failure of meiosis during the formation of the gametes. The resulting gametes are $2n$. Fertilization can then be either $n + 2n = 3n$ or $2n + 2n = 4n$. Polyploidy is always lethal in humans although it is often found in fish and plants. Finally, *chromosome breakage* might be induced by environmental factors or mutagenic drugs.

Gene Mutations

As discussed in the chapter on cellular and molecular biology, there might be changes in the base sequence of DNA that result in changes in single genes, changing one or more base pairs and the protein produced by reading the gene.

Mutagenic Agents

Mutagenic agents induce mutations. For example, *uv* light, X-rays, radioactivity, and some chemicals will cause mutations by damaging DNA. Such agents are also typically carcinogenic.

This concludes our chapter on classical genetics. With the knowledge you have gained here, you should be able to look at your own family and study the heritability of certain traits. If you have blue eyes, a recessive trait, you must be homozygous for that trait; each of your parents gave you an allele for blue eyes. If one of your parents has brown eyes, he or she must be a heterozygote, possessing alleles for both blue and brown eyes.

Now let's see how you do on some quiz questions.

The Two Faces of Mutations

Mutations may be beneficial (enabling a population to develop new adaptations to the changing environment and leading to evolution of a species). Yet they can also damage the organisms in which they occur, sometimes fatally. Mutations are the root cause of diseases like cancer. The vast majority of mutations are probably neutral or negative.

Classical Genetics Quiz

1. Breeding animals closely related in a pedigree is known as

   (A) inbreeding
   (B) codominance
   (C) linkage analysis
   (D) selective breeding
   (E) test breeding

2. A process that CANNOT take place in haploid cells is

   (A) mitosis
   (B) meiosis
   (C) cell division
   (D) growth
   (E) digestion

3.
   A ——— B C ——————— D ———

   If the diagram above represents genes on a chromosome, which genes would have the highest frequency of recombination between them?

   (A) *A* and *B*
   (B) *A* and *D*
   (C) *B* and *C*
   (D) *B* and *D*
   (E) the frequencies are the same for all crossovers

4. Laboratory mice are to be classified based on genes *A*, *B*, and *C*. How many genetically different gametes can be formed by a mouse that is genotypically *AaBbCc*? (Assume that none of these is a lethal gene.)

   (A) 3
   (B) 6
   (C) 8
   (D) 9
   (E) 12

5. An individual with type O blood must have which of the following?

   (A) multiple alleles at the blood type gene
   (B) several different genes that control bloodtype
   (C) homozygous recessive alleles
   (D) selection against other types that delete the gene
   (E) spontaneous mutations

6. The gene for red-green color blindness is located on the *X* chromosome. The offspring of a man suffering from red-green color blindness would have which of the following characteristics if he married a normal homozygous female?

   (A) 50% of the females would be carriers; 100% of the males would be affected.
   (B) 100% of the females would be normal; 50% of the males would be affected.
   (C) 100% of the females would be carriers; 100% of the males would be normal.
   (D) 50% of the females would be affected; 100% of the males would be affected.
   (E) 100% of the females would be normal; 50% of the males would be carriers.

GO ON TO THE NEXT PAGE

7.   A mutation in a gene in a somatic cell is deleterious because

    (A)   it will affect gamete formation
    (B)   it will be dominant
    (C)   it may be passed on to subsequent generations
    (D)   it may lead to a tumor in that tissue
    (E)   none of the above

8.   Which of the following genetic mutations will NEVER affect the protein produced?

    (A)   point
    (B)   silent
    (C)   insertion
    (D)   frame shift
    (E)   all of the above

9.   Tall is dominant over short in a certain plant. A tall plant was crossed with a short plant, and both tall and short offspring were produced. This demonstrates

    (A)   the law of segregation
    (B)   incomplete dominance
    (C)   linkage
    (D)   mutation
    (E)   the law of independent assortment

10.   A typical human gamete

    (A)   contains a haploid number of genes
    (B)   always contains an X or Y chromosome
    (C)   is a result of the meiotic process
    (D)   has genetic material that has undergone recombination
    (E)   all of the above

11.   Spermatogenesis and oogenesis differ in that

    (A)   spermatogenesis is mitotic while oogenesis is meiotic
    (B)   oogenesis is mitotic while spermatogenesis is meiotic
    (C)   spermatogenesis produces gametes while oogenesis does not
    (D)   spermatogenesis produces four haploid sperm cells while oogenesis produces one egg cell and more than one polar body
    (E)   spermatogenesis involves unequal division of cytoplasm

12.   If a male with blood type A marries a female with blood type B, which of the following types would be impossible for a first generation child?

    (A)   type B
    (B)   type A
    (C)   type O
    (D)   type AB
    (E)   all types are possible

GO ON TO THE NEXT PAGE

13. Polar bodies are formed during

   (A) male mitosis
   (B) female mitosis
   (C) male meiosis
   (D) female meiosis
   (E) two of the above

14. Red is dominant over white in a certain flower. To test whether a red offspring is homozygous or heterozygous in this flower, one would

   (A) cross it with a red plant that had a white parents
   (B) cross it with a red plant that had two red parents
   (C) cross it with a white plant
   (D) Two of the above will work.
   (E) None of the above will work.

15. Green (*Y*) is dominant over yellow (*y*) in peas, and the smooth allele (*W*) is dominant over wrinkled (*w*). Which cross must produce all green, smooth peas?

   (A) *YyWw* × *YyWw*
   (B) *Yyww* × *YYWw*
   (C) *YyWW* × *yyWW*
   (D) *YyWw* × *YYWW*
   (E) none of the above

16. Unequal division of the cytoplasm occurs in

   (A) production of sperm cells
   (B) production of egg cells
   (C) mitosis of an epidermal cell
   (D) binary fission in bacteria
   (E) none of the above

17. Disjunction is the process whereby

   (A) homologous chromosomes separate into two cells
   (B) homologous pairs of chromosomes recombine
   (C) the spindle apparatus is formed from the centrioles
   (D) the cell membrane invaginates to form two daughter cells
   (E) none of the above

STOP

Answers and Explanations to the Classical Genetics Quiz

1. **(A)** Inbreeding occurs when animals that are closely related are bred to create progeny. (B) Codominance is a pattern trait of expression. (C) Linkage analysis is the study of recombination between genes to map their positions. Selective breeding (D) is defined as the creation of certain strains of specific traits through controlled breeding, while test breeding (E) is the breeding of an organism with a homozygous recessive in order to determine whether that organism is homozygous dominant or heterozygous dominant for a given trait.

2. **(B)** A cell that is n (haploid) cannot undergo meiosis to become $\frac{1}{2}n$. (A), (C), and (D) are incorrect because there are a number of organisms that are haploid. These organisms undergo mitosis, divide, and grow. Meanwhile, (E) is incorrect because an organism, whether it is diploid or haploid, must be able to digest to maintain life.

3. **(B)** Homologous recombination occurs during metaphase I tetrad formation. The farther apart two genes are, the more likely a homologous recombination will occur between them. Therefore, the genes that are farthest apart are also those most likely to to have recombination occur between them. It is also important to note that the farther away from the centromere genes are, the more likely they are to recombine.

4. **(C)** The gametes that can be formed by the mouse in this case are ABC, ABc, AbC, Abc, aBC, aBc, abC, and abc. To calculate, multiply the number of alleles at each gene: $2 \times 2 \times 2 \times 2 = 8$ possible gametes.

5. **(C)** Type O blood is found in people with two recessive alleles that express neither the A or B antigen.

6. **(C)** A male affected with red-green color blindness would have a genotype of $X_{cb}Y$. If he mated with a normal female, XX, all their female offspring would be $X_{cb}X$, receiving one good copy of the X chromosome from their mother and the color blindness gene from their father. Since color blindness is a recessive trait, all the female offspring would be carriers. All the male offspring of this mating would be XY, receiving one good copy of the X chromosome from their mother and the Y chromosome from their father.

7. **(D)** Mutations in somatic cells (cells of the body, not germ tissue) affect only the individuals involved. They cannot be passed on to the next generation and will not affect gamete formation. These mutations are typically recessive, although there are some instances of dominant negative mutations. The major concern about mutations in somatic cells is that they are linked to the development of tumors; these are produced by proteins that have lost their functions due to somatic mutation.

8. **(B)** A silent mutation is a point mutation that either occurs in a noncoding region or does not change the amino acid sequence, due to degeneracy of the genetic code. Therefore, silent mutations by definition do not affect the protein produced. Point mutations, meanwhile, occur when a single nucleotide base is substituted for another nucleotide base which can change a codon to a different amino acid. A frameshift mutation is either an insertion or deletion of a number of nucleotides. These mutations have serious effects on the protein coded for, since nucleotides are read as series of triplets. The addition or loss of nucleotides (except in multiples of three) will change the reading frame of the mRNA.

9. **(A)** The law of segregation states that when gametes are formed, the two alleles for a particular trait will separate or segregate into the gametes, so that each of the gametes only contains one of the alleles for a given trait. So if tall is dominant over short, and both tall and short offspring were produced, then the tall plant is a hybrid or heterozygous plant. This means that the genotype of the tall plant contains both one tall allele and one short allele. The short plant contains two short alleles. When the gametes are formed for this mating, the two alleles in the tall plant, the tall and short, will segregate into the gametes, forming tall-containing gametes and short-containing gametes. When these meet and fertilize, the short-containing gametes from the other plant, half the offspring produced will be tall because they are the result of the tall gamete's fertilization of a short gamete, and the other half will be the result of the short gamete's fertilization of the other short gamete.

In (B), incomplete dominance, or blending, occurs when two individuals mate and the resulting offspring is a phenotype that appears to be midway between the two phenotypes of the two parents. For example, if a tall and short plant were crossed in this case and they produced an offspring of medium height, the result would be incomplete dominance. As for (C), linkage refers to genes or alleles that travel with each other on the same chromosome. Hence linked genes cannot segregate into two separate gametes because they are on the same chromosome. Genes or alleles which separate must, by definition, be on separate chromosomes. (D) is not correct either; mutation refers to changes in the DNA sequence of a chromosome, and there is no evidence of mutation occurring in this question. Finally, in (E) the law of independent assortment states that when we are dealing with more than one trait at a time and these traits are not linked, they are carried on different chromosomes, and the inheritance of these traits is not connected.

10. **(E)** During meiosis, the gamete reduces its genetic component from $2n$ to n, resulting in a haploid cell with half the normal chromosome number. When a haploid egg and sperm unite, they form a diploid organism known as a zygote. All ova will contain an X chromosome, and all sperm will contain either an X or a Y chromosome. These gametes are formed during the two reductional divisions of meiosis. During Metaphase I of Meiosis I, tetrads form and sister chromatids undergo a homologous recombination known as crossing over.

11. **(D)** Spermatogenesis and oogenesis are both forms of gametogenesis in which haploid gametes are produced through reductional division (meiosis) of diploid cells. Both processes occur in the gonads. They differ, however, in that in spermatogenesis, the cytoplasm is equally divided during meiosis, and four viable sperm are produced form one diploid cell. In oogenesis, the cytoplasm is divided unequally and only one ovum, containing the bulk of the cytoplasm, is produced, along with two or three inert polar bodies.

12. **(E)** The Type A man can be either AA or Ai and the type B woman can be either BB or Bi. A and B blood groups are codominant over blood antigen O. Therefore, if a man heterozygous for blood type A (AO) married a woman heterozygous for blood type D (BO), they could have children with the possible genotypes AO, BO, OO, or AB. Therefore, all blood types are possible in this mating.

13. **(D)** Polar bodies are nonfunctional, gametelike cells that are formed during female meiosis. Recall that meiosis is a two-stage process. There are a total of four haploid cells formed from each original diploid germ cell. In the case of sperm cells, four functional haploid gamete sperm cells are formed. In the case of egg cells, the first meiotic division involves an unequal division of cytoplasm, resulting in the formation of one large cell and one small cell. The large cell will go on to divide again, while the smaller cell is known as the first polar body. This small cell polar body may then divide again to form two other polar bodies. The large cell, meanwhile, undergoes a second meiotic division, again unequally dividing the cytoplasm into one larger resulting cell, the final haploid ovum, and another small cell, which again becomes a polar body. Hence it is possible that during a female meiotic division, one large ovum or functional egg cell and three polar bodies may be formed. Polar bodies are not formed in mitosis, which is how all other cell divisions occur in the body; in other words, equal cell divisions do not form sex cells.

14. (C) This question illustrates a test cross. A test cross is performed to determine if a particular dominant individual's phenotype is caused by a homozygous or heterozygous genotype. In this case, there are two possible red genotypes: *RR*, the pure homozygous red, and *Rr*, the hybrid heterozygous red. These two individuals would have the same phenotype. In order to determine the genotype, the unknown red plant would be mated with a recessive, white plant. If the red plant and the white plant produce only red offspring, then it can be assumed that the original red plant is homozygous. If the mating of the unknown red organism and the white organism produces any white offspring at all, then we know that the original unknown red was heterozygous. This is because white offspring can be produced only through the production of one white gamete by each parent.

15. (D) Both green and smooth are dominant phenotypes. The goal in this question is to produce only green smooth peas, so we want only dominant phenotype offspring. Therefore, we must avoid any crossing that may result in the mating or combining of two recessive alleles. In (A), crossing *Yy* and *Yy* could result in approximately one quarter of the offspring turning out yellow. Similarly, in (B), *ww* crossed with *Ww* would produce offspring of which approximately half would possess a wrinkled phenotype. In (C), *Yy* crossed with *yy* is likely to produce offspring which are approximately half yellow. (D) is correct because one of the parents is a double dominant, meaning that all offspring will have the dominant phenotype, regardless of the genotype of the other parent.

16. (B) Unequal cytoplasm division occurs when egg cells are produced during the meiotic process of oogenesis. In the first stage of egg production, the precursor diploid cell produces two daughter cells, but one of the daughter cells receives almost the entire amount of cytoplasm, while the other becomes a nonfunctioning polar body. In the second division of oogenesis, the large daughter cell divides again, and once again one of the new daughter cells receives almost all of the cytoplasm, and the other becomes a small nonfunctional polar body. The original first polar body may also divide to form two nonfunctional polar bodies. The final result is a potential four haploid cells, but only one of them—the one that received a greater amount of cytoplasm during each meiotic division—becomes a functional egg cell. (A) is incorrect; during spermatogenesis, one diploid precursor cell forms four functional haploid sperm cells. In this case, both divisions are equal, and all sperm cells are equal in amount of cytoplasm. As for (C), during mitosis of epidermal cells, cytoplasm is distributed equally. Likewise, in binary fission of bacteria, in which bacterial cells are dividing as a means of reproduction, cytoplasmic division will be equal.

17. (A) Disjunction is the separation of homologous chromosomes during meiosis. Each tetrad is separated into two halves. One of each pair of chromosomes (each containing two chromatids) is pulled to opposite ends of the cell. Note that some of the maternal chromosomes can go to one end and some to the other end of the cell. The distribution of homologous chromosomes between the two resultant nuclei is random.

CHAPTER SIX

ECOLOGY

To understand how organisms live, biologists study molecules, cells, tissues and organs, breaking organisms down into their fundamental units. Organisms from bacteria to humans do not live on Earth in an isolated state, however. All organisms, including humans, live by interacting with other organisms and with the nonliving (*abiotic*) environment. Life on earth is a network of interacting organisms that depend on each other for survival. Ecology is the study of the interactions between organisms and their environment and how these shape both the organisms and the environments they live in.

Populations in the Environment

Since ecology seeks to understand life at a broader level than the organism, it is the population rather than the individual that is the basic unit of study in this discipline. A *population* is a group of individuals that interbreed and share the same gene pool, the same definition used in population genetics. Every environment will include many different interacting populations. There are properties of populations that are not present in individuals, such as population growth and maximal population size, that are not properties of individuals. Also, populations are concerned with the maintenance of the population, and are not as concerned with individual members of the population. These distinct properties of populations are important for ecosystems.

Patterns of Population Growth

One of the key characteristics of a population is its rate of population growth. At any given time a population can grow, stay the same, or shrink in size. The birth rate, the death rate, and the population size determine the rate of growth, with the birth rate and death rate influenced by the environment. If the birth rate is high and the death rate is low, as in an environment where resources are unlimited, a population will grow rapidly. If a population of mice starts with a male and female mouse, breeds once every three months, producing six male and female offspring in each generation, the population will have over two million mice in two years. A sin-

The Big Picture

Let's take a moment to assess the links between ecological levels of biological organization and the cellular biology we discussed in an earlier chapter. From smallest to largest, the levels of biological organization found on Earth are as follows:

1) Biological chemistry
2) Cell
3) Tissue
4) Organ
5) Organism
6) Population
7) Community
8) Ecosystem
9) Biosphere

gle bacteria reproducing by binary fission every thirty minutes can produce 8 million bacteria in 12 hours. This form of population growth produces a curve with rapidly increasing slope and is termed *exponential growth*, since every generation increases the population size in an exponential manner.

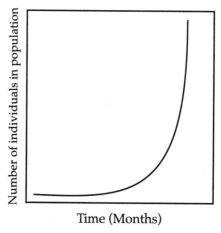

Exponential Population Growth

In nature, this rapid population growth may be observed when a population first encounters a favorable new environment, such as a rich growth medium inoculated with a small number of bacteria, or a fertile empty field invaded by a weed. Exponential growth cannot be maintained forever, though. A population of mice growing exponentially in a field of wheat will soon eat so much of the available food that starvation will occur; growth will slow and then halt. Bacteria reproducing without check would in a few days weigh more than the mass of the earth. Limitations of the environment prevent exponential growth from proceeding indefinitely. Reasons for a slowdown in the growth rate include a lack of food, competition for other resources, predation, disease, accumulation of waste, or lack of space. All of these factors act more strongly to slow growth as the population becomes denser. Under these conditions, the growth curve may appear sigmoidal, as in the figure on the next page, with rapid exponential growth at first, followed by a slowing and leveling off of growth. In this curve, the population size at the point where the growth curve is flat is the maximum sustainable number of individuals, called the *carrying capacity*, and is observed when the birth rate and death rate are equal.

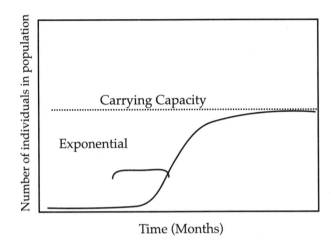

Sigmoidal Population Growth

In natural populations, the environment is constantly changing, and the carrying capacity varies with it. For example, the carrying capacity for rabbits in a grassy plain will be greater in a year of plentiful rain and lush vegetation growth than in a year of drought. Populations often have regular fluctuations in size, suggesting that the carrying capacity changes in a periodic manner. If a population of rabbits consumes all available vegetation, the carrying capacity will be reduced and the population size will fall until the vegetation regrows and the carrying capacity for rabbits is increased once again. An example that is often used is the size of hare and lynx populations in Canada. In this environment, the primary food of the lynx is the hare. The hare population size regularly cycles up and then crashes, perhaps due to the rapid spread of disease in crowded conditions. The lynx population size cycles along with that of the hare, crashing in size after the hare population crashes, then building in size once again after the hare population rebounds.

Reproductive Strategies and Population Growth

Different strategies of reproduction can produce different patterns of population growth. An example is the choice between sexual and asexual reproduction. Sexual reproduction occurs in most complex multicellular organisms, and helps to create and maintain diversity in the gene pool of a population. Asexual reproduction can allow a population to grow very rapidly, as occurs in some plant species that reproduce from shoots rather than seeds. The benefit of asexual reproduction is a reduced cost to produce new organisms. For example, it takes less energy for a plant to produce a shoot and reproduce asexually than for it to flower or to produce fruit and seeds. By reproducing through *parthenogenesis* (in which an egg divides in the absence of fertilization by a male), only female offspring are produced, all of which produce their own young, increasing by two-fold the rate of population growth. The cost of asexual reproduction is a reduction in genetic variability. Some species will reproduce asexually in times of abundant resources to maximize the opportunity for rapid growth, and reproduce sexually when resources are limited, perhaps generating the genetic diversity required to survive a changing environment.

Species also use two different strategies that affect the number of offspring produced in each generation and the amount of care they receive. In the first strategy, offspring reach sexual maturity very rapidly, and produce a large number of young in each generation that receive little or no parental care. Insects and plants often reproduce in this way, producing large numbers of eggs or seeds that are left in the environment to fend for themselves. In colonizing a new environment, a species with this reproduction pattern will grow exponentially to rapidly exploit this opportunity. The lack of parental care for a species with this strategy can cause a high death rate early in life when these species are in a competitive environment. For example, some marine species release their young as large number of eggs that develop without care. As a result, many are consumed early in life. The large number of young ensure survival despite the high mortality rate. Species with this reproductive strategy are also prone to rapid crashes in population size as the environment rapidly becomes depleted of resources after a period of exponential growth.

The other reproductive strategy is to delay sexual maturity, have few young, and put a great deal of parental care into offspring. Large mammals often fall in this category. These species have long life spans and are highly adapted to compete for resources in a competitive environment that is at, or near, the carrying capacity. Since these organisms invest so much into their young, they have a lower mortality rate. These species also have difficulty recovering from catastrophic decreases in their population size; examples include California condors or whooping cranes.

The Role of the Abiotic Environment in Population Growth

The size and growth of a population are affected both by the biotic (living) and abiotic (nonliving) portions of the environment. The abiotic portions include the air, water, soil, light and temperature that living organisms require. Not only are organisms dependent on the abiotic environment, but they in turn modify it. Plants create shade that alters the light environment for other plants, preserve water in the soil, consume carbon dioxide, and produce oxygen. The modifications of the environment by a population affect the types of species the population lives in.

Water is essential to all life and is a major component of all living things. Our bodies are made mostly of water. Animals must regulate their water content to ensure the proper volume and salt concentrations inside and outside of the cell.

Don't Mix These Up on Test Day

The *abiotic environment* encompasses:

- Temperature
- Light
- Water
- Oxygen supply
- Soil

The *biotic* (living) *environment* encompasses:

- Organisms and their relationships with other organisms

Osmoregulation

Osmoregulation may be defined as the ways in which organisms regulate the volume and salt content of their internal fluids. Saltwater fish, for example, live in a hyperosmotic environment that causes them to lose water and take in salt. In constant danger of dehydration, they must compensate by constantly drinking and actively excreting salt across their gills. Freshwater fish, in contrast, live in a hypo-osmotic environment that causes intake of excess water and excessive salt loss. These fish correct this condition by drinking infrequently, absorbing salts through the gills, and excreting dilute urine.

On land, insects excrete solid uric acid crystals in order to conserve water, while desert animals possess adaptations for avoidance of desiccation. For example, the horned toad has a thick, scaly skin. Other desert animals burrow in the sand during the day and search for food at night, thereby avoiding the intense heat that causes water loss.

As for plants, nondesert land plants possess waxy cuticles on leaf surfaces and stomates on their lower leaf surfaces only, and shed leaves in winter to avoid water loss. The water-conserving adaptations of desert plants include extensive root systems, fleshy stems to store water, spiny leaves to limit water loss, extra thick cuticles, and small numbers of stomates.

Thermoregulation

The external temperature is part of the abiotic environment that life contends with and adapts to. Temperature affects organisms' rate of metabolic activity and rate of water loss. Extremes of temperature retard most life, although there are organisms that live only in boiling hot springs or at subfreezing temperatures.

Organisms must also develop ways to regulate heat. Cellular respiration transfers only some of the energy derived from the oxidation of carbohydrates, fats, and proteins into the high-energy bonds of ATP. Roughly 60 percent of the total energy is not captured; most of this is transformed to heat. The vast majority of animals are *cold-blooded*, or *ectothermic*—most of their heat energy escapes to the environment. Consequently, the body temperature of ectotherms, also known as poikilotherms, is very close to that of their surroundings. Since an organism's metabolism is closely tied to its body temperature, the activity of ectothermic animals such as snakes is radically affected by environmental temperature changes. As the temperature rises (within limits, since very high temperatures would be lethal), these organisms become more active; as temperatures fall, they become sluggish.

Some animals, notably mammals and birds, are endotherms; they are *warm-blooded*, or *homeothermic*. They have evolved physical mechanisms that allow them to make use of the heat produced as a consequence of respiration. Physical adaptations like fat, hair, and feathers actually retard heat loss. Homeotherms maintain constant body temperatures higher than the

Why the Camel Has a Hump

The camel can tolerate a wide range of body temperatures because of its hump. This protective fat layer is strategically located on the animal's back, an area especially vulnerable to solar radiation.

Don't Mix These Up on Test Day

Osmoregulation is the regulation of water and salt levels within an organism.

Thermoregulation is the regulation of an organism's temperature.

environment around them. Hence they are less dependent upon environmental temperature than poikilothermic animals, and are able to inhabit a comparatively greater range of variable conditions as a result.

Sunlight

Sunlight serves as the ultimate source of energy for almost all organisms. Green plants must compete for sunlight in forests. To this end, they develop adaptations to capture as much sunlight as possible (including broad leaves, branching, greater height, and vine growth). In water, the *photic zone*—the top layer through which light can penetrate—is where all photosynthetic activity takes place. In the *aphotic zone*, only animal life and other heterotrophic life exist.

Oxygen Supply

Oxygen supply poses no problem for terrestrial life, since air is composed of approximately 20 percent oxygen. Aquatic plants and animals utilize oxygen dissolved in water, where oxygen is present only in parts per million. Pollution can significantly lower oxygen content in water, threatening aquatic life. It can also benefit certain organisms at the expense of others.

Substratum (Soil or Rock)

Substratum determines the nature of plant and animal life in the soil. Some soil factors include:

- *Acidity (pH).* Acid rain may make soil pH too low for most plant growth. Rhododendrons and pines, however, are more well suited to acidic soil.

- *Texture of soil and clay content.* These determine the quantity of water the soil can hold. Most plants grow well in loams that contain high percentages of each type of soil.

- *Minerals.* Nitrates, phosphates, and other minerals determine the type of vegetation soil will support. Beach sand has been leached of all minerals and is unable to support plant life.

- *Humus quantity.* This is determined by the amount of decaying plant and animal life in the soil.

Plants That *Like* Pollution

Duckweed grows rapidly when exposed to pollution, "choking" other plant and animal life by utilizing all the available oxygen.

Chemical Cycles

Also included in the abiotic environment are inorganic chemicals required for life such as carbon and nitrogen. The movement of these essential elements between the biotic and abiotic environment form cycles that are central to all life on earth. Some organisms take the simple inorganic starting chemicals up from the soil and air and convert them into a biologically useful form. After material passes through the biological community, respiration and decay organisms return these chemicals to their inorganic state to begin the cycle again.

Carbon Cycle. The carbon cycle commences as gaseous CO_2 enters the living world when plants take it in and use it to produce glucose via photosynthesis. Plants use energy stored in glucose to make starch, proteins, and fat.

Next, animals eat plants and use the digested nutrients to form carbohydrates, fats, and proteins characteristic of the species. Part of these organic compounds is used as fuel in respiration in plants and animals. The metabolically produced CO_2 is then released to the air. Aside from expelled wastes, the rest of the organic carbon remains locked within an organism until its death, at which time decaying processes return the CO_2 to the air.

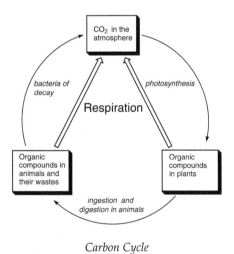

Carbon Cycle

Waste Not, Want Not

Because raw materials are finite in quantity here on Earth, nutrients like carbon and nitrogen must continually be recycled.

Nitrogen Cycle. Nitrogen is an essential element of amino acids and nucleic acids, which are the building blocks for all living things. Since there is a finite amount of nitrogen on the earth, it is important that it be recovered and reused.

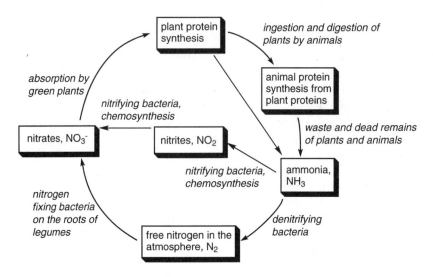

Nitrogen Cycle

Well Hidden

Pull up the root of a legume, and you'll see many large nodules. These serve as the site of nitrogen fixation by symbiotic bacteria.

The following bullets refer to the figure of the nitrogen cycle above.

- Elemental (free) nitrogen, at the bottom of the figure, is chemically inert and cannot be used by most organisms. Lightning and nitrogen-fixing bacteria in the roots of legumes change the nitrogen to usable, soluble nitrates.

- The nitrates are absorbed by plants and are used to synthesize nucleic acids and plant proteins.

- Animals eat the plants and synthesize specific animal proteins from the plant proteins. Both plants and animals give off wastes and, eventually, die.

- The nitrogen locked up in the wastes and dead tissue is released by the process of decay, which converts the proteins into ammonia.

- Two fates await the ammonia (NH_3): Part of it is nitrified to nitrites by chemosynthetic bacteria and then to usable nitrates by nitrifying bacteria. The rest of the ammonia is denitrified. This means that the ammonia is broken down to release free nitrogen, which returns us to the beginning of the cycle.

- Note that there are four kinds of bacteria: decaying, nitrifying, denitrifying, and nitrogen-fixing. The bacteria have no use for the excretory ammonia, nitrites, nitrates, and nitrogen they produce. These materials are essential, however, for the existence of other living organisms.

Other Material Cycles

Other material cycles include the water, oxygen, and phosphorous cycles. These substances are available in limited amounts, and they are used by almost all living things. Like nitrogen, they must be returned by the biotic community to the environment in such a way that they can be reused. In the oxygen cycle, for example, humans utilize oxygen and exhale carbon dioxide. This carbon dioxide is used by plants in photosynthesis in order to release oxygen into the atmosphere, which is once again utilized by humans.

Populations in Communities and Ecosystems

The next level of biological organization beyond a population is a *community*, which is all the interacting populations living together in an environment. The populations within a community interact with each other in a variety of ways, including *predation*, *competition*, or *symbiosis*. These interactions affect the number of individuals in each population in the community and the number of different species in the community. The living community combined with the abiotic environment, the interactions between populations, and the flow of energy and molecules within the system define an *ecosystem*.

Predation

Predation is the consumption of one organism by another, usually resulting in the death of the organism that is eaten. Both carnivores that consume meat, and herbivores, consuming plants only, are types of predators. Predation includes a zebra eating grass, a lion eating a zebra, a whale eating plankton, a paramecium eating yeast, or a Venus flytrap eating a housefly. Predators often select weak or sick members of the prey population, removing alleles with poor fitness, and driving evolution in the prey toward more effective means of escaping predation. Predator and prey often coevolve, with the predator evolving to become more effective as the prey evolves to escape predation. Predator-prey relationships between populations in a community can influence the carrying capacity of prey populations involved and tend to achieve a balance such that the predator is effective enough to maintain its own population without decimating the prey it is dependent on. Predation can cause a community to maintain a greater diversity of species—without predation, one prey species will often predominate.

Competition and the Niche

A competitive relationship between populations in a community exists when different populations in the same location use a limiting resource. Competition can be interspecific (between species) or intraspecific (between organisms of the same species). Integral in understanding interspecific com-

Survival of the Fittest

Two species occupying similar niches will do one of the following:

• Compete until one species is driven to extinction

• Evolve in divergent directions

petition is the idea of the ecological *niche*. If the habitat is the physical environment in which the population lives, the niche is the way it lives within the habitat, including what it eats, where it lives, how it reproduces, and all other aspects of the species that define the role it plays in the ecosystem. The niche occupied by each species is unique to that species and can in part define that species. Another way to understand the niche is to say that if the habitat is the address of a population, the niche is its profession.

Interspecific Competition

When two populations have overlap in their niches, such as by eating the same insects or occupying the same nesting sites, there is competition between the populations. The more the niches overlap, the greater the competition. Generally, when two populations compete, one will compete more effectively than the other and grow more rapidly. Competition can drive the less efficient population out of the community, with the "winner" occupying the niche on its own. Another result of competition for a niche can be that evolution drives the two populations to occupy niches that overlap less, reducing the competition. For example, if two species of related birds compete for the same nesting site, then they may evolve to reduce competition by using different nesting sites (see the figure below). Even in an environment with several different herbivores, their niches are unique since they evolve to have different heights, different sizes, different teeth and digestive tracts to avoid competition for the same plants. Several closely related species of birds can live in the same tree and eat similar food, and yet occupy distinct niches by living in different part of the tree, with some near the crown, others in the middle, and still others close to the ground.

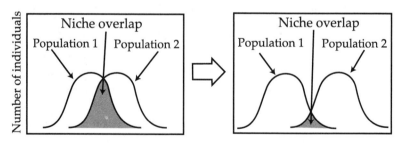

Height of nesting site in apple trees

Evolution Drives Reduced Niche Overlap

Symbiosis

Symbionts live together in an intimate, often permanent association that may or may not be beneficial to them. Some symbiotic relationships are obligatory—that is, one or both organisms cannot survive without the other. Types of symbiotic relationships are generally classified according to the benefits the symbionts receive. Symbiotic relationships include commensalism, mutualism, and parasitism.

Commensalism. In this relationship, one organism is benefited by the association and the other is not affected (this is symbolized as +/0). The host neither discourages nor fosters the relationship. The remora (shark-sucker), for example, attaches itself to the underside of a shark. Through this association, the remora obtains the food the shark discards, wide geographic dispersal, and protection from enemies. The shark is totally indifferent to the association. A similar association links the barnacle and the whale. The barnacle is a sessile crustacean that attaches to the whale and obtains wider feeding opportunities through its host's migrations.

Mutualism. This is a symbiotic relationship from which both organisms derive some benefit (+/+). In the instance of the tick bird and rhinoceros, the rhinoceros aids the bird through the provision of food in the form of parasites on its skin. The bird in its turn aids the rhinoceros by removing the parasites and by warning the rhinoceros of danger when it suddenly flies away.

A more intimate mutualistic association exists between a fungus and an algae in the form of the lichen. Lichens are found on rocks and tree barks. The green algae produces food for itself and the fungus by photosynthesis. Meshes of fungal threads support the algae and conserve rain water. Thus, the fungus provides water, respiratory carbon dioxide, and nitrogenous wastes for the algae, all of which are needed for photosynthesis and protein synthesis. Lichens are significant in that they were the first organism capable of establishing a terrestrial existence; they are pioneer organisms in the order of ecological succession on bare rock.

Nitrogen-fixing bacteria and legumes also engage in mutualism. Nitrogen-fixing bacteria invade the roots of legumes and infected cells grow to form root nodules. In the nodule, the legume provides nutrients for the bacteria, and the bacteria fixes nitrogen (by changing it to soluble nitrate, a mineral essential for protein synthesis by the plant). These bacteria are a major source of usable nitrogen, which is needed by all plants and animals.

Protozoa and termites work together in a similar fashion. Termites chew and ingest wood, but are unable to digest its cellulose. Protozoa in the digestive tract of the termite secrete an enzyme that is capable of digesting cellulose, and both organisms share the carbohydrates. In this manner, the protozoa are guaranteed protection and a steady food supply, while the termite obtains nourishment from the ingested wood. Likewise, in the case of intestinal bacteria and humans, bacteria utilize some of the food material not fully digested by humans and, in turn, manufacture vitamin B_{12}.

Parasitism. A parasite takes from the host but gives nothing in return; thus, the parasite benefits at the expense of the host (+/−). Examples of parasites include leeches, ticks, and sea lampreys. Parasitism exists when competition for food is most intense. Few autotrophs (green plants) exist as parasites (mistletoe is an exception).

Don't Mix These Up on Test Day

The three types of symbiosis are:

- *Commensalism* is a +/0 relationship in which one organism benefits and the other is unaffected.

- *Mutualism* is a +/+ relationship in which both organisms benefit.

- *Parasitism* is a +/− relationship in which one organism benefits and the other is harmed.

Parasites Exercise Restraint

It is not to a parasite's advantage to maximize its food intake and to severely injure its host. If the host dies, the parasite will lose its free ride and perish, since it is adapted to a parasitic lifestyle.

Ectoparasites cling to the exterior surface of the host with suckers or clamps, bore through the skin, and suck out juices. Endoparasites, on the other hand, live within the host. In order to gain entry into the host, they must break down formidable defenses, including skin, digestive juices, antibodies, and white blood cells. Parasites possess special adaptations to overcome these defenses.

Parasitism is advantageous and efficient, since the parasite lives with a minimum expenditure of energy. Parasites may even be parasitic on other parasites. Thus, a mammal may have parasitic worms, which in turn are parasitized by bacteria, which in turn are victims of bacteriophages.

A prominent example of a parasitic relationship is that between the virus and its host cell. All viruses are parasites. They contain nucleic acids surrounded by a protein coat, and are nonfunctional outside their host cells. As viral nucleic acid enters the host, the virus takes over the host cell functions and redirects them into replication of the virus.

The tapeworm-human relationship is a particularly good example of parasitism. Tapeworms can live inside their hosts' intestines for many years, growing longer and longer. It is interesting to note that successful parasites do not kill their hosts, as this would, counterproductively, lead to the death of the parasite itself. The more dangerous the parasite is to its host, the less chance it has of ultimate survival.

Intraspecific Competition

Competition is not restricted to interspecific interactions. Individuals belonging to the same species utilize the same resources; if a particular resource is limited, these organisms must compete with one another. Members of the same species compete, but they must also cooperate. Intraspecific cooperation may be extensive (as in the formation of societies in animal species) or may be nearly nonexistent. Hence, within a species, relationships between individuals are influenced by both disruptive and cohesive forces. Competition (for food or a mate, for example) is the chief disruptive force, while cohesive forces include reproduction, protection from predators, and destructive weather.

Community Structure

Producers

The populations within a community are organized in many different ways. Within the community, each population plays a different role depending on the source of energy for that population. Producers are *autotrophs*, organisms that get energy from the environment (the sun or inorganic molecules) and use this energy along with simple molecules (carbon dioxide, water and minerals) to drive the biosynthesis of their own proteins, carbohydrates, and lipids. The energy a producer such as a plant gets from the sun is stored in chemical bonds in the biological molecules it produces. Producers form the foundation of any community, passing on their energy to other organisms. In a terrestrial environment, green plants, photosynthetic bacteria, or mosses are producers, using the energy of sunlight to produce biosynthetic energy through photosynthesis. In marine environments, green plants or algae are the main producers. There are even marine ecosystems at deep, dark ocean geothermal vents at which the entire community is based not on photosynthetic producers but on chemosynthetic bacteria that use the energy of inorganic molecules released from the volcanic vent to drive biosynthesis.

Consumers

Consumers get the energy to drive their own biosynthesis and to maintain life by ingesting and oxidizing the complex molecules synthesized by other organisms. Since they get their energy by consuming other organisms, they are called *heterotrophs*. Herbivores (plant eaters), carnivores (meat eaters), and omnivores (eating both plants and animals) are all consumers. The adaptations of each consumer depend on the type of food it eats. Herbivores tend to have teeth for grinding and long digestive tracts that allow for the growth of symbiotic bacteria to digest cellulose found in plants. Carnivores are more likely to have pointed, fanglike teeth for catching and tearing prey and shorter digestive tracts than herbivores.

Primary consumers are animals that consume producers like green plants—for example, the cow, the grasshopper, and the elephant. *Secondary consumers*, meanwhile, are carnivorous and consume primary consumers—for example, frogs, tigers, and dragonflies. Finally, *tertiary consumers* feed on secondary consumers; examples include snakes that eat frogs.

Decay Organisms (Saprophytes)

Decay organisms, also called saprophytes or decomposers, are heterotrophs, since they derive their energy from oxidizing complex biological molecules, but they do not consume living organisms. Decay organisms get energy from the biological organic molecules they encounter left as waste by producers and consumers, or the debris of dead organisms. They perform respiration to derive energy, and return carbon dioxide, nitrogen, phosphorous and other inorganic compounds to the environment to renew the cycles of these materials between the biotic and physical environments. Bacteria and fungi are the primary examples of decay organisms. Scavengers such as hyenas or vultures play a similar role, living on the stored chemical energy found in dead organisms.

The Food Web

The term *food chain* is often used to describe a community, depicting a simple linear relationship between a series of species, with one eating the other. For example, a food chain might contain grass as the producer, mice as the

The Chain of Life

Producers are consumed by primary consumers, who are consumed by secondary consumers, who are consumed by tertiary consumers—and all of these organisms are broken down by decomposers.

Good Thing Someone's Cleaning This Place Up!

If the earth had no saprophytes to consume decaying matter, we'd be climbing over the carcasses of dinosaurs and other animals that died millions of years ago.

Eat or Be Eaten

Organisms play different roles in the food chain at different times. Where in the food chain would you be if you ate a plate of french fries? How about if you ate a hamburger? Or if a lion ate you?

primary consumer, snakes as the secondary consumer, and hawks as the tertiary consumer (see the figure below). The different levels in the food chain, such as producers and primary consumers, are sometimes called *trophic levels*. A more realistic depiction of the relationships within the community is a *food web,* in which every population interacts not with one other population, but several other populations. An animal in an ecosystem is often preyed on by several different predators, and predators commonly have a diet of several different prey, not just one. The greater the number of potential interactions in a community food web, the more stable the system will be, and the better able it will be to withstand and rebound from external pressures such as disease or weather.

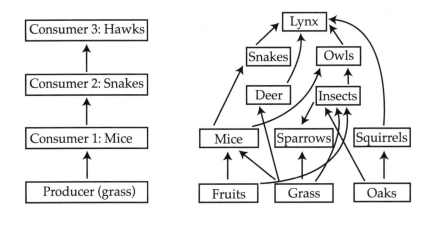

Food Chain   *Food Web*

Energy Flow in Communities

Each tropic level in a food web contains different quantities of stored chemical energy in the populations it contains. When consumers eat producers and secondary consumers eat primary consumers, some energy is lost in each transfer from one level to another. As producers get energy from the sun, not all of the energy is converted into stored energy in chemical bonds. Some of the energy is lost at that level to the metabolic energy an organism requires to maintain its life. Plants consume some of the energy they produce in respiration to support their own metabolic activities. The total chemical energy generated by producers is *gross primary productivity*, and the total with losses to respiration subtracted is the *net primary productivity*.

At the next level of the energy pyramid, herbivores consume primary producers, incorporating about 10 percent of the energy consumed into their own stored chemical energy. The remainder is lost through respiration. Only about 10 percent of the stored chemical energy is present in the next higher trophic level at every stage (see the figure below). The energy contained in a community can be visualized as a *pyramid*, with the most energy in the producers and less energy at successive levels of consumers (see figure). The efficiency of energy transfer between levels can differ greatly from 10 percent, but the pyramid will always have the most energy in the producer level, with less in each level of consumers. A similar pyramid is observed if one compares the biomass or numbers of individuals in a community, with each successive level about 10 percent the size of the level beneath it. In terms of numbers, the shape of the pyramid can often vary, with a single large producer like a tree supporting a large number of primary consumers like birds or insects.

The Higher You Go, the More You Lose

As a general rule, energy, mass, and numbers are lost as a food pyramid is ascended.

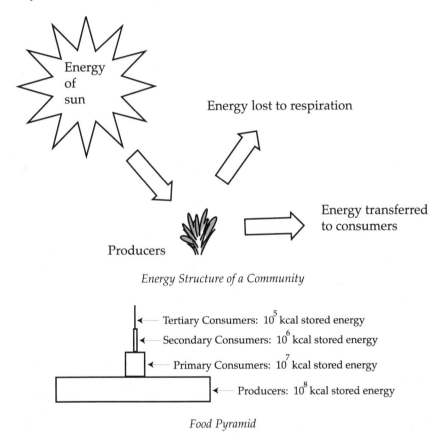

Energy Structure of a Community

Tertiary Consumers: 10^5 kcal stored energy
Secondary Consumers: 10^6 kcal stored energy
Primary Consumers: 10^7 kcal stored energy
Producers: 10^8 kcal stored energy

Food Pyramid

Community Diversity

The number of species within a community is termed the community diversity. The types of interactions between populations within a community affect the number of species in the community, as well as the physical environment. Predation has been observed as one factor that increases the diversity of species in a community, and competition may do the same through selective pressure driving populations into distinct niches. Warm environments like the tropics with very high productivity have the greatest diversity, and colder environments have less community diversity. Topographic diversity increases community diversity, perhaps by creating a greater number of niches in the environment. Larger land masses or ecosystems also have a greater community diversity.

Changes in Community over Time: Succession

Communities can change in composition over time, either as a result of a changing physical environment such as the climate or as the result of changes created by the populations that live in the community. When a population changes the environment it lives in, it may make the environment more favorable for some populations and less favorable for others, including itself. When a community changes as a result of organisms that live in the community, this is termed *succession*. For example, a grassland may provide abundant sunlight and rich soil that lead to colonization by trees, followed by others trees that grow best in the shade of the pioneer trees. Successive communities are composed of populations best able to exist in each new set of conditions, both biotic and abiotic.

The community will continue to change until it arrives at combination of populations that do not change the environment any further, leaving the community the same over time in what is called the *climax community*. The climax community is stable over time, with each generation leaving the environment it resides in the same, and it will remain in place unless it is disturbed by climate change, fire, humans, or other catastrophes. If the climax community is disturbed, the series of community successions will begin again until a climax community is achieved once again. The type of climax community that is present in an environment depends on the abiotic factors of the ecosystem, including rainfall, temperature, soil, and sunlight.

Consider, for example, a rocky area in the northeastern United States, barren, perhaps, as a result of a severe forest fire. Lichen would be a good candidate to be the first or pioneer organism to resettle this virgin area. Recall that a lichen is an association between an alga and a fungus that can live on a rocky surface. Acids produced by the lichen attack rock, breaking it down to form the first layers of soil. Since lichens thrive only on a solid surface, conditions at this stage are worse for the lichen but better for mosses. Airborne spores of mosses land on the soil and germinate. The result is a new *sere* (a series of ecological communities formed in ecological succession), with moss supplanting lichen as the dominant species in the community.

As the remains of the moss build up the soil still more, annual grasses and then perennial grasses with deeper roots become the dominant species. As time marches on, we find shrubs and then trees. The first trees will be the sun-loving gray birch and poplar. As more and more trees compete for the sun, the birch and poplar will be replaced by white pine and, finally, maples and beeches, which grow in deep shade.

The growth of maples and beeches produces the same conditions that originally favored their appearance. And so this community remains for a thousand years. In the final maple-beech community, you would also find foxes,

Nature's Irony

In nature's irony, each level in the succession of a community changes the physical environment, making it a better place for organisms in the next stage of succession to live, but pushing out its own organisms in the process.

deer, chipmunks, and plant-eating insects. These are all animals that would not have been found in the original barren rock terrain.

To summarize this example of ecological succession:

Lichen —> mosses —> annual grasses —> perennial grasses —> shrubs —> sun-loving trees (poplar) —> thick shade trees (hemlock, beech, maple)

Time elapsed: About 1,000 years

Here's an example of the progression of a climax community in an aquatic environment. This community starts with a pond:

- *Step 1—pond.* This pond contains plants such as algae and pondweed and animals such as protozoa, water insects, and small fish.

- *Step 2—shallow water.* The pond begins to fill in with reeds, cattails, and water lilies.

- *Step 3—moist land.* The former pond area is now filled with grass, herbs, shrubs, willow trees, frogs, and snakes.

- *Step 4—woodland.* Pine or oak becomes the dominant tree of the climax community.

It is important to remember that the dominant species of the climax community is determined by such physical factors as temperature, nature of the soil, and rainfall. Thus the climax community at higher elevations in New York state is hemlock-beech-maple, while at lower elevations, it is more often oak-hickory. In cold Maine, the climax community is dominated by the pine; in the wet areas of Wisconsin, by cypress; in sandy New Jersey, by pine; in Georgia, by oak, hickory, and pine; and on a cold, windy mountain top, by scrub oak.

Biomes

The conditions in a particular terrestrial and climatic region select plants and animals possessing suitable adaptations for that particular region. Each geographic region is inhabited by a distinct community called a *biome*.

Terrestrial Biomes

Land biomes are characterized and named according to the climax vegetation of the region in which they are found. The climax vegetation is the vegetation that becomes dominant and stable after years of evolutionary development. Since plants are important as food producers, they determine the nature of the inhabiting animal population; hence the climax vegetation determines the climax animal population. There are eight types of terrestrial biomes that can be formed as a result of all these factors:

Tropical Forests

Tropical forests are characterized by high temperatures and, in tropical rain forests, by high levels of rainfall. The climax community includes a dense growth of vegetation that does not shed its leaves. Vegetation like vines and epiphytes (plants growing on the other plants) and animals like monkeys, lizards, snakes, and birds inhabit the typical tropical forest, or rain forest. Trees grow closely together in a dense canopy high above the ground; sunlight barely reaches the forest floor. The floor is inhabited by saprophytes living off dead organic matter. Tropical rain forests are found in central Africa, Central America, the Amazon basin, and southeast Asia, and are one of the most productive and diverse communities.

Savanna

The savanna (grassland) is characterized by low rainfall (usually 10–30 inches per year), although it gets considerably more rain than the desert biomes do. Grassland has few trees and provides little protection for herbivorous mammals (such as bison, antelopes, cattle, and zebras) from carnivorous predators. That is why animals that do inhabit the savanna have generally developed long legs and hoofs, enabling them to run fast. Examples of savanna include the prairies east of the Rockies, the Steppes of the Ukraine, and the Pampas of Argentina.

Desert

The desert receives less than ten inches of rain per year, and this rain is concentrated within a few heavy cloudbursts. The growing season in the desert is restricted to those days after rain falls. Generally, small plants and animals inhabit the desert. Most desert plants (for example, cactus, sagebrush, and mesquite) conserve water actively and avoid extreme heat, often by being nocturnal. Desert animals like the lizard, meanwhile, live in burrows. Birds and mammals found in the deserts also have developed adaptations for maintaining constant body temperatures. Examples of desert biomes include the Sahara in Africa, the Mojave in the United States, and the Gobi in Asia.

Temperate Deciduous Forest

Temperate deciduous forests have cold winters, warm summers, and a moderate rainfall. Trees such as beech, maple, oaks, and willows shed their leaves during the cold winter months. Animals found in temperate deciduous forests include the deer, fox, woodchuck, and squirrel. The forest floor is a rich soil of decaying matter, inhabited by worms and fungi. Temperate deciduous forests are located in the northeastern and central eastern United States and in central Europe.

You Won't Sweat Here

We generally think of deserts as very hot and dry. However, cold deserts also exist. Deserts develop in regions in which less than 30 cm of rain falls per year; temperature does not play any role in their formation.

Northern Coniferous Forest

Northern coniferous forests are cold, dry, and inhabited by fir, pine, and spruce trees. Much of the vegetation here has evolved adaptations for water conservation—that is, needle-shaped leaves. These forests are found in the extreme northern part of the United States and in Canada. The forest floor is dry and contains a layer of needles with fungi, moss, and lichens. Common animals include (in North America) moose, deer, black bears, hares, wolves, and porcupines.

Taiga

The taiga receives less rainfall than the temperate forests, has long, cold winters, and is inhabited by a single type of coniferous tree, the spruce. The forest floors in the taiga contain moss and lichens. Birds are the most common animal; however, the black bear, the wolf, and the moose are also found here. Taiga exists in the northern parts of Canada and Russia.

Tundra

Tundra is a treeless, frozen plain located between the taiga and the northern icesheets. Although the ground is always frozen, the surface can melt during summer. It has a very short summer and a very short growing season, during which time the ground becomes wet and marshy. Lichens, moss, polar bears, musk oxen, and arctic hens make their homes here.

Polar Region

The polar region is a frozen area with very few types of vegetation or terrestrial animals. Animals that do inhabit polar regions, such as seals, walruses, and penguins, generally live near the polar oceans, surviving through preying on marine life.

Terrestrial Biomes and Altitude

The sequence of biomes between the equator and the pole is comparable to the sequence of regions on mountains. The nature of those regions is determined by the same decisive factors—temperature and rainfall. The base of the mountain, for example, would resemble the biome of a temperate deciduous area. As one ascends the mountain, one would pass through a coniferous-like biome, then taigalike, tundralike, and polarlike biomes.

Aquatic Biomes

In addition to the eight terrestrial biomes, there are aquatic biomes, each with its own characteristic plants and animals. More than 70 percent of the earth's surface is covered by water, and most of the earth's plant and animal life is found there. As much as 90 percent of the earth's food and oxygen production (photosynthesis) takes place in the water. Aquatic biomes are classified according to criteria quite different from the criteria used to

Two Tundras

There are two types of tundra: arctic tundra, which is found just south of the North Pole, and alpine tundra, located above the tree line on high mountains all over the world.

classify terrestrial biomes. Plants have little controlling influence in communities of aquatic biomes, as compared to their role in terrestrial biomes.

Aquatic areas are also the most stable ecosystems on Earth. The conditions affecting temperature, amount of available oxygen and carbon dioxide, and amount of suspended or dissolved materials are stable over very large areas, and show little tendency to change. For these reasons, aquatic food webs and aquatic communities tend to be balanced. There are two types of major aquatic biomes: marine and freshwater.

Marine Biomes

The oceans connect to form one continuous body of water that controls the earth's temperature by absorbing solar heat. Water has the distinctive ability to absorb large amounts of heat without undergoing a great temperature change. Marine biomes contain a relatively constant amount of nutrient materials and dissolved salts. Although ocean conditions are more uniform than those on land, distinct zones in the marine biomes do exist, including the intertidal zone, littoral zone, and pelagic zone.

Intertidal Zone. The intertidal zone is a region exposed at low tide that undergoes variations in temperature and periods of dryness. Populations in the intertidal zone include algae, sponges, clams, snails, sea urchins, sea stars (starfish), and crabs.

Littoral Zone. The littoral zone is a region on the continental shelf that contains ocean area with depths of up to 600 feet, and extends several hundred miles from the shores. Populations in littoral zone regions include algae, crabs, crustacea, and many different species of fish.

Pelagic Zone. The pelagic zone is typical of the open seas and can be divided into photic and aphotic zones. The photic zone is the sunlit layer of the open sea extending to a depth of 250–600 feet. It contains plankton—passively drifting masses of microscopic photosynthetic and heterotrophic organisms—and nekton—active swimmers such as fish, sharks, and whales that feed on plankton and smaller fish. The chief autotroph is the diatom, an alga.

Meanwhile, the *aphotic zone* may be defined as the region beneath the photic zone with no sunlight and no photosynthesis; only heterotrophs can survive here. Deep-sea organisms in this zone have adaptations that enable them to survive in very cold water, high pressure, and complete darkness. The zone contains nekton and benthos—the crawling and sessile organisms. Some are scavengers and some are predators. The habitat of the aphotic zone is fiercely competitive.

Not to Be Underestimated

Since nearly three quarters of the earth is covered by the oceans, marine biomes are extremely important. Ocean temperatures affect the planet's climate and wind patterns, while marine algae supply us with a large proportion of our oxygen.

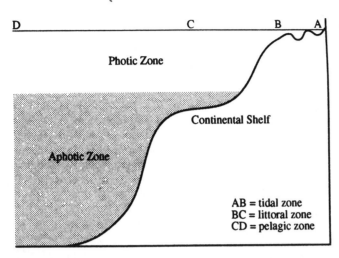

Aquatic Biomes

Freshwater Biomes

Rivers, lakes, ponds, and marshes—the links between the oceans and land—contain freshwater. Rivers are the routes by which ancient marine organisms reached land and evolved terrestrial adaptations. Many forms failed to adapt to land and developed adaptations for freshwater. Others developed special adaptations suitable for both land and freshwater. As in marine biomes, factors affecting life in freshwater include temperature, transparency (illumination due to suspended mud particles), depth of water, available CO_2 and O_2, and most importantly, salt concentration.

Freshwater biomes differ from salt water biomes in three basic ways:

- Freshwater has a lower concentration of salt (greater concentration of water) than the cell, creating a diffusion gradient that results in the passage of water into the cell. Freshwater organisms have homeostatic mechanisms to maintain water balance by the regular removal of excess water, such as the contractile vacuole of protozoa and excretory systems of fish.

- In rivers and streams, strong, swift currents have caused fish to develop strong muscles and plants to develop rootlike holdfasts.

- Freshwater biomes (except for very large lakes) are affected by variations in climate and weather. They might freeze, dry up, or have mud from their floors stirred up by storms. Temperatures of freshwater bodies vary considerably.

In this chapter, we've covered the basic concepts of ecology. By this point, you should be familiar with topics like energy flow, nutrient cycles, levels of biological organization, the physical environment, the ecosystem, and biomes. Now it's time to try your hand at the following quiz.

Stopping the Flood

Freshwater plant cells have rigid cell walls, building up cell pressure (cell turgor) as waste passes in. This pressure counteracts the gradient pressure, stops the influx of water, and, as a result, establishes a water balance.

Ecology Quiz

1. In a pond community, the greatest mass present would consist of

    (A) algae
    (B) insects
    (C) frogs
    (D) fish
    (E) fungi

2. Mutualism is exemplified by

    (A) lichens
    (B) tapeworms
    (C) bread mold
    (D) nematodes
    (E) epiphytes

3. A climax community

    (A) consists of only one species of life
    (B) is populated mainly by so-called pioneer organisms
    (C) is stable within a given climate
    (D) is independent of the environment
    (E) consists of decaying organic matter

4. In an ecosystem, the greatest amount of stored chemical bond energy is generally found in

    (A) primary producers
    (B) secondary producers
    (C) primary consumers
    (D) secondary consumers
    (E) tertiary consumers

5. Living in a close nutritional relationship with another organism in which one organism benefits while the other is neither harmed nor benefited is best defined as

    (A) symbiosis
    (B) mutualism
    (C) saprophytism
    (D) commensalism
    (E) parasitism

6. Digestion of cellulose by protozoans found in a termite's gut is an example of

    (A) mutualism
    (B) parasitism
    (C) saprophytism
    (D) commensalism
    (E) autotrophism

7. Which of the following fixes atmospheric N_2?

    (A) photosynthesis
    (B) symbiotic bacteria on the roots of legumes
    (C) *uv* light
    (D) decay organisms
    (E) autotrophs

GO ON TO THE NEXT PAGE

8. Denitrifying bacteria

    (A) turn ammonia into NO_2 (nitrites)
    (B) turn ammonia into N_2
    (C) turn ammonia into NO_3 (nitrates)
    (D) do not use nitrogen in their life cycle
    (E) none of the above

9. A stable ecosystem

    (A) requires a constant energy source
    (B) requires a living system
    (C) is self-sustaining
    (D) requires cycling of materials between the living system and the environment
    (E) all of the above

10. Which of the following is a marine zone?

    (A) intertidal zone
    (B) littoral zone
    (C) pelagic zone
    (D) all of the above
    (E) two of the above

11. A biome characterized by permafrost and located immediately north of the timberline and south of the permanent ice sheet would best be characterized as

    (A) taiga
    (B) arctic tundra
    (C) alpine tundra
    (D) coniferous forest
    (E) spruce/moose forest

12. In a food chain of grass —> prairie dog —> snake —> owl, the secondary consumer would be

    (A) grass
    (B) prairie dog
    (C) snake
    (D) owl
    (E) microbes of decay

13. In the northeastern United States, the final succession climax stage of a freshwater lake will be a

    (A) marsh
    (B) grassland
    (C) bog
    (D) estuary
    (E) deciduous forest

STOP

Answers and Explanations to the Ecology Quiz

1. **(A)** In an ecology pyramid, the primary producers (photosynthetic or chemosynthetic organisms, such as algae) are always the largest population. (B) and (E) are primary consumers, while (D) and (C) are secondary consumers.

2. **(A)** Mutualism is a close nutritional relationship between two species in which both benefit. Lichen is the result of a mutualistic relationship between fungi and algae. The algae attach to a rock via rootlets that the fungi produce. Through these rootlets, the fungi receives water, nutrients, and an attachment, the lichen. The alga in its turn receives an attachment (lichen) to the rock, and can produce carbohydrates through photosynthetically utilizing the water and nutrients from the fungi's rootlets. This is considered a +/+ situation. Tapeworms (B) are parasites. The tapeworm gets nutrition from the host as the host loses it, a +/– situation. Saprophytes, such as bread mold (C), are organisms that feed on dead and decaying material, while nematodes (D) are either free-living saprophytes or parasites. Finally, (E) epiphytes may be defined as plants that live on the branches of other plants. In this manner they receive greater exposure to sunlight than they would normally have access to, exemplifying commensalism, a +/0 relationship.

3. **(C)** A climax community is the final community in a particular biome's succession. In the northeastern part of the United States, the climax community is the deciduous forest, while in the midwest it is the grasslands. (A) is incorrect because many species will live in a community. Meanwhile, pioneer species (B) are the species that colonize a biome, such as lichen on rocks, and are therefore the earliest species in a succession. (D) The climax community is dependent on the environment, especially the climate; factors such as type of soil and amount of rainfall will determine what organisms will survive and thrive there. As for (E), dead and decaying organisms are present in all communities, but are not the sole inhabitants of any of them.

4. **(A)** In an ecosystem, the greatest amount of energy is always found among primary producers. Primary producers are either plants or photosynthetic bacteria. Energy is lost with each level in the pyramid, as it is utilized for maintenance of the organism, movement, and warmth. Only a fraction of the energy produced becomes new tissue that can be harvested by the next level up in the pyramid. This should remind you of the general chemistry principle that with any exchange of energy, some is lost. Primary consumers are herbivores, while carnivores that ingest the herbivores are known as secondary consumers. They in their turn are preyed upon by tertiary consumers.

5. **(D)** In a commensal relationship, which is a form of a symbiotic relationship, two organisms live in close association with each other. One benefits from this association, while the other is neither harmed nor benefited (in what is sometimes described as a "+/0" relationship). An example of a commensal relationship is the epiphyte plant, which lives on the branches of rainforest trees, gaining the advantage of being closer to sunlight. Symbiosis (A) is a general term describing close nutritional relationships of all types, including mutualism, commensalism, and parasitism. For explanations of the concepts of mutualism (B), saprophytism (C), and parasitism (E), see explanation (2) above.

6. **(A)** Both the termite and the protozoans benefit. Termites cannot actually digest cellulose, but protozoans in their digestive systems can. In return, these tiny organisms receive a home and food and water. The terms mutualism, parasitism (B), saprophytism (C), and commensalism (D) are all defined in the explanation to question 2 above. (E) Autotropism, meanwhile, describes self-feeders such as photosynthetic or chemosynthetic organisms.

7. **(B)** Elemental nitrogen (N_2) is chemically inert and cannot be used by most organisms. Lightning and nitrogen-fixing bacteria in the roots of legumes change the nitrogen to usable, soluble nitrates.

8. **(B)** Denitrifying bacteria break down NH_3 into N_2. Nitrifying bacteria turn ammonia into NO_2 (nitrites), while nitrogen-fixing bacteria turn N_2 into NO_3 (nitrates).

9. **(E)** A stable ecosystem is self-sustaining and will therefore remain stable in the presence of a relatively stable physical environment (abiotic factors) and a relatively stable biotic community. A stable ecosystem requires a constant energy source, a living system incorporating this energy into organic compounds, and a cycling of materials between the living system and the environment.

10. **(D)** All of the zones listed may be classified as types of marine biomes. The intertidal zone (A) is a region exposed at low tides that undergoes variations in temperature and periods of dryness. Populations in the intertidal zones include algae, sponges, clams, snails, sea urchins, starfish, and crabs. Meanwhile, the littoral zone (B) is the term used for a region on the continental shelf that contains ocean area with depths of up to 600 feet and can extend many miles from the shore. Populations in this zone include algae, crabs, crustacea, and many different species of fish. Finally, the pelagic zone, typical of the open sea, is divided into the photic zone (the sunlit layer containing plankton and fish, sharks, or whales) and the aphotic zone (the sunless zone containing the crawling and sessile organisms).

11. **(B)** The arctic tundra is a treeless, frozen plain found between the taiga lands and the northern ice sheets, as well as south of the permanent ice sheet. It has a very short summer and growing season. Lichens, moss, polar bears, musk oxen, and arctic hens are found here. The taiga (A), on the other hand, receive less rainfall than the temperate forests and have long, cold winters. These forests are made up entirely of spruce trees, and the most common animals are the moose, the black bear, the wolf, and birds.

12. **(C)** The snake, a secondary consumer, consumes the prairie dog, a primary consumer. For definitions of the terms named in this question, see explanation (4).

13. **(E)** The northeastern part of the United States would have the climax community of a temperate deciduous forest. These forests have cold winters, warm summers, and moderate rainfall. They are populated by deer, fox, woodchucks, and squirrels, and their trees (beech, maple, oaks, and willows) shed their leaves during the cold winter months.

CHAPTER SEVEN

EVOLUTION AND DIVERSITY

On your SAT II exam, you'll probably be faced with quite a few questions dealing with the evolution and classification of the millions of species on Earth. This chapter will give you the background you need to ace these questions. We'll be covering topics ranging from types of evidence for evolution to the taxonomic classification of various common species.

Evidence of Evolution

Evolution provides a sweeping framework for the understanding of the diversity of life on earth. Living systems, from the cell to the organism to the ecosystem, arose through a long process through geologic time, selecting solutions out of diverse possibilities. What is the evidence that supports the evolutionary view of life? The evidence takes several forms.

The Fossil Record

Fossils are preserved impressions or remains in rocks of living organisms from the past. Fossils provide some of the most direct and compelling evidence of evolutionary change and are generally found in sedimentary rock. When animals settle in sediments after death, their remains can be embedded in the sediment. These sediments then might be covered over with additional layers of sediment that turn to rock through heat and pressure over many millions of years. The embedded remains turn to stone, replaced with minerals that preserve an impression of the form of the organism, often in a quite detailed state. Most fossils are of the hard bony parts of animals, since these are preserved the most easily. Fossils of soft body parts or of invertebrates are much more unusual, probably since these parts usually decay before fossil formation can occur. In some cases, however it appears that animals died in anaerobic sediments that resisted decay to provide soft-body fossils.

One of the questions about fossils when they are discovered is their age, to place the fossil in correctly in the timeline of life on earth. One way to place the date is to compare the location of the fossil sediment to other sedimentary rock formations in which the age is already known. Dating using

How about a Different Kind of Engagement Ring?

Specimens of amber that contain insect fossils are highly prized by jewellers.

Don't Mix These Up on Test Day

Homologous structures share a common ancestry.

Analogous structures are not inherited from a common ancestor, but perform similar functions.

radioactive decay is also very useful. Carbon dating is frequently used for material that is only a few thousand years old, but cannot be used for older material since the decay rate of carbon is too rapid.

The conditions for fossil formation are relatively particular, especially for the preservation of invertebrates or soft body parts. Scientists locate fossils by luck, and overall can only look at a tiny percentage of possible fossil locations. They have over time located a great variety of fossils, including fossils that provide a clear story for the evolution of modern species. Archaeopteryx is an example of a feathered dinosaur that was probably an intermediate species in the evolution of birds. Changes in fossils over time have revealed a great deal about evolution and insight into the evolutionary paths that resulted in modern species including horses, whales, and humans. Any of the so-called "gaps" in the fossil record are probably the result of scarcity of fossils and difficulty in finding them, and is not evidence that evolution did not occur.

Comparative Anatomy

One way to find the evolutionary relationship between organisms is by examining their external and internal anatomy. Animals that evolved from a common ancestor might be expected to have anatomical features in common that they share with their common ancestor. Alternatively, two organisms might share features that look the same but evolved from different ancestors and resulted in similar structures as a result of similar functions. When we compare the anatomies of two or more living organisms, we can not only form hypotheses about their common ancestors, but we can also glean clues that shed light upon the selective pressures that led to the development of certain adaptations, such as the ability to fly. Comparative anatomists study *homologous* and *analogous structures* in organisms.

Homologous structures. Homologous structures have the same basic anatomical features and evolutionary origins. They demonstrate similar evolutionary patterns with late divergence of form due to differences in exposure to evolutionary forces. Examples of homologous structures include the wings of a bat, the flippers of a whale, the forelegs of a horse, and the arms of a human. These structures were all derived from a common ancestor but diverged to perform different functions in what is termed *divergent evolution.*

Analogous Structures. Analogous structures have similar functions but may have different evolutionary origins and entirely different patterns of development. The wings of a fly (membranous) and the wings of a bird (bony and covered in feathers) are analogous structures that have evolved to perform a similar function—to fly. The wings of flies and birds might look the same but this does not indicate that they share a winged ancestor. The evolution of structures that look the same for a common function but are not derived from a common ancestor is called *convergent evolution.* Analogous organs demonstrate superficial resemblances that cannot be used as a basis for classification.

Comparative Embryology

Comparing the anatomy of adult organisms is one method used to derive evolutionary relationships, and comparison of embryonic structures and routes of embryo development is another way to derive these relationships. The development of the human embryo is very similar to the development of other vertebrate embryos. Adult tunicates (sea squirts) and amphibians lack a notochord, one of the key traits of the chordate phylum, but their embryos both possess notochords during development, indicating these animals are in fact vertebrates with a common evolutionary ancestor even though the adults do not resemble each other. The earlier that embryonic development diverges, the more dissimilar the mature organisms are. Thus, it is difficult to differentiate between the embryo of a human and that of an ape until relatively late in the development of each embryo, while human and flatworm embryos diverge much earlier.

Other embryonic evidence of evolution includes such characteristics as teeth in an avian embryo (recalling the reptile stage); the resemblance of the larvae of some mollusks (shellfish) to annelids (segmented worms), and the tail of the human embryo (indicating relationships to other mammals).

Molecular Evolution

If organisms are derived from a common ancestor, this should be evident not just at the anatomical level but also at the molecular level. The traits that distinguish one organism from another are ultimately derived from differences in genes. With the advent of molecular biology, the genes and proteins of organisms can be compared to determine their evolutionary relationship. The closer the genetic sequences of organisms are to each other, the more closely related they are in evolution and the more recently they diverged from a common ancestor.

Some genes change rapidly in evolution while others have changed extremely slowly. The rate of change in a gene over time is called the *molecular clock*. The rate of change in a gene's sequence is probably a function of the tolerance of the gene to changes without disrupting its function. Genes that change very slowly over extremely long periods of time probably do not tolerate change very well and play key roles in the life of cells and organisms. The large ribosomal RNA has changed slowly enough that it can be used to compare organisms all the way back to the divergence of eukaryotes, bacteria, and archaebacteria. The enzymes of glycolysis play an essential role in energy production for all life, and also evolve very slowly, allowing comparison of their gene sequences to illuminate evolutionary relationships over billions of years. Fossil genes are not known, but using computers to compare the gene sequences of many organisms allows researchers to determine how long ago organisms evolved from a common ancestor. More recently evolved genes and genes that evolve more rapidly can be used to compare more recent evolutionary events.

When the Appendix Makes Its Presence Known

Although useless in present-day humans, the appendix can cause a lot of trouble if left unattended when it becomes inflamed. If it bursts, it can cause illness and even death.

Vestigial Structures

Vestigial structures are structures that appear to be useless in the context of a particular modern-day organism's behavior and environment. It is apparent, however, that these structures used to have some function in an earlier stage of a particular organism's evolution. They serve as evidence of an organism's evolution over time, and can help scientists to trace its evolutionary path.

There are many examples of vestigial structures in humans, other animals, and plants. The appendix—small and useless in humans—assists digestion of cellulose in herbivores, indicating human's vegetarian ancestry, while the animal-like tail in humans is reduced to a few useless bones (coccyx) at the base of the spine. The splints on the legs of a horse are vestigial remains of the two side toes of the eohippus. Finally, adult pythons have legs that are reduced to useless bones embedded in their sides, as do whales.

Mechanisms of Evolution

The Population as the Basic Unit of Evolution

Evolution is the change in a species over time. These changes are the result of changes in the gene pool of a population of organisms. Evolution does not happen in one individual, but in a population of a organisms. What is a population? A *population* is a group of individuals of a species that interbreed. In classical genetics, it is observed that the genotype of organisms produces their phenotype, the physical expression of inherited traits. A population of organisms includes individuals with a range of phenotypes and genotypes. It is possible, however, to describe a population not by their individual characteristics, but by certain traits of the group as a whole, including the abundance of alleles within the whole population. The sum total of all alleles in a population is called the *gene pool* and the frequency of a specific allele in the gene pool is called the *allele frequency*. Each individual receives their specific set of alleles from the gene pool, and not every individual receives the same alleles, leading to individual variation in genotypes and phenotypes.

How is the allele frequency calculated and used? For a diploid organism, the total number of alleles for a specific gene in a population is the number of individuals multiplied by 2. If there are 1,000 rabbits in a population, and they are diploid, with two copies of every gene, then the gene pool of the rabbit population will include 2,000 alleles for genes. If the genotype of every rabbit is known, then the allele frequency in the population can be determined by adding up how many copies of each allele are found in the population. If 100 rabbits are homozygous for an allele (both gene copies are the same) and another 200 rabbits in the population are heterozygous for the same allele, then the allele frequency = $(2(100) + 1(200))/2,000$. Allele frequency is the decimal fraction representing the presence of an allele for all members of a population that have this particular gene locus. The letter

p is used for the frequency of the dominant allele of a particular gene locus. The letter q represents the frequency of the recessive allele. For a given gene locus, $p + q = 1$. The total allele frequency for a gene must always equal one.

Sexual reproduction constantly shifts alleles around in a population, mixing and remixing them in new combinations through meiotic recombination, independent segregation of chromosomes during meiosis, and random matching of alleles from parents during mating and fertilization. All of these allow for mixing of alleles in a population to create variation in individual genotypes and phenotypes. Mutation in a population can create new alleles. Evolution is caused by changes in the gene pool of a population over time, as a result of changes that occur to individuals in the population caused by their phenotype and the alleles they carry.

Hardy-Weinberg and Population Changes

The allele frequencies in the gene pool of a population determine how many individuals in a population get each allele and this in turn determines the phenotypes of individuals. If nothing changes the allele frequencies, then every generation will get the same alleles in the same proportions, and the population will not change over time. This idea is the foundation of population genetics and the central idea of *Hardy-Weinberg equilibrium* in population genetics.

According to the Hardy-Weinberg principle, allele frequencies in a population remain constant from generation to generation and a population is maintained in equilibrium as long as certain assumptions are met. If the assumptions are met, and the allele frequencies in the gene pool of a population are constant over time, the population does not change and evolution does not occur. If the assumptions are not true, then the allele frequencies of the population will change and the population will evolve. The assumptions for Hardy-Weinberg equilibrium to be maintained are:

- Random mating (no isolation) must occur, so that no particular trait is favored. There can be no assortative mating (in other words, no organisms may select mating partners that resemble themselves).

- Migration—that is, immigration or emigration—cannot take place.

- There must be no total mutations.

- Large populations are required. As in all cases of probability, large samplings are needed to provide an accurate approximation of the expected occurence. In addition, it must be highly unlikely that chance alone could significantly alter gene frequencies.

- Natural selection does not occur.

The Pouch Holds Its Ground

Pouched animals are primitive, and have generally died out in areas in which they were put in direct competition with more evolutionarily advanced placental animals. Opossums, however, did manage to cross the Bering Strait land bridge and compete successfully with placental mammals.

Systematics

Creating phylogenetic trees is part of systematics, the field of biology that investigates the diversity of life.

Not Completely Wrong

Although Lamarck's theory of evolution has been disproved, he did correctly observe that evolution should not be envisioned as a ladder, but instead as a pathway with many branches.

Under the above conditions, there is a free flow of alleles between members of the same species, while the total content of the gene pool is continually being shuffled. A constant gene pool is nevertheless maintained for the entire species. The constancy of the gene pool is always threatened by changes in the environment (which would favor certain genes), mutations, migrations (new genes introduced), or reproductive isolation (lack of random mating favors certain genes).

Mathematical Demonstration of the Hardy-Weinberg Equilibrium Principle. We can cross two individuals to demonstrate mathematically that the gene pool frequencies remain constant generation after generation. Let us assume for the original gene pool that the gene frequency of the dominant allele for tallness, p, is $0.80T$, and the gene frequency of the recessive allele, q, is $0.20t$. Thus $p = 0.80$ and $q = 0.20$. The parents are crossed and their offspring frequencies are shown with a Punnett square.

The resulting gene frequency in the F_1 generation is: 64 percent TT, 16 percent + 16 percent or 32 percent Tt, and 4 percent tt.

Possible Sperm

		$0.80T$	$0.20t$
Possible Eggs	$0.80T$	$0.64TT$ ($p^2 = 0.64$)	$0.16Tt$ ($pq = 0.16$)
	$0.20t$	$0.16Tt$ ($pq = 0.16$)	$0.04tt$ ($q^2 = 0.04$)

If there are no mutations, no migrations, and no decrease in population size, the frequencies of the above F_1 generation are applicable in calculating the frequencies for the F_2 generation. The alleles are reshuffled between individuals between generations but they are not lost.

Some Mathematical Applications of the Hardy-Weinberg Principle. In working out population genetics problems, we must recall that p = gene frequency of the dominant allele, q = frequency of the recessive allele, and $p + q = 1$, since frequencies of the dominant and recessive alleles total 100 percent. In a population that includes alleles p and q, the alleles will produce offspring with the following types and frequencies:

Possible Sperm

		p	q
Possible Eggs	p	$p\,p$	$p\,q$
	q	$p\,q$	$q\,q$

Thus, when we cross $p + q$, we obtain $p^2 + 2pq + q^2 = 1$ (the "1" indicates that the total is 100 percent of the offspring). Note that the result is scientific confirmation of an obvious algebraic identity: $(p + q)(p + q) = p^2 + 2pq + q^2$. The official key for working out the following problem is as follows:

p = frequency of dominant allele

q = frequency of recessive allele

p^2 = frequency of homozygous dominant individuals

$2pq$ = frequency of heterozygous individuals

q^2 = frequency of homozygous recessive individuals

Another mathematical application of the Hardy-Weinberg principle is explained below:

Problem: In a certain population, the frequency of homozygous curly hair (CC) is 64 percent. What percentage of the population has curly hair?

Solution: According to the key, p represents the frequency of the dominant allele (C), while q represents the frequency of the recessive allele (c). We are told that the CC frequency is 64 percent. This means that $p^2 = 0.64$ or $p = 0.8$. Since $p + q = 1$, $q = 1 - 0.8$, or 0.2. An individual with curly hair may be either CC or Cc. The frequency of each genotype = $p^2 + 2pq + q^2$.

$p^2 = 0.64$ or 64 percent homozygous curly hair

$2pq = 2(0.8)(0.2)$ or 32 percent heterozygous curly hair

$q^2 = 0.04$ or 4 percent homozygous straight hair

Therefore, the percentage of the population possessing curly hair is as follows:

$p^2 + 2pq = 64$ percent + 32 percent = 96 percent

Disruption of Hardy-Weinberg Equilibrium in Evolution

The Hardy-Weinberg principle describes the stability of the gene pool. However, no population stays in Hardy-Weinberg equilibrium for very long, because the stable, ideal conditions needed to maintain it do not exist. The assumptions required for equilibrium cannot be met in the real world.

As conditions change, the gene pool changes and the population changes. Changes in the gene pool caused by breaking the assumptions are the basis of evolution.

Mutation

If the gene pool is not going to change, then there can be no new alleles that appear in the population. Mutations may be infrequent in a population as a result of the great accuracy of DNA replication, but DNA replication is never perfect and some mutations will occur at least infrequently. Radiation from the environment and environmental mutagens also contribute a low by inescapable level of mutation in any population. The mutations will not form a large part of allele frequency, but they do form an important com-

Test Strategy

All you need to know to solve any Hardy-Weinberg problem is the value of p (or p^2) and q (or q^2). Once you know these, you can calculate everything else using the formulas $p + q = 1$ and $p^2 + 2pq + q^2 = 1$.

Adapting to Poison

DDT is an insecticide used to kill insects. In the past, when it has been introduced into insect populations, a favorable environmental change has been created for DDT-resistant mutants, and a fatal environmental change for the rest of the insect population. Conditions have selected for the survival of DDT-resistant organisms, resulting in the evolution of a new DDT-resistant species.

ponent, as a source of variation in a population. Most mutations are harmful, but a small minority may confer a selective advantage in some way. Phenotypes are the material that evolution acts on in a population and mutations are the only source of truly new alleles that will result in truly new phenotypes.

Migration

If two populations are separated from each other and do not interbreed, then the allele frequencies in their gene pools may be different from each other. If individuals move between the populations however, carrying their alleles with them, this creates gene flow, and will alter the frequency of alleles in both of the populations involved.

Population Size

One of the assumptions for the maintenance of Hardy-Weinberg equilibrium is that a population is large. Small populations are subject to random events that can statistically alter the gene pool. Changes in the gene pool caused by random events in a small population are called *genetic drift*. One example is a *population bottleneck*. If an event like a flood suddenly and dramatically reduces the size of a population, the allele frequencies of the survivors are not necessarily the same as the allele frequencies in the original population. When the population grows in size again, the allele frequencies in the new gene pool will represent the frequencies in the small bottleneck population, not the population before the reduction in size. A similar phenomena called the *founder effect* occurs in the colonization of a new habitat. When a new island forms, it might be colonized by a very small number of individuals from another population. Since the new population is founded by a small number of individuals, it is unlikely statistically that the island population will represent the same allele frequencies as the population they were derived from.

Nonrandom Mating

If a population is going to maintain constant allele frequencies, then alleles must be matched randomly in each new generation. This requires that individuals mate with each other without any preference for specific traits or individuals. If the phenotype of individuals influences mating, this will change allele frequencies and disrupt Hardy-Weinberg equilibrium. Most species are quite discerning in mate selection, however, blocking maintenance of Hardy-Weinberg equilibrium.

Natural Selection

Within a population of organisms, individuals are non-identical. Mutation is a source of new alleles, and sexual reproduction leads to constant shuffling of alleles in new genotypes. The variety of genotypes created in a population in this way creates a variety of phenotypes. If individuals have different phenotypes, then these individuals probably interact with their envi-

Instability in the Gene Pool

The following factors cause instability in the gene pool, resulting in the loss of Hardy-Weinberg equilibrium:

- Mutations

- Genetic drift

- Migration

- Natural selection

- Nonrandom mating

ronment with differing degrees of success in escaping predators, finding food, avoiding disease, and reproducing. The differential survival and reproduction of individuals based on inherited traits is *natural selection* as described first by Charles Darwin.

Fitness is a quantitative measure of the ability to contribute alleles and traits to offspring and future generations. The key to fitness is reproduction and survival of offspring. Avoiding predators, finding food, resistance to disease and other factors that improve survival are likely to improve fitness but only to the extent that they lead to more offspring and more of the alleles involved in the future gene pool. Finding a mate, mating, successful fertilization, and caring for offspring are factors that can improve fitness as well. There are different strategies for improving fitness. For example, some animals have lots of offspring but provide little parental care, while other animals have few offspring but provide lots of care for each of them.

None of the other factors that alter Hardy-Weinberg equilibrium alter it in a directed fashion. Genetic drift, mutation, and migration are all random in their effects on the gene pool. Natural selection, however, increases the prevalence of alleles in a population that increase survival and reproduction. Alleles that increase fitness will over time increase in their allele frequency in the gene pool, and increase the abundance of the associated phenotype as well. This effect will change the population in a directed manner over many generations, creating a population that is better adapted to its environment.

Different types of natural selection can occur, including *stabilizing selection*, *disruptive selection* and *directional selection*. Traits in a population such as the height of humans are often distributed according to a bell-shaped curve. The type of selection that occurs can affect the average value for the trait or it can alter the shape of the curve around the average. *Stabilizing selection* does not change the average, but makes the curve around the average sharper, so that values in the population lie closer to the average. For example if both very small fish and very large fish tend to get eaten, then stabilizing selection may not alter the average fish size, but is likely to cause future generations to be closer to average.

Disruptive selection is the opposite, in which the peak value is selected against, selecting for either extreme in a trait, so that a single peak for a trait in a population tends to be split into two peaks. *Directional selection* alters the average value for a trait, such as selecting for dark wings in a population of moths in an industrial area.

Natural selection acts on an individual and its direct descendants. In some cases natural selection can also act on closely related organisms that share many of the same alleles. This type of natural selection, called *kin selection*, occurs in organisms that display social behavior. The key to fitness is that an organism's alleles are contributed to the next generation. Contribution of alleles can happen by an individual or by close relatives like siblings, aunts, uncles, etc, who share many of the same alleles. The evolution of

Fitness

Fitness is defined as the ability of an organism to contribute its alleles and therefore its phenotypic traits to future generations.

When Different Species Interbreed

Members of different species can interbreed under certain circumstances, but the offspring they produce is frequently infertile. Hence the mule, which is the offspring of two different species, the horse and the ass, is usually sterile.

social organisms is the result of the increased fitness that social behavior provides. Described cases of altruistic behavior in animals is probably the result of kin selection at work, in which an animal might sacrifice its own safety to allow relatives to survive, thereby increasing the fitness of itself and the whole social group it shares alleles with.

Speciation

A species is a group of organisms that is able to successfully interbreed with each other and not with other organisms. The key to defining a species is not external appearance. Within a species, there can be great phenotypic variation, as in the domestic dog. What defines a species is reproductive isolation, an inability to interbreed and create fertile offspring. Actual interbreeding is not necessary to make organisms the same species. Two groups of animals may live in different locations and never contact each other to interbreed, but if a researcher transports some of the animals and they create fertile offspring, they are part of the same species. Horses and donkey can interbreed and create offspring, the mule. The mule, however, is sterile, meaning the horses and donkeys are two different species.

Speciation, the creation of a new species, occurs when the gene pool for a group of organisms becomes reproductively isolated. At this point, evolution can act on that group that shares a gene pool separately from all other life on earth. Two species can be derived from a single common ancestor species. In many cases this occurs when two populations of a species are separated geographically through a process known as *allopatric speciation*.

Separation of a widely distributed population by emerging geographic barriers causes each population to evolve specific adaptations for the environment in which it lives, in addition to the accumulation of neutral (random, non-adaptive) changes. These adaptations will remain unique to the population in which they evolve, provided that interbreeding is prevented by the barrier. In time, genetic differences will reach the point where interbreeding becomes impossible and reproductive isolation would be maintained if the barrier were removed. In this manner, geographic barriers promote evolution.

Adaptive radiation is the production of a number of different species from a single ancestral species. Radiation refers to a branching out; adaptive refers to the hereditary change that allows a species to be more successful in its environment or to be successful in a new environment. Whenever two or more closely related populations occur together, natural selection favors evolution of different living habits. This results in the occupation of different niches by each population (this process is discussed in detail in our chapter on ecology). This divergent evolution through adaptive radiation has been an extremely frequent occurrence, as demonstrated by the famous example of Darwin's finches.

Lamarckian Evolution

Until it was supplanted by Darwin's ideas, the scientist Lamarck's theory was one of the more widely accepted explanations of the mechanisms of evolution. The cornerstone of Lamarck's hypothesis was the principle of use and disuse. He asserted that organisms developed new organs, or changed their existing ones, in order to meet their changing needs. The amount of change that occurred was thought to be based on how much or little the organ in question was actually used.

Unfortunately for Lamarck, this theory of use and disuse was based upon a fallacious understanding of genetics. Any useful characteristic acquired in one generation was thought to be transmitted to the next. An oft-cited example was that of early giraffes, which stretched their necks to reach for leaves on higher branches. The offspring were believed to inherit the valuable trait of longer necks as a result of their parents' excessive use of their necks. Modern genetics has disproved this concept of acquired characteristics.

It has now been established that changes in the DNA of sex cells are the only types of changes that can be inherited; because acquired changes are changes in the characteristics and organization of somatic cells, they cannot be inherited.

Classification and Diversity

Evolution has created a great diversity of organisms on earth, but these organisms are related to each other through common ancestors they shared in the history of life. By examining organisms for common features and common ancestors, it should be possible to make sense of the diversity of life by grouping organisms into categories together. The science of classifying living things and using a system of nomenclature to name them is called *taxonomy*. Carolus Linnaeus invented modern taxonomy in the 1700s, grouping organisms and naming them according to a hierarchical system.

A modern classification system seeks to group organisms on the basis of evolutionary relationships. The bat, whale, horse, and human are placed in the same class of animals (mammals) because they are believed to have descended from a common ancestor. The taxonomist classifies all species known to have descended from the same common ancestor within the same taxonomic group.

Since much about early evolutionary history is not understood, there is some disagreement among biologists as to the best classification to employ, particularly with regard to groups of unicellular organisms. Taxonomic organization proceeds from the largest, broadest group to the smaller, more specific subgroups. The largest group, or kingdom, is broken down into smaller and smaller subdivisions. Each smaller group has more specific characteristics in common. Furthermore, each subgroup is distinguishable from the next. The naming system is subject to discussion and revision as research yields new insights over time into the relationship between organisms. Some classifications are clearer than others.

Viruses are obligate intracellular parasites that cannot conduct metabolic activities or replicate on their own. As such, they are not generally considered living, although they are certainly important to living systems. They are not classified within this taxonomy however.

Classification and Subdivisions

Each *kingdom* has several major phyla. A *phylum* or division has several *subphyla* or subdivisions, which are further divided into *classes*. Each class consists of many *orders*, and these orders are subdivided into *families*. Each family is made up of many *genera*. Finally, the *species* is the smallest subdivision.

Hence the order of classificatory divisions is as follows:

KINGDOM —> PHYLUM —> SUBPHYLUM —> CLASS —> ORDER —> FAMILY —> GENUS —> SPECIES

The complete classification of humans is:

Kingdom:	Animalia
Phylum:	Chordata
Subphylum:	Vertebrata
Class:	Mammalia
Order:	Primates
Family:	Hominidae
Genus:	*Homo*
Species:	*Sapiens*

Assignment of Scientific Names

All organisms are assigned a scientific name consisting of the genus and species names of that organism. Thus, humans are *Homo sapiens*, and the common housecat is *Felis domestica*.

One of the primary groupings of all living organisms separates *prokaryotes* from *eukaryotic* organisms. The prokaryotes include bacteria, and another type of organisms called archaebacteria. Like the bacteria, archae have no organelles and have a simple circular DNA genome. Archae were relatively unknown until recently, and tend to inhabit harsh environments like hot springs that might resemble the early earth. Archae are distinct from bacteria in many ways such as the composition of their membrane lipids, and in some ways appear to be related more closely to eukaryotes than prokaryotes. For this reason, more recent classification schemes break all living things into three *domains*, groups at a higher level than kingdom: bacteria, archaebacteria, and eukaryotes.

What Came First?

Heterotrophs were the first form of life to develop. As their need for nutrients surpassed the rate at which these nutrients were being spontaneously formed, autotrophs developed.

Bacteria

The ubiquitous bacteria are single-celled, lack true nuclei, lack a cytoskeleton and contain double-stranded circular chromosomal DNA that is not enclosed by a nuclear membrane. These creatures nourish themselves heterotrophically—either saprophytically or parasitically—or autotrophically, depending upon the species. Bacteria are classified by their morphological appearance: cocci (round), bacilli (rods), and spirilla (spiral). Some forms are duplexes (diplococci), clusters (staphylococci), or chains (streptococci).

Bacteria have cell walls made of peptidoglycan, a specialized matrix of carbohydrates and peptides. A method of staining bacteria called Gram staining separates them into two groups according to the strength of the staining of their cell wall: Gram positive and Gram negative. Gram positive cells have a thick peptidoglycan cell wall that stains strongly while Gram negative cells have a thin cell wall and an outer membrane that stains poorly.

The Protist Kingdom

The simplest eukaryotic organisms are the *protists*. Protists probably represent the evolution between prokaryotes and the rest of the eukaryotic kingdoms, including fungi, plants and animals. Most, but not all, protists are unicellular eukaryotes. One way to define the protists is that this group includes organisms that are eukaryotes but are not plants, animals, or fungi. The protists include heterotrophs like *amoeba* and *paramecium*, photosynthetic autotrophs like *euglena* and *algae*, and fungilike organisms like *slime molds*. Some protists are mobile through the use of flagella, cilia, or amoeboid motion. Protists use sexual reproduction in some cases and asexual reproduction in others.

One of the best known types of protists are the *amoebas*. Amoebas are large single-celled organisms that do not have a specific body shape. They move and change their shape through changes in their cytoskeleton and streaming of cytoplasm within the cell into extensions called pseudopods. Amoebas are heterotrophs that feed by engulfing a food source through phagocytosis, internalizing the food to digest it in vacuoles in the cytoplasm.

Ciliates are another well-known group of protists and are complex single-cell organisms, including paramecium. The surface of ciliates is covered with cilia that beat in a coordinated fashion to move the cell through water. Ciliates have a defined shape and contractile vacuoles that visibly beat under the microscope to remove excess water from the cell that enters through osmosis. Food (yeast cells in the case of paramecium) is internalized through an oral groove where it is internalized in digestive vacuoles. Paramecia reproduce mitotically, but also have a mechanism called conjugation for exchange of genetic material between cells.

Slime molds are an interesting group of organisms that in some cases are grouped with the fungi kingdom. They are heterotrophs, with some slime

Study Tip

Know the following broad categories of organisms by heart when the time comes to take your SAT II: Biology E/M exam:

- Heterotrophic aerobes (amoebae, earthworms, humans)

- Heterotrophic anaerobes (yeast)

- Autotrophic aerobes (green plants)

- Autotrophic anaerobes (chemosynthetic bacteria)

molds spending some of their time as independent cells, but at other times gathering together to form multicellular forms that produce spores.

Algae are an important group of photosynthetic protists, mostly unicellular. Algae include diatoms, single celled organisms with intricate silica shells; dinoflagellates, with flagella; and brown algae. Algae can reproduce sexually and with alternation of generations between diploid sporophytes and haploid gametophytes, as occurs in plants. The algae include large multicellular forms like giant kelp that might be grouped with the protists since they are an algae, but are also grouped with plants by others. It is likely that the plants evolved from one group of algae, the green algae.

The Fungi Kingdom

Fungi are heterotrophs that absorb nutrients from the environment. Fungi are often saprophytic, feeding off of dead material as their nutrition source, and are important along with bacteria in the decay of material in ecosystems. Absorptive nutrition involves the secretion of enzymes that digest material in the extracellular environment, followed by absorption of the digested material back into the cell. One of the distinguishing features of fungi is their cell wall made of chitin, unlike the cellulose found in plants. Fungi often form long slender filaments called hyphae. Mushrooms, molds and yeasts are all examples of fungi.

Most fungi reproduce both sexually and asexually. Asexual reproduction can occur through the production of haploid spores or through splitting of a piece of fungus that grows mitotically into a new organism. Sexual reproduction in fungi does not involve distinct male or female sexes, but multiple mating types that are not distinct in their morphology. Fungi spend most of their life cycle in a haploid form. Fertilization of haploid gametes forms a diploid zygote that usually quickly enters meiosis to produce haploid spores that can grow mitotically into mature haploid fungi.

The Plant Kingdom

Plants are multicellular eukaryotes that produce energy through photosynthesis in chloroplasts, using the energy of the sun to drive the production of glucose. A cell wall of cellulose is a common feature of all plants, along with a life cycle featuring alternation of generations between haploid gametophytes and diploid sporophytes. The sporophyte is a diploid form that makes haploid spores that grow into a complete haploid form, the gametophyte. The gametophyte in turn is a haploid form that produces haploid gametes that unite through fertilization during sexual reproduction. The resulting diploid zygote grows into the mature sporophyte once again.

Plants are distinct from animals in that plants are usually nonmotile while animals are heterotrophic and move. Plant structure is adapted for maximum exposure to light, air, and soil by extensive branching; animals, on the other hand, are adapted usually in compact structures for minimum surface exposure and maximum motility. Animals have much more centralization in their physiology while plants often exhibit delocalized control of processes and growth.

The evolution of plants has included the ongoing increase in the ability to conquer land through resistance to gravity and ability to tolerate drier conditions. The first plants probably evolved from green algae in or near shallow water. These first plants were nontracheophytes, without water transport systems called vascular systems. Nontracheophytes include mosses. They also lack woody stems. Their lack of a vascular system restricts their size and generally restricts their range to very moist environments. In nontracheophytes, the sporophyte is larger than the gametophyte.

The evolution of vascular systems was a major adaptation in plants. The first vascular plants, tracheophytes that did not produce seeds, included ferns and horsetails, plants with cells called tracheids that form tubes for the movement of fluid in the plant tissue called xylem. This vascular system also helps to provide rigid stems that plants need to live on land. These plants colonized land about 400 million years ago, making it possible for animals like arthropods to colonize land soon after. The non-seed tracheophytes dominated the land for 200 million years before plants with seeds appeared. Ferns form large sporophytes, which release haploid spores. These spores grow into gametophytes that produce haploid gametes. Fertilization in ferns requires sperm to swim through water, restricting most ferns to moist environments.

The evolution of the seed was the next major event in plant evolution, found first in the gymnosperms and later in the flowering plants, the angiosperms. The seed is a young sporophyte that becomes dormant early in development. The embryo is usually well-protected in the seed and able to survive unfavorable conditions by remaining dormant until conditions become more favorable again and the embryo begins to grow again, sprouting. In some cases seeds can remain viable for many years, waiting for the right conditions for the sporophyte to grow. This increases the ability of plants to deal with the variable conditions found on land. Seeds are produced as the result of fertilization of male and female gametes produced by male and female gametophytes. In the seed plants, the male and female gametophytes are small.

The conifers are the most abundant gymnosperms today. About 200 million years ago gymnosperms replaced the nonseed vascular plants like ferns as the predominant plant forms on land. Pines and other conifers are large diploid sporophytes. Gymnosperms like conifers produce male and female spores in separate cones. The male cones make the male gametophytes, which are pollen grains. In conifers, pollen grains are usually dispersed by the wind to find female cones. Unlike ferns, conifers do not require male gametes to swim in water to find the female gametes to fertilize, an adaptation that allows these plants to live in drier environments. When the pollen grain finds the female cone, they grow pollen tubes from the pollen

Mnemonic

The following sentence will help you remember the order of classificatory divisions:

King **P**hillip **S**wiftly **C**ame **O**ver **F**or **G**ood **S**ushi.

Versatile Viruses

Viruses possess the characteristics of life only when they've infected a living host cell. Some examples of the wide range of diseases caused by viruses are:

- Chicken pox (caused by varicella zoster)
- AIDS (HIV)
- Colds (rhinoviruses)
- Cold sores (herpes simplex virus I)

grain to the female gametophyte that contains the eggs of the plant. Male gametes swim through the pollen tube to the eggs to fertilize them and create the diploid zygote that will grow into a seed and eventually another mature plant sporophyte.

Following the evolution of the seed, the next big innovation in plant evolution was the flower. The angiosperms represent the flowering plants and are today the predominant plant group in many ecosystems. Like the gymnosperms, angiosperms produce seeds. The seeds of gymnosperms are "naked," growing without a large amount of nutritional material or protective tissues. Angiosperms produce flowers for fertilization and produce seeds that are surrounded by nutritional tissues. Angiosperm development involves a double fertilization. One of the fertilizations involves the fertilization of an egg by one sperm that grows into the embryo. The other fertilization involves the fertilization of two female nuclei by one sperm to form a triploid tissue that grows to form the nutritive component of the seed, the endosperm. When the seed embryo germinates, it first gains nutrition from the endosperm.

The Animal Kingdom

Animals are fairly easy to recognize: animals are all multicellular heterotrophs. The evolution of animals has included the evolution of a variety of body plans to solve problems like getting food, avoiding predators, and reproducing. Over time, animals have tended to become larger in size, and more complex, with greater specialization of tissues. Another trend in the animal kingdom has been the evolution of increasingly complex nervous systems to enable complex behaviors in response to the environment.

Different groups of animals have evolved different body shapes, reflecting their different life styles. Animals with *radial symmetry* are organized with their body in a circular shape radiating outward. The echinoderms like sea stars and the cnidarians like jellyfish are examples of animals with radial symmetry. Another common body plan is *bilateral symmetry*, in which the body has a left side and a right side that are mirror images of each other. Humans are a good example of bilateral symmetry, in which a plane drawn vertically through the body splits the body into left and right sides that look the same. The front of the body, where the head is located, is the anterior, and the rear of the animal is the posterior. The back of the animal, where the backbone is located in vertebrates, is the dorsal side (like the dorsal fin) and the front of the animal is the ventral side.

The method used to capture food is intimately tied to their body shape. Some animals that do not move are called *sessile*. These animals gather food by filtering it from the environment. Examples of sessile filter feeders include sponges and cnidarians. This is a highly successful life style that requires little energy to gather food, waiting instead for the food to come to you, but animals with this lifestyle are at the mercy of their environment and must compete for space and resources. Animals with more active life

The Life of a Parasite

The *plasmodium*, which causes malaria, begins its life cycle when an infected mosquito bites a human. Sporozoites enter the human's bloodstream and infect the red blood cells. These red blood cells eventually lyse, causing severe anemia.

styles have evolved increasingly complex nervous systems and motor systems to enable them to navigate their environment.

During the early stages of animal development, immediately after fertilization, the embryo enters into several rapid cycles of cell division that split the zygote into increasing smaller cells. In some animals called *protostomes* the cells in the early embryo divide in a spiral pattern while in *deuterostome* animals, cells are cleaved in a radial pattern. The protostomes include annelids, arthropods, mollusks, and roundworms, while the deuterostomes include the echinoderms and chordates. These divisions reflect one of the major evolutionary divisions in the animal kingdom.

The body cavity in animals has evolved over the history of animals. The body cavity is the area between the gut and the exterior of the animal. Early animals like flatworms have only solid tissue between the gut wall and the exterior surface. Other animals like annelids and chordates have evolved a cavity called the *coelom* between the gut and the exterior wall. The coelom is lined with muscle both around the gut and the interior wall. The organs of the chest and abdomen in chordates reside with the coelom. The coelom in annelids makes coordinated motion with a hydrostatic skeleton possible.

The phyla that are described here do not include all of the animal phyla, but most of the more abundant and important phyla.

Phylum Porifera (sponges). Animals probably evolved from simple colonial heterotrophic protists with groups of cells living together and starting to specialize for different functions. These simple animals, probably representing the first evolutionary step between protists and animals, are the *sponges, phylum porifera.* Sponges resemble a colonial organism, with only a small amount of specialization of cells within the animal, no organs, and distributed function. Sponges usually only have a few different types of cells, no nervous system, and if broken apart can reassemble into new sponges. With a saclike structure, sponges have flagellated cells that move water into the animal through pores into a central cavity and back out again. Cells lining the cavity capture food from water as it moves past.

Sponge

Phylum Cnidaria (hydra, sea anemone, jelly fishes). Cnidarians, also called coelenterates, have radial symmetry, with tentacles arranged around a simple gut opening. Their gut has only one opening to the environment through which food passes in and wastes pass out. They are aquatic animals and represent one of the earliest phyla of animals in evolution, with only two cell layers, the endoderm and ectoderm. With only two cell layers, cnidarians do not need circulatory or respiratory systems. These animals have a simple nerve net to respond to the environment, a decentralized system for simple responses to the environment such as retraction of tentacles or swimming motions with the body. The tentacles contain one of the trademarks of cnidarians, stinging cells called *nematocysts* that have a harpoon-like structure toxins to capture prey. The life cycle of cnidarians can include a polyp stage and a medusa stage. The polyp is settled on a solid surface, with the mouth opening pointed upward while the medusa is a swimming form, with the mouth opening pointed downward. Polyps are asexual. Sea anemones on a rock in a cluster are often clones of each other that have reproduced by budding, competing for space with other clones. Sexual reproduction occurs in medusa, where sperm and eggs are produced and released into the environment for fertilization.

Cnidarian

Phylum Platylhelminthes (flatworms). *Flatworms* are ribbonlike with bilateral symmetry. They possess three layers of cells, including a solid mesoderm but lack a circulatory system. Their nervous system consists of simple light detection organs, an anterior brain ganglion, and a pair of longitudinal nerve cords. Their digestive system is a cavity with a single opening, and they lack a coelom. These animals are not swift moving, using cilia to move over surfaces. A common flatworm is the planaria, famous for its regeneration. The shape of the worm, elongated and without appendages, has evolved in many phyla, as a compact structure that is well designed for movement. Planaria are free-living but many flatworms are internal parasites, including flukes and tapeworms, deriving their nutrition by direct absorption into their cells from the host.

Flatworm

Phylum Nematoda (roundworms). *Nematodes* are roundworms, with three cell layers, including mesoderm, a complete digestive tract with two openings, a mouth and an anus, and a body cavity called a *pseudocoelom* around the gut. The pseudocoelom has muscle lining the interior body wall but not around the gut. This allows for very active movement but more of a wiggling motion than movement in a specific direction. Nematodes do not have respiratory or circulatory systems, exchanging gases directly with the environment. Roundworms are one of the most abundant animal groups, including huge numbers of free-living, scavenging species as well as parasites. The species *C. elegans* is a simple organism with only 950 cells that has made it popular in modern biology for studies of cell differentiation and genetics.

Phylum Annelida (segmented worms). The earthworm and leaches are examples of annelid worms, commonly called *segmented worms*. The annelids are worms with segmented bodies and a coelom body cavity. The division of the body into segments and the presence of the coelom body cavity filled with water creates a hydrostatic skeleton that allows annelids complex, sophisticated movement, coordinated by a nervous system. Each segment has local control by a ganglion of nerves but these nerves are coordinated by a ventral nerve cord and a larger nerve collection that might be called a brain in the front of the worm. Annelids exchange gases directly with their environment through their skin, an important reason why they have moist skin and live in moist environments. Each segment has a twin set of excretory organs called *nephridia*. Annelids have a complete digestive tract with some specialization into organs along the tract and they also have a closed circulatory system with five pairs of hearts.

Annelids

Phylum Arthropoda. Arthropods have jointed appendages, exoskeletons of chitin, and open circulatory systems. With an exoskeleton, the coelom of arthropods is reduced and less important in movement. The three most important classes of arthropods are insects, arachnids, and crustaceans. The exoskeleton of arthropods has muscles attached to their interior for movement. The exoskeleton provides protection, and has a variety of specialized appendages. The exoskeleton prevents gas exchange between the skin and exterior, however, as occurs in annelids, requiring the evolution of a respiratory system. Insects possess three pair of legs, spiracles, and tracheal tubes designed for breathing outside of an aquatic environment. Arachnids have four pair of legs and "book lungs"; examples include the scorpion and

Evolutionary Pathway

The evolutionary pathway to humans is as follows:

Porifera
Radiata
Acoelomates
Pseudocoelomates
Protostones
Annelida
Arthopoda
Deuterostomes
Chordata
Vertebrata
Mammalia
Homo sapiens

the spider. Most arthropods have complex sensory organs, including compound eyes, that provide information about the environment to their increasingly complex nervous systems. Crustaceans have segmented bodies with a variable number of appendages. Crustaceans like the lobster, crayfish, and shrimp possess gills for gas exchange. The exoskeleton of arthropods allowed them to colonize land and become the first winged organisms as well. Arthropods, particularly insects, remain one of the most abundant and varied groups of organisms on earth.

Phylum Mollusca. The *mollusks* include animals like clams, squid, and snails. In their body shape these animals do not resemble each other very much but they do share a few basic traits that lead biologists to classify them together as mollusks. These shared mollusk traits include a muscular foot, a mantle that secretes a shell, and a rasping tongue called the *radula*. Most mollusks are covered by a hard protective shell secreted by the mantle. Mollusks are mostly aquatic and use gills for respiration that are enclosed in a space created by the mantle, the mantle cavity. The gills are also involved in feeding, and move water over their surface with the beating of cilia.

Mollusk

Phylum Echinodermata. The *echinoderms*, which include sea stars and sea urchins, are spiny and have radial symmetry, contain a water-vascular system, and possess the capacity for regeneration. The echinoderms may not resemble the vertebrates but they share in common with chordates that they are deuterostomes. The water vascular system is a unique adaptation of the echinoderms, with a network of vessels that carry water to extensions called tube feet. The tube feet are the small suckerlike extensions in sea stars, sea urchins and sand dollars that allow the animals to adhere and to move. Echinoderms also have a hard internal skeleton formed from calcium deposits that assists in protection and locomotion.

Phylum Chordata. The chordates have a stiff, solid dorsal rod called the *notochord* at some stage of their embryologic development, as well as paired gill slits. Chordata have dorsal hollow nerve cords, tails extending beyond the anus at some point in their development, and a ventral heart. These adaptations may not sound impressive but they paved the way for the evolution of the vertebrates, a major subphylum of chordates.

The chordates probably originated from animals like tunicates, commonly called sea squirts. Adult tunicates are sessile filter feeders that do not resemble vertebrates at all. Tunicate larvae however are free-swimming, with a notochord and a dorsal nerve cord, and resemble tadpoles.

The *vertebrates* are a subphylum of the chordates that includes fish, amphibians, reptiles, birds and mammals. In vertebrates the notochord is present during embryogenesis but is replaced during development by a bony, segmented vertebral column that protects the dorsal spinal cord and provides anchorage for muscles. Vertebrates have bony or cartilaginous endoskeletons, chambered hearts for circulation, and increasingly complex nervous systems. The vertebrate internal organs are contained in a coelom body cavity.

The first vertebrates were probably filter-feeding organisms that evolved into swimming jawless fishes that were still filter feeders. Jawless fish such as lampreys still exist today. The evolution of fish with jaws led to the development of the cartilaginous and bony fishes that are dominant today. These fish use gills for respiration, and move water over the gills through paired gill slits. The jaw allows fish to adopt new life styles other than filter feeding, grabbing food with their jaws. Cartilaginous fish (*class Chondrichthyes*) like sharks and rays have an endoskeleton that is made entirely of cartilage rather than hard, calcified bone. Bony fishes (*class Osteichthyes*) have swim bladders for the regulation of buoyancy in water.

Two adaptations were important to set the stage for vertebrates to colonize the land. One was the presence of air-sacs that allowed some fish in shallow water to absorb oxygen from air for breath periods. The other adaptation was a change in the structure of fins to have lobes that allowed some degree of movement on land. Fish with these features evolved into the *amphibians* about 350 million years ago. Most amphibians like frogs and salamanders still live in close association with water and have only simple lungs or gills supplemented by oxygen absorbed through the skin. Another reason that amphibians are mostly associated with water is that amphibian eggs lack hard shells and will dry out on land. Amphibian larvae often live in water and then metamorphose into the adult form.

Reptiles became independent of water for reproduction through the evolution of hard-shelled eggs that do not dry out on land. The egg shell protects the developing embryo but still allows gas exchange with the environment. Reptiles also evolved more effective lungs and heart and thicker dry skins to allow them a greater metabolic activity than amphibians and the ability to survive on land.

Birds evolved from reptilian relatives of dinosaurs with the development of wings, feathers, and light bones for flight. Birds also have four-chambered hearts and uniquely adapted lungs to supply the intense metabolic needs of flight. Birds have hard-shelled eggs and usually provide a great deal of parental care during embryonic development and maturation after hatching. A famous evolutionary intermediate from the fossil record is *Archaeopteryx*, which is dinosaurlike in some respects, but had feathers and wings.

Mammals are the remaining major class of vertebrates. Mammals have hair, sweat glands, mammary glands, and four-chambered hearts. The fossil record indicates that mammals evolved 200 million years ago and coexisted with the dinosaurs up until the major extinction 65 million years ago. At this time, mammals diversified to occupy many environmental niches and become the dominant terrestrial vertebrates in many ecosystems. Mammals are highly effective in regulation of their body temperature, and most mammals provide extensive care for their young. One small group of mammals, the Monotremes (for example, the duck-billed platypus), lay eggs. Other mammals gestate their embryos internally and give birth to young. Marsupial mammals give birth after a short time and complete development of young in an external pouch. Placental mammals gestate their young to a more mature state, providing nutritition to the embryo with the exchange of material in the placenta. Marsupial mammals were once widespread across the globe, but were replaced in most cases by placental mammals. Australia being isolated was a haven for marsupial mammals until the present day.

Among the mammals, the primates have opposable thumbs and stereoscopic vision for depth perception, adaptations for life in the trees and traits that have been important factors in the evolution of humans. Many primates have complex social structures. The ancestors of humans included *australopithecines*. Fossils indicate these ancestors were able to walk upright on two legs on the ground. Fossil remains of hominids such as *Homo habilus* from 2-3 million years ago display an increasing size of the cortex. *Homo habilus* probably used tools, setting the stage for modern humans, *Homo sapiens*.

Congratulations! You've made it through the chapter, and you should now be well versed in topics evolution, speciation, the origin of early life, and classification and diversity. See just how much you've learned as you tackle the quiz on the following pages.

Evolution and Diversity Quiz

1. Nematocysts are characteristic of

   (A) Porifera
   (B) Protozoa
   (C) Cnidarians
   (D) Annelida
   (E) Echinodermata

2. According to the modern theory of evolution, which of the following evolved first?

   (A) the Krebs cycle
   (B) anaerobic respiration
   (C) autotrophic nutrition
   (D) photosynthesis
   (E) chemosynthesis

3. Which of the following is an INCORRECT association?

   (A) Porifera: sessile
   (B) Echinodermata: radial symmetry
   (C) Annelida: coelom
   (D) Platyhelminthes: anus
   (E) insects: tracheal tubes

4. Intestinal nematodes evolved from a free-living to a parasitic form through developing

   (A) special reproductive segments
   (B) a symbiotic relationship with intestinal bacteria
   (C) an external cuticle resistant to digestive enzymes
   (D) a long digestive tube
   (E) an intricate nervous system

5. Which statement about the phylum Echinodermata is false?

   (A) The phylum includes starfish and sea urchins.
   (B) Echinoderms reproduce sexually.
   (C) The phylum includes crayfish.
   (D) Echinoderms are heterotrophs.
   (E) Echinoderms are invertebrates.

6. Echinoderms are regarded as being closely related to chordates due to their

   (A) bilateral symmetry
   (B) form of circulatory system
   (C) endoskeleton
   (D) early embryonic development
   (E) possession of a notochord

7. Which of the following is NOT a distinction between plants and animals?

   (A) Plants have cellulose cell walls.
   (B) Plants have intermediate larval stages.
   (C) Plants are extensively branched.
   (D) Animals are heterotrophic.
   (E) All of the above.

8. Which of the following organisms is a chordate but NOT a vertebrate?

   (A) shark
   (B) lamprey eel
   (C) turtle
   (D) tunicate
   (E) none of the above

GO ON TO THE NEXT PAGE

9. The notochord is

   (A) present in all adult chordates
   (B) present in all echinoderms
   (C) present in chordates during embryological development
   (D) always a vestigial organ in chordates
   (E) part of the nervous system of all vertebrates

10. Which order of classificatory divisions is correct?

   (A) kingdom, subphylum, genus, family
   (B) phylum, class, order, genus
   (C) species, order, family, phylum
   (D) subphylum, kingdom, family, order
   (E) genus, order, species, phylum

11. The hypothesis that chloroplasts and mitochondria were originally prokaryotic organisms living within eukaryotic hosts is supported by the fact that mitochondria and chloroplasts

   (A) possess protein synthetic capability
   (B) possess genetic material
   (C) possess a plasma membrane
   (D) possess characteristic ribosomes
   (E) all of the above

12. Which of the following is a member of the protist kingdom?

   (A) archaebacteria
   (B) unicellular green alga
   (C) anaerobic bacteria
   (D) fungi
   (E) mosses

13. Which of the following statement about viruses is NOT true?

   (A) Their genetic material may be DNA or RNA.
   (B) The virus may replicate in a bacterial host.
   (C) The virus may replicate in a eukaryotic host.
   (D) The virus may replicate autonomously in the absence of a host.
   (E) The protein coat of the virus does not enter a host bacterial cell.

GO ON TO THE NEXT PAGE

14. Which of the following has a chitinous exoskeleton?

    (A) oriole
    (B) starfish
    (C) clam
    (D) honeybee
    (E) earthworm

15. In the speculation concerning the origins of life, one theory states that purines, pyrimidines, sugars, and phosphates combined to form

    (A) nucleotides
    (B) nucleosides
    (C) carbohydrates
    (D) fats
    (E) proteins

16. Compound eyes are found in

    (A) Porifera
    (B) Coelenterata
    (C) Mollusca
    (D) Arthropoda
    (E) Annelida

17. Reptiles

    (A) are homeothermic
    (B) respire with gills
    (C) must live in water at some stage of their life cycle
    (D) possess notochords as adults
    (E) lay leathery eggs

18. All viruses

    (A) carry DNA
    (B) carry RNA
    (C) lack protein
    (D) have chromosomes
    (E) cannot reproduce outside of cells

19. Which of the following men might have explained the auk's loss of the ability to fly with the following hypothesis?

    "Since the auk stopped using its wings, the wings became smaller, and this acquired trait was passed on to the offspring."

    (A) Darwin
    (B) Mendel
    (C) De Vries
    (D) Lamarck
    (E) Morgan

GO ON TO THE NEXT PAGE

20. All arthropods

    (A) have a body consisting of three parts (the head, the thorax, and the abdomen)
    (B) breathe through spiracles that lead to the tracheal tubes
    (C) have a calcerous shell with a soft mantle
    (D) have radial symmetry
    (E) have jointed appendages and exoskeletons

21. Which of the following will NOT affect the frequency of a gene in an ideal population?

    (A) environmental selective pressure
    (B) mutation
    (C) random breeding
    (D) nonrandom matings
    (E) selective emigration

STOP

Answers and Explanations for the Evolution and Diversity Quiz

1. **(C)** Nematocysts are stinging cells found in coelenterates; they are used to immobilize prey. Coelenterates are organisms like hydra and jellyfish that are radially symmetrical and have one opening to their digestive tract that acts as both a mouth and an anus. Of the remaining choices, (A) Porifera are organisms such as sponges that have no defined tissue or organs, (B) Protozoa are one-celled organisms such as paramecium and amoebae, and (D) annelids are segmented invertebrates (such as earthworms) with a true body cavity (coelem) and two openings in their digestive tract. Lastly, echinoderms (E) are described in detail in the explanation to question 5 below.

2. **(B)** According to the evolutionary theory, large numbers of spontaneously formed organic molecules became simple organisms able to respire in the simplest way—anaerobically. These anaerobic heterotrophs developed first, followed by chemosynthetic bacteria. Next, photosynthetic bacteria developed and released O_2 as a byproduct, allowing the development of aerobic heterotrophs and the Krebs cycle.

3. **(D)** Platyhelminthes are ribbonlike, bilaterally symmetrical organisms with three layers of cells, including a solid mesoderm. They do not have a circulatory system and their nervous system consists of eyes, an anterior brain ganglion, and a pair of longitudinal nerve cords. They also have a primitive excretory system containing flame cells. They do not have an anus. Meanwhile, (A) porifera are sessile sponges that have two layers of cells, pores, and a low degree of cellular specialization. Echinoderms (B) are spiny, radially symmetrical, contain a water-vascular system, and possess the capacity for regeneration of parts, and (C) annelids are segmented worms which possess a true body cavity, the coelem, contained in the mesoderm. Annelids have well-defined nervous, circulatory, and excretory systems. As for (E), insects are arthropods and possess jointed appendages, chitinous exoskeletons, and open circulatory systems. They have three pairs of legs, spiracles, and tracheal tubes designed for breathing outside of an aquatic environment.

4. **(C)** This development was the most important evolutionary development for internal intestinal parasites, enabling them to avoid being digested by the host's acidic environment in the gut. (D) is false; as nematodes (roundworms) consume the already digested food of their host, they do not need developed digestive tracts. (B) and (E) have nothing to do with parasitism; nematodes actually have quite primitive nervous systems. And (A) is a characteristic of tapeworms, not roundworms. Roundworms actually reproduce by releasing eggs, so they can act as intestinal parasites without this ability.

5. **(C)** Echinoderms are the invertebrate predecessors of the chordates; because of this, they are a favorite question topic on the Biology E/M test. Examples include the starfish, the sea urchin, and the sea cucumber. They are characterized by having a primitive vascular system known as the water vascular system. The adult echinoderm is radially symmetrical, while larvae are bilaterally symmetrical. They move with structures known as tube feet, have no backbone, and do not make their own food—in other words, they are heterotrophic. Crayfish are members of the phylum Arthropoda, not Echinodermata; their phylum includes insects, which are characterized by segmented bodies covered in a chitinous exoskeleton and jointed appendages.

6. **(D)** Echinoderms are considered our invertebrate predecessors because they are the only other organism besides chordates that are deuterostomes. In other words, the echinoderm's mouth is formed after the anus is formed from the blastopore, rather than before the anus is formed. In all other organisms, the blastopore forms the mouth first. As for (C), the endoskeleton of the echinoderm is made up of calcareous plates, unlike that of a chordate, and (E) echinoderms do not have a notochord in any

stage of their development. See explanation (5) above for more characteristics of the echinoderm.

7. **(B)** Animals often go through a larval stage—a developmental stage between the fertilized egg and the adult (examples include the metamorphosis of tadpoles and caterpillars). Plants do not pass through intermediate larval stages. Another distinction between plants and animals is that plants are typically photosynthetic and sessile, while animals are generally heterotrophic and motile. Also, plant structure is adapted for maximum exposure to light, air, and soil by extensive branching; animals, meanwhile, are adapted for minimum surface exposure. They are compact. Finally, plant cells contain cell walls composed of dellulose, while animal cells do not.

8. **(D)** Members of the phylum Chordata have a dorsal notochord, while members of the subphylum Vertebrata have a backbone. Nonvertebrate chordates include the tunicate, also called the sea squirts. (A) and (B) are vertebrates with cartilaginous skeletons, while (C) has a bony skeleton.

9. **(C)** The notochord is a semirigid chord in the dorsal part of all chordates during a certain stage of embryonic development. In lower chordates, this chord remains as a semirigid chord, although in higher chordates it is seen only in the embryo and not as a vestigial organ. As for the other alternatives, (B) echinoderms do not possess a notochord, (D) the notochord remains in the lower chordates (such as the amphioxus and the tunicate worm), and (E) the notochord is not part of the nervous system.

10. **(B)** The correct order of classificatory divisions is: Kingdom, phylum, subphylum, class, order, family, genus, and species.

11. **(E)** The endosymbiotic hypothesis states that blue-green algae entered into a symbiotic arrangement with early eukaryotic plants to develop into chloroplasts, while bacteria entered into a similar arrangement with eukaryotic animal cells to become mitochondria. Chloroplasts and mitochondria have a plasma membrane (their inner membrane) and the ability to produce their own proteins without utilizing the cell's machinery (C). In addition, they have circular DNA, and their rRNA subunits are characteristic of prokaryotes.

12. **(B)** Protists include some single-celled eukaryotic photosynthetic organisms, including some algae (including green algae). Archaebacteria and anaerobic bacteria are prokaryotes, mosses are plants, and fungi have their own kingdom.

13. **(D)** A virus is a simple nonliving organism that takes on living characteristics when it enters cells that support its replication. The virus is made up of genetic material, either DNA or RNA, and a protein coat, and it is incapable of replicating autonomously; it must replicate within its host cell. A bacteriophage, for example, will inject its DNA into the bacterium, leaving its protein coat on the cell surface. In eukaryotes, however, the whole virus may enter the cell and not become unencapsulated until it enters the cytoplasm.

14. **(D)** Insects like the honeybee are characterized by having chitinous exoskeletons with three segmented bodies made up of a head, thorax, and abdomen. They have three pairs of jointed legs, two pairs of wings, and breathe through tracheal tubes. (A) Orioles, meanwhile, are vertebrates with bony endoskeletons, and (B) starfish are echinoderms whose endoskeletons are made up of calcareous plates. (C) Clams are mollusks characterized by having a mantle secrete a calcium-mineral shell, and (E) earthworms are considered to have a hydrostatic skeleton of fluid surrounded by a layer of muscle.

15. **(A)** Nucleotides are made up of a nitrogenous base; guanine, uracil, adenine, cytosine, or thymidine; a sugar, either ribose or deoxyribose; and a phosphate group. Adenine and guanine are purines, while thymidine, cytosine, and uracil are pyrimidines. (B) Nucleosides do not contain the phosphate group, only the nitrogenous base and the sugar. Meanwhile, (C) carbohydrates are only sugars, while (D) fats are a combination of glycerols and three fatty acids. (E) Proteins are only linked amino acids and do not contain anything found in a nucleotide.

16. (D) Compound eyes are large groups of individual simple eyes, forming a type of composite eye characteristic of insects (phylum Arthropoda). Arthropods have jointed legs and an exoskeleton normally containing chitin. (A) The porifera, however, are sponges, extremely simple animal organisms that do not have eyes. (B) The coelenterates, including hydra and jellyfish, are radially symmetrical organisms which do not have eyes. Likewise, (C), the mollusks (snails, slugs, clams, and other shellfish) do not have compound eyes. Lastly, (E) annelids are segmented worms, such as the earthworm. They are invertebrate, have a fluid vascular skeleton that repeats segments, and do not have eyes.

17. (E) Reptiles lay soft, leathery eggs; they are water resistant, which implies that they tend to retain water on the inside. This is due to the fact that reptiles are terrestrial organisms. Reptiles are not homeothermic, or warm-blooded; they are poikilothermic or ectothermic, in other words, cold-blooded. Their internal temperature is greatly dependent on the external environment. Reptiles do not respire with gills; fish and some larval amphibians, on the other hand, do. Reptiles also have functional lungs. They do not need to live in water at any stage of their life cycle; (C) refers to amphibians. Finally, reptiles do not possess notochords as adults; notochords are cartilaginous chords that form in the chordates during embryological development. In the chordates known as vertebrates, these are not present in the adult stage.

18. (E) Viruses must have a host cell in order to replicate, although this cell can be either a eukaryote or a prokaryote. (D) is incorrect because the genetic material of the virus does not take the form of a chromosome (the chromosome is actually a complex of nucleic acids and proteins known as a histone). Refer to the explanation for question 13 for more characteristics of the virus.

19. (D) Lamarck's theory of evolution states that new organs or changes in existing ones were believed to arise because of needs of the organism. The amount of change was thought to be based on the use or disuse of the organ. This theory was based on a fallacious understanding of genetics. Any useful characteristic acquired in one generation was thought to be transmitted to the next, but in actuality, only changes in the DNA of sex cells can be inherited. As for (A), Darwin's theory of natural selection states that pressures in the environment select the organism most fit to survive and reproduce. Darwin's theory incorporates a number of factors and stages, including overpopulation, variation, competition, natural selection, inheritance of variations, and evolution of a new species. (B) Mendel, meanwhile, defined modern genetics through experiments with pea plants, describing dominance, segregation, and independent assortment. (C) De Vries confirmed Mendel's observations with results from experiments using a variety of different plant species. Finally, (E) Morgan induced mutation in Drosophila and studied the inheritance of these mutations. He also described sex-linked inheritance.

20. (E) Arthropods such as crustaceans, insects, spiders, and scorpions are characterized by a bilateral symmetry of jointed appendages, chitinous exoskeletons, and open circulatory systems. (A) and (B) refer specifically to insects, a class of arthropods. They have a body consisting of a head, thorax, and an abdomen and a respiratory system consisting of spiracles and tracheal tubes. (C) Mollusks, such as snails, slugs, and clams, are characterized by a calcerous shell with a soft mantle; arthropods are not. Finally, coelenterates, not arthropods, are radially symmetrical organisms with a top and a bottom but no left and right (D).

21. (C) The Hardy-Weinberg Law states that for a population to have stability in its gene frequencies, it must be large, with no migration, no mutation, and random breeding. Only if these criteria are met will gene frequencies not change. Only (C) is one of these criteria; the other answer choices will all affect gene frequency.

READY, SET, GO!

READY, SET, GO!

CHAPTER EIGHT

STRESS MANAGEMENT

The countdown has begun. Your date with THE TEST is looming on the horizon. Anxiety is on the rise. The butterflies in your stomach have gone ballistic. Perhaps you feel as if the last thing you ate has turned into a lead ball. Your thinking is getting cloudy. Maybe you think you won't be ready. Maybe you already know your stuff, but you're going into panic mode anyway. Worst of all, you're not sure of what to do about it.

Don't freak! It is possible to tame that anxiety and stress—before and during the test. We'll show you how. You won't believe how quickly and easily you can deal with that killer anxiety.

Make the Most of Your Prep Time

Lack of control is one of the prime causes of stress. A ton of research shows that if you don't have a sense of control over what's happening in your life, you can easily end up feeling helpless and hopeless. So, just having concrete things to do and to think about—taking control—will help reduce your stress. This section shows you how to take control during the days leading up to taking the test.

Identify the Sources of Stress

In the space provided, jot down anything you identify as a source of your test-related stress. The idea is to pin down that free-floating anxiety so that you can take control of it. Here are some common examples to help get you started:

- I always freeze up on tests.

- I'm nervous about trig (or functions, or geometry, etcetera).

- I need a good/great score to go to Acme College.

- My older brother/sister/best friend/girl- or boyfriend did really well. I must match their scores or do better.

Avoid Must-y Thinking

Let go of "must-y" thoughts, those notions that you must do something in a certain way—for example, "I must get a great score, or else!" "I must meet Mom and Dad's expectations."

Don't Do It in Bed

Don't study on your bed, especially if you have problems with insomnia. Your mind might start to associate the bed with work, and make it even harder for you to fall asleep.

Think Good Thoughts

Create a set of positive but brief affirmations and mentally repeat them to yourself just before you fall asleep at night. (That's when your mind is very open to suggestion.) You'll find yourself feeling a lot more positive in the morning. Periodically repeating your affirmations during the day makes them more effective.

- My parents, who are paying for school, will be really disappointed if I don't test well.

- I'm afraid of losing my focus and concentration.

- I'm afraid I'm not spending enough time preparing.

- I study like crazy, but nothing seems to stick in my mind.

- I always run out of time and get panicky.

- I feel as though thinking is becoming like wading through thick mud.

Sources of Stress

_____   _____

_____   _____

_____   _____

_____   _____

Take a few minutes to think about the things you've just written down. Then rewrite them in some sort of order. List the statements you most associate with your stress and anxiety first, and put the least disturbing items last. Chances are, the top of the list is a fairly accurate description of exactly how you react to test anxiety, both physically and mentally. The later items usually describe your fears (disappointing Mom and Dad, looking bad, etcetera). As you write the list, you're forming a hierarchy of items so you can deal first with the anxiety provokers that bug you most. Very often, taking care of the major items from the top of the list goes a long way toward relieving overall testing anxiety. You probably won't have to bother with the stuff you placed last.

Strengths and Weaknesses

Take one minute to list the areas of the test that you are good at. They can be general ("physiology") or specific ("the immune system"). Put down as many as you can think of, and if possible, time yourself. Write for the entire time; don't stop writing until you've reached the one-minute stopping point.

Strong Test Subjects

_____   _____

_____   _____

_____   _____

_____   _____

Next, take one minute to list areas of the test you're not so good at, just plain bad at, have failed at, or keep failing at. Again, keep it to one minute, and continue writing until you reach the cutoff. Don't be afraid to identify and write down your weak spots! In all probability, as you do both lists, you'll find you are strong in some areas and not so strong in others. Taking stock of your assets and liabilities lets you know the areas you don't have to worry about, and the ones that will demand extra attention and effort.

Weak Test Subjects

_____   _____

_____   _____

_____   _____

_____   _____

Facing your weak spots gives you some distinct advantages. It helps a lot to find out where you need to spend extra effort. Increased exposure to tough material makes it more familiar and less intimidating. (After all, we mostly fear what we don't know and are probably afraid to face.) You'll feel better about yourself because you're dealing directly with areas of the test that bring on your anxiety. You can't help feeling more confident when you know you're actively strengthening your chances of earning a higher over-all test score.

Now, go back to the "good" list, and expand it for two minutes. Take the general items on that first list and make them more specific; take the specific items and expand them into more general conclusions. Naturally, if anything new comes to mind, jot it down. Focus all of your attention and effort on your strengths. Don't underestimate yourself or your abilities. Give yourself full credit. At the same time, don't list strengths you don't really have; you'll only be fooling yourself.

Very Superstitious

Stress expert Stephen Sideroff, Ph.D., tells of a client who always stressed out before, during, and even after taking tests. Yet, she always got outstanding scores. It became obvious that she was thinking superstitiously—sub-consciously believing that the great scores were a result of her worrying. She didn't trust herself, and believed that if she didn't worry she wouldn't study hard enough. Sideroff convinced her to take a risk and work on relaxing before her next test. She did, and her test results were still as good as ever—which broke her cycle of superstitious thinking.

Get It Together

Don't work in a messy or cramped area. Before you sit down to study, clear yourself a nice, open space. And, make sure you have books, paper, pencils—whatever tools you will need—within easy reach before you sit down to study.

Expanding from general to specific might go as follows. If you listed "algebra" as a broad topic you feel strong in, you would then narrow your focus to include areas of this subject about which you are particularly knowledgeable. Your areas of strength might include multiplying polynomials, working with exponents, factoring, solving simultaneous equations, etcetera.

Whatever you know comfortably goes on your "good" list. Okay. You've got the picture. Now, get ready, check your starting time, and start writing down items on your expanded "good" list.

Strong Test Subjects: An Expanded List

_____	_____
_____	_____
_____	_____
_____	_____

After you've stopped, check your time. Did you find yourself going beyond the two minutes allotted? Did you write down more things than you thought you knew? Is it possible you know more than you've given yourself credit for? Could that mean you've found a number of areas in which you feel strong?

You just took an active step toward helping yourself. Notice any increased feelings of confidence? Enjoy them.

Here's another way to think about your writing exercise. Every area of strength and confidence you can identify is much like having a reserve of solid gold at Fort Knox. You'll be able to draw on your reserves as you need them. You can use your reserves to solve difficult questions, maintain confidence, and keep test stress and anxiety at a distance. The encouraging thing is that every time you recognize another area of strength, succeed at coming up with a solution, or get a good score on a test, you increase your reserves. And, there is absolutely no limit to how much self-confidence you can have or how good you can feel about yourself.

Imagine Yourself Succeeding

This next little group of exercises is both physical and mental. It's a natural follow-up to what you've just accomplished with your lists.

First, get yourself into a comfortable sitting position in a quiet setting. Wear loose clothes. If you wear glasses, take them off. Then, close your eyes and breathe in a deep, satisfying breath of air. Really fill your lungs until your rib cage is fully expanded and you can't take in any more. Then, exhale the air completely. Imagine you're blowing out a candle with your last little puff of air. Do this two or three more times, filling your lungs to their maximum and emptying them totally. Keep your eyes closed, comfortably but not tightly. Let your body sink deeper into the chair as you become even more comfortable.

With your eyes shut you can notice something very interesting. You're no longer dealing with the worrisome stuff going on in the world outside of you. Now you can concentrate on what happens *inside* you. The more you recognize your own physical reactions to stress and anxiety, the more you can do about them. You might not realize it, but you've begun to regain a sense of being in control.

Let images begin to form on the "viewing screens" on the back of your eyelids. You're experiencing visualizations from the place in your mind that makes pictures. Allow the images to come easily and naturally; don't force them. Imagine yourself in a relaxing situation. It might be in a special place you've visited before or one you've read about. It can be a fictional location that you create in your imagination, but a real-life memory of a place or situation you know is usually better. Make it as detailed as possible, and notice as much as you can.

Stay focused on the images as you sink farther back into your chair. Breathe easily and naturally. You might have the sensations of any stress or tension draining from your muscles and flowing downward, out your feet and away from you.

Take a moment to check how you're feeling. Notice how comfortable you've become. Imagine how much easier it would be if you could take the test feeling this relaxed and in this state of ease. You've coupled the images of your special place with sensations of comfort and relaxation. You've also found a way to become relaxed simply by visualizing your own safe, special place.

Now, close your eyes and start remembering a real-life situation in which you did well on a test. If you can't come up with one, remember a situation in which you did something (academic or otherwise) that you were really proud of—a genuine accomplishment. Make the memory as detailed as possible. Think about the sights, the sounds, the smells, even the tastes associated with this remembered experience. Remember how confident you felt as you accomplished your goal. Now start thinking about the upcoming test. Keep your thoughts and feelings in line with that successful experience. Don't make comparisons between them. Just imagine taking the upcoming test with the same feelings of confidence and relaxed control.

This exercise is a great way to bring the test down to Earth. You should practice this exercise often, especially when the prospect of taking the exam starts to bum you out. The more you practice it, the more effective the exercise will be for you.

Ocean Dumping

Visualize a beautiful beach, with white sand, blue skies, sparkling water, a warm sun, and seagulls. See yourself walking on the beach, carrying a small plastic pail. Stop at a good spot and put your worries and whatever may be bugging you into the pail. Drop it at the water's edge and watch it drift out to sea. When the pail is out of sight, walk on.

Counseling

Don't forget that your school probably has counseling available. If you can't conquer test stress on your own, make an appointment at the counseling center. That's what counselors are there for.

Take a Hike, Pal

When you're in the middle of studying and hit a wall, take a short, brisk walk. Breathe deeply and swing your arms as you walk. Clear your mind. (And, don't forget to look for flowers that grow in the cracks of the sidewalk.)

Play the Music

If you want to play music, keep it low and in the background. Music with a regular, mathematical rhythm—reggae, for example—aids the learning process. A recording of ocean waves is also soothing.

Exercise Your Frustrations Away

Whether it is jogging, walking, biking, mild aerobics, pushups, or a pickup basketball game, physical exercise is a very effective way to stimulate both your mind and body and to improve your ability to think and concentrate. A surprising number of students get out of the habit of regular exercise, ironically because they're spending so much time prepping for exams. Also, sedentary people—this is a medical fact—get less oxygen to the blood and hence to the head than active people. You can live fine with a little less oxygen; you just can't think as well.

Any big test is a bit like a race. Thinking clearly at the end is just as important as having a quick mind early on. If you can't sustain your energy level in the last sections of the exam, there's too good a chance you could blow it. You need a fit body that can weather the demands any big exam puts on you. Along with a good diet and adequate sleep, exercise is an important part of keeping yourself in fighting shape and thinking clearly for the long haul.

There's another thing that happens when students don't make exercise an integral part of their test preparation. Like any organism in nature, you operate best if all your "energy systems" are in balance. Studying uses a lot of energy, but it's all mental. When you take a study break, do something active instead of raiding the fridge or vegging out in front of the TV. Take a 5- to 10-minute activity break for every 50 or 60 minutes that you study. The physical exertion gets your body into the act, which helps to keep your mind and body in sync. Then, when you finish studying for the night and hit the sack, you won't lie there, tense and unable to sleep because your head is overtired and your body wants to pump iron or run a marathon.

One warning about exercise, however: It's not a good idea to exercise vigorously right before you go to bed. This could easily cause sleep onset problems. For the same reason, it's also not a good idea to study right up to bedtime. Make time for a "buffer period" before you go to bed: For 30 to 60 minutes, just take a hot shower, meditate, simply veg out.

The Dangers of Drugs

Using drugs (prescription or recreational) specifically to prepare for and take a big test is definitely self-defeating. (And if they're illegal drugs, you can end up with a bigger problem than the SAT II Bio test on your hands.) Except for the drugs that occur naturally in your brain, every drug has major drawbacks—and a false sense of security is only one of them.

You may have heard that popping uppers helps you study by keeping you alert. If they're illegal, definitely forget about it. They wouldn't really work anyway, since amphetamines make it hard to retain information. Mild stimulants, such as coffee, cola, or over-the-counter caffeine pills can sometimes help as you study, since they keep you alert. On the down side, they can also lead to agitation, restlessness, and insomnia. Some people can drink a pot of high-octane coffee and sleep like a baby. Others have one cup and start to vibrate. It all depends on your tolerance for caffeine. Remember, a little anxiety is a good thing. The adrenaline that gets pumped into your bloodstream helps you stay alert and think more clearly. But, too much anxiety and you can't think straight at all.

Instead, go for endorphins—the "natural morphine." Endorphins have no side effects and they're free—you've already got them in your brain. It just takes some exercise to release them. Running around on the basketball court, bicycling, swimming, aerobics, power walking—these activities cause endorphins to occupy certain spots in your brain's neural synapses. In addition, exercise develops staying power and increases the oxygen transfer to your brain. Go into the test naturally.

Take a Deep Breath . . .

Here's another natural route to relaxation and invigoration. It's a classic isometric exercise that you can do whenever you get stressed out—just before the test begins, even *during* the test. It's very simple and takes just a few minutes.

Close your eyes. Starting with your eyes and—without holding your breath—gradually tighten every muscle in your body (but not to the point of pain) in the following sequence:

1. Close your eyes tightly.

2. Squeeze your nose and mouth together so that your whole face is scrunched up. (If it makes you self-conscious to do this in the test room, skip the face-scrunching part.)

3. Pull your chin into your chest, and pull your shoulders together.

4. Tighten your arms to your body, then clench your hands into tight fists.

5. Pull in your stomach.

6. Squeeze your thighs and buttocks together, and tighten your calves.

7. Stretch your feet, then curl your toes (watch out for cramping in this part).

At this point, every muscle should be tightened. Now, relax your body, one part at a time, *in reverse order*, starting with your toes. Let the tension drop out of each muscle. The entire process might take five minutes from start to finish (maybe a couple of minutes during the test). This clenching and unclenching exercise should help you to feel very relaxed.

Cyberstress

If you spend a lot of time in cyberspace anyway, do a search for the phrase *stress management*. There's a ton of stress advice on the Net, including material specifically for students.

Nutrition and Stress: The Dos and Don'ts

Do eat:

- Fruits and vegetables (raw is best, or just lightly steamed or nuked)
- Low-fat protein such as fish, skinless poultry, beans, and legumes (like lentils)
- Whole grains such as brown rice, whole wheat bread, and pastas (no bleached flour)

Don't eat:

- Refined sugar; sweet, high-fat snacks (simple carbohydrates like sugar make stress worse and fatty foods lower your immunity)
- Salty foods (they can deplete potassium, which you need for nerve functions)

The Relaxation Paradox

Forcing relaxation is like asking yourself to flap your arms and fly. You can't do it, and every push and prod only gets you more frustrated. Relaxation is something you don't work at. You simply let it happen. Think about it. When was the last time you tried to force yourself to go to sleep, and it worked?

Enlightenment

A lamp with a 75-watt bulb is optimal for studying. But don't put it so close to your study material that you create a glare.

And Keep Breathing

Conscious attention to breathing is an excellent way of managing test stress (or any stress, for that matter). The majority of people who get into trouble during tests take shallow breaths. They breathe using only their upper chests and shoulder muscles, and may even hold their breath for long periods of time. Conversely, the test taker who by accident or design keeps breathing normally and rhythmically is likely to be more relaxed and in better control during the entire test experience.

So, now is the time to get into the habit of relaxed breathing. Do the next exercise to learn to breathe in a natural, easy rhythm. By the way, this is another technique you can use during the test to collect your thoughts and ward off excess stress. The entire exercise should take no more than three to five minutes.

With your eyes still closed, breathe in slowly and *deeply* through your nose. Hold the breath for a bit, and then release it through your mouth. The key is to breathe slowly and deeply by using your diaphragm (the big band of muscle that spans your body just above your waist) to draw air in and out naturally and effortlessly. Breathing with your diaphragm encourages relaxation and helps minimize tension. Try it and notice how relaxed and comfortable you feel.

CHAPTER NINE

THE FINAL COUNTDOWN

Quick Tips for the Days Just Before the Exam

- The best test takers do less and less as the test approaches. Taper off your study schedule and take it easy on yourself. You want to be relaxed and ready on the day of the test. Give yourself time off, especially the evening before the exam. By then, if you've studied well, everything you need to know is firmly stored in your memory banks.

- Positive self-talk can be extremely liberating and invigorating, especially as the test looms closer. Tell yourself things such as, "I choose to take this test" rather than "I have to"; "I will do well" rather than "I hope things go well"; "I can" rather than "I cannot." Be aware of negative, self-defeating thoughts and images and immediately counter any you become aware of. Replace them with affirming statements that encourage your self-esteem and confidence. Create and practice visualizations that build on your positive statements.

- Get your act together sooner rather than later. Have everything (including choice of clothing) laid out days in advance. Most important, know where the test will be held and the easiest, quickest way to get there. You will gain great peace of mind if you know that all the little details—gas in the car, directions, etcetera—are firmly in your control before the day of the test.

- Experience the test site a few days in advance. This is very helpful if you are especially anxious. If at all possible, find out what room your part of the alphabet is assigned to, and try to sit there (by yourself) for a while. Better yet, bring some practice material and do at least a section or two, if not an entire practice test, in that room. In this situation, familiarity doesn't breed contempt, it generates comfort and confidence.

Dress for Success

On the day of the test, wear loose layers. That way, you'll be prepared no matter what the temperature of the room is. (An uncomfortable temperature will just distract you from the job at hand.) And, if you have an item of clothing that you tend to feel "lucky" or confident in—a shirt, a pair of jeans, whatever—wear it. A little totem couldn't hurt.

What Are "Signs of a Winner," Alex?

Here's some advice from a Kaplan instructor who won big on *Jeopardy!*™ In the green room before the show, he noticed that the contestants who were quiet and "within themselves" were the ones who did great on the show. The contestants who did not perform as well were the ones who were fact-cramming, talking a lot, and generally being manic before the show. Lesson: Spend the final hours leading up to the test getting sleep, meditating, and generally relaxing.

• Forego any practice on the day before the test. It's in your best interest to marshal your physical and psychological resources for 24 hours or so. Even race horses are kept in the paddock and treated like princes the day before a race. Keep the upcoming test out of your consciousness; go to a movie, take a pleasant hike, or just relax. Don't eat junk food or tons of sugar. And—of course—get plenty of rest the night before. Just don't go to bed too early. It's hard to fall asleep earlier than you're used to, and you don't want to lie there thinking about the test.

Handling Stress During the Test

The biggest stress monster will be the test itself. Fear not; there are methods of quelling your stress during the test.

• Keep moving forward instead of getting bogged down in a difficult question. You don't have to get everything right to achieve a fine score. The best test takers skip difficult material temporarily in search of the easier stuff. They mark the ones that require extra time and thought. This strategy buys time and builds confidence so you can handle the tough stuff later.

• Don't be thrown if other test takers seem to be working more furiously than you are. Continue to spend your time patiently thinking through your answers; it's going to lead to better results. Don't mistake the other people's sheer activity as signs of progress and higher scores.

• *Keep breathing!* Weak test takers tend to forget to breathe properly as the test proceeds. They start holding their breath without realizing it, or they breathe erratically or arrhythmically. Improper breathing interferes with clear thinking.

• Some quick isometrics during the test—especially if concentration is wandering or energy is waning—can help. Try this: Put your palms together and press intensely for a few seconds. Concentrate on the tension you feel through your palms, wrists, forearms, and up into your biceps and shoulders. Then, quickly release the pressure. Feel the difference as you let go. Focus on the warm relaxation that floods through the muscles. Now you're ready to return to the task.

• Here's another isometric that will relieve tension in both your neck and eye muscles. Slowly rotate your head from side to side, turning your head and eyes to look as far back over each shoulder as you can. Feel the muscles stretch on one side of your neck as they contract on the other. Repeat five times in each direction.

With what you've just learned here, you're armed and ready to do battle with the test. This book and your studies will give you the information you'll need to answer the questions. It's all firmly planted in your mind. You also know how to deal with any excess tension that might come along, both when you're studying for and taking the exam. You've experienced everything you need to tame your test anxiety and stress. You're going to get a great score.

KAPLAN PRACTICE TESTS

Instructions for the
Biology E/M Tests

- The two tests that follow offer realistic practice for the SAT II: Biology E/M Subject Test. Test Two contains the kinds of questions you will encounter if you select the Ecological-option of the Biology E/M Test, and Test Three reflects the range of questions found on the Molecular-option of the exam. To get the most out of these practice tests, you should take them under testlike conditions.

- Take the tests in a quiet room with no distractions. Bring some No. 2 pencils.

- Time yourself. You should spend no more than one hour on the 80 questions on each test.

- Use the answer sheets provided on the pages before each test to mark your answers.

- Answers and explanations follow each test.

- Scoring instructions are in the "Compute Your Test Score" sections immediately following the answer keys of each test.

ANSWER SHEET FOR BIOLOGY E/M
TEST TWO: E-OPTION

1 (A) (B) (C) (D) (E)	21 (A) (B) (C) (D) (E)	41 (A) (B) (C) (D) (E)	61 (A) (B) (C) (D) (E)
2 (A) (B) (C) (D) (E)	22 (A) (B) (C) (D) (E)	42 (A) (B) (C) (D) (E)	62 (A) (B) (C) (D) (E)
3 (A) (B) (C) (D) (E)	23 (A) (B) (C) (D) (E)	43 (A) (B) (C) (D) (E)	63 (A) (B) (C) (D) (E)
4 (A) (B) (C) (D) (E)	24 (A) (B) (C) (D) (E)	44 (A) (B) (C) (D) (E)	64 (A) (B) (C) (D) (E)
5 (A) (B) (C) (D) (E)	25 (A) (B) (C) (D) (E)	45 (A) (B) (C) (D) (E)	65 (A) (B) (C) (D) (E)
6 (A) (B) (C) (D) (E)	26 (A) (B) (C) (D) (E)	46 (A) (B) (C) (D) (E)	66 (A) (B) (C) (D) (E)
7 (A) (B) (C) (D) (E)	27 (A) (B) (C) (D) (E)	47 (A) (B) (C) (D) (E)	67 (A) (B) (C) (D) (E)
8 (A) (B) (C) (D) (E)	28 (A) (B) (C) (D) (E)	48 (A) (B) (C) (D) (E)	68 (A) (B) (C) (D) (E)
9 (A) (B) (C) (D) (E)	29 (A) (B) (C) (D) (E)	49 (A) (B) (C) (D) (E)	69 (A) (B) (C) (D) (E)
10 (A) (B) (C) (D) (E)	30 (A) (B) (C) (D) (E)	50 (A) (B) (C) (D) (E)	70 (A) (B) (C) (D) (E)
11 (A) (B) (C) (D) (E)	31 (A) (B) (C) (D) (E)	51 (A) (B) (C) (D) (E)	71 (A) (B) (C) (D) (E)
12 (A) (B) (C) (D) (E)	32 (A) (B) (C) (D) (E)	52 (A) (B) (C) (D) (E)	72 (A) (B) (C) (D) (E)
13 (A) (B) (C) (D) (E)	33 (A) (B) (C) (D) (E)	53 (A) (B) (C) (D) (E)	73 (A) (B) (C) (D) (E)
14 (A) (B) (C) (D) (E)	34 (A) (B) (C) (D) (E)	54 (A) (B) (C) (D) (E)	74 (A) (B) (C) (D) (E)
15 (A) (B) (C) (D) (E)	35 (A) (B) (C) (D) (E)	55 (A) (B) (C) (D) (E)	75 (A) (B) (C) (D) (E)
16 (A) (B) (C) (D) (E)	36 (A) (B) (C) (D) (E)	56 (A) (B) (C) (D) (E)	76 (A) (B) (C) (D) (E)
17 (A) (B) (C) (D) (E)	37 (A) (B) (C) (D) (E)	57 (A) (B) (C) (D) (E)	77 (A) (B) (C) (D) (E)
18 (A) (B) (C) (D) (E)	38 (A) (B) (C) (D) (E)	58 (A) (B) (C) (D) (E)	78 (A) (B) (C) (D) (E)
19 (A) (B) (C) (D) (E)	39 (A) (B) (C) (D) (E)	59 (A) (B) (C) (D) (E)	79 (A) (B) (C) (D) (E)
20 (A) (B) (C) (D) (E)	40 (A) (B) (C) (D) (E)	60 (A) (B) (C) (D) (E)	80 (A) (B) (C) (D) (E)

Use the answer key following the test to count up the number of questions you got right and the number you got wrong. (Remember not to count omitted questions as wrong.) The "Compute Your Score" section following the Answer Key will show you how to find your score.

Remove this answer sheet and use it to complete the Practice Test.

BIOLOGY E/M
TEST TWO: E-OPTION

Part A

Directions: Each question or incomplete statement below is followed by five possible answers or completions, lettered A–E. Choose the answer that is the best in each case. Fill in the corresponding oval on your answer sheet.

1. $6CO_2 + 6H_2O \longrightarrow C_6H_{12}O_6 + 6O_2$

This process is completed

(A)   in the cytoplasm
(B)   in the area of the cell membrane
(C)   in the chloroplast
(D)   in the mitochondria
(E)   in the area around the ribosomes

2.   Which of the following is a feature in fetal development?

   I.   ductus arteriosus
   II.   foramen ovale
   III.   fetal hemoglobin with a higher affinity for oxygen than adult hemoglobin

(A)   I only
(B)   II only
(C)   III only
(D)   I and III
(E)   I, II, and III

3.   Which of the following associations most closely corresponds to that of ectoderm: endoderm?

(A)   heart: stomach
(B)   retina: lungs
(C)   skeletal muscles: liver
(D)   skin: stomach muscle
(E)   taste buds: uterus

4.   A movement of muscle that bends a joint to a more acute angle is termed

(A)   flexion
(B)   insertion
(C)   tonus
(D)   diastole
(E)   extension

5.   Which hormone triggers your body to retain NaCl, especially during periods of excessive heat?

(A)   aldosterone
(B)   progesterone
(C)   ACTH
(D)   epinephrine
(E)   testosterone

GO ON TO THE NEXT PAGE

6. If one parent is homozygous dominant and the other is homozygous recessive, which of the following might appear in an F_2 generation, but not in an F_1 generation?

    I. heterozygous genotype
    II. dominant phenotype
    III. recessive phenotype

    (A) I only
    (B) II only
    (C) III only
    (D) I and II
    (E) II and III

7. Which of the following statements about oxidative phosphorylation is NOT correct?

    (A) It occurs in the inner membrane of the mitochondrion.
    (B) It involves O_2 as the final electron acceptor.
    (C) It produces 2 ATPs for each $FADH_2$.
    (D) It can occur under anaerobic conditions.
    (E) It involves a proton gradient.

8. The climate with the shortest growing season would be located in the

    (A) taiga
    (B) tropical rain forest
    (C) deciduous forest
    (D) savanna
    (E) steppe

9. A factor that tends to keep the gene pool constant is

    I. nonrandom mating
    II. freedom to migrate
    III. no net mutations
    IV. large populations

    (A) I and II
    (B) III and IV
    (C) I, III, and IV
    (D) II, III, and IV
    (E) I, II, III, and IV

10. A person takes a large overdose of antacid. As a result, the activity of which of the following enzymes would be most affected?

    (A) maltase
    (B) lactase
    (C) lipase
    (D) pepsin
    (E) sucrase

11. Which of the following is not a characteristic of the kingdom Protista?

    (A) Members can be photosynthetic.
    (B) Members can be free living.
    (C) Some members move via flagella.
    (D) Some members are shaped like rods and termed bacilli.
    (E) Some members spend part of their life cycle inside insects.

GO ON TO THE NEXT PAGE

12. Which enzyme digests disaccharides to monosaccharides?

(A) lactase
(B) kinase
(C) zymogen
(D) lipase
(E) phosphorylase

13. The vessel with the LEAST oxygenated blood is the

(A) pulmonary vein
(B) aorta
(C) renal artery
(D) pulmonary artery
(E) superior vena cava

14. Which of the following has a chambered heart and breathes through gills?

(A) cardinal
(B) sea urchin
(C) snail
(D) praying mantis
(E) earthworm

15. A single nondisjunction may cause all of the following except

(A) spontaneous miscarriage of the fetus
(B) 47 chromosomes per somatic cell
(C) 45 chromosomes per somatic cell
(D) congenital disorders such as Down's syndrome
(E) breakage near the centromere

16. A person's anterior pituitary gland is removed by surgery. Plasma concentrations of which of the following hormones would be LEAST affected?

(A) GH
(B) LH
(C) FSH
(D) insulin
(E) ACTH

17. How many different types of gametes would be produced by an organism of genotype *AabbCcDD* if all of the genes were to assort independently?

(A) 4
(B) 6
(C) 8
(D) 10
(E) 12

18. Ribosomes function in aggregates called

(A) histones
(B) nucleoli
(C) endoplasmic reticulum
(D) the Golgi complex
(E) polysomes

GO ON TO THE NEXT PAGE

19. Salmon return to their specific home stream to spawn. This is an example of

   (A) pheromones
   (B) a reflex
   (C) imprinting
   (D) classical conditioning
   (E) circadian rhythms

20. If two animals mate and produce viable, fertile offspring under natural conditions, we can conclude that

   (A) they both have diploid somatic cells
   (B) they are both from the same species
   (C) for any given gene, they both have the same allele
   (D) their blood types are compatible
   (E) none of the above

21. In an emergency, an individual with type AB antigen in his red blood cells may receive a transfusion of

   I.   type O blood
   II. type A blood
   III. type B blood

   (A) I only
   (B) II only
   (C) II and III
   (D) III only
   (E) I, II, and III

22. The gene for color blindness is X-linked. If normal parents have a color-blind son, what is the probability that he inherited the gene for color blindness from his mother?

   (A) 0%
   (B) 25%
   (C) 50%
   (D) 75%
   (E) 100%

23. Which of the following is a correct association?

   (A) mitochondria: transport of materials form the nucleus to the cytoplasm
   (B) Golgi apparatus: modification and glycosylation of proteins
   (C) endoplasmic reticulum: selective barrier for the cell
   (D) ribosomes: digestive enzymes most active at acidic pH
   (E) lysosomes: membrane-bound organelles that convert fat into sugars

GO ON TO THE NEXT PAGE

24. Spermatogenesis and oogenesis differ in that

   I. meiosis proceeds continually without pausing in spermatogenesis, while oogenesis involves a meiotic pause

   II. spermatogenesis only occurs at puberty

   III. spermatogenesis produces four haploid sperm cells from each diploid precursor cell, while oogenesis produces one egg cell and two or more polar bodies

   (A) I only
   (B) II only
   (C) III only
   (D) I and III
   (E) I, II, and III

25. Which of the following foods contains the greatest amount of energy per gram?

   (A) sugar
   (B) starch
   (C) fat
   (D) proteins
   (E) vitamins

26. Which statement about human gamete production is false?

   (A) In the testes, sperm develop in the seminiferous tubules.
   (B) In the ovaries, eggs develop in the ovarian follicles.
   (C) FSH stimulates gamete production in both sexes.
   (D) Gametes arise via meiosis.
   (E) The result of meiosis in females is the production of four egg cells from each diploid precursor cell.

27. The hormone progesterone

   (A) stimulates follicle growth
   (B) stimulates FSH production
   (C) is solely responsible for the maintenance of secondary sex characteristics
   (D) is produced by the anterior pituitary
   (E) readies the uterus for implantation

28. In a particular population, for a trait with two alleles, the frequency of the recessive allele is 0.6. What is the frequency of individuals expressing the dominant phenotype?

   (A) 0.16
   (B) 0.36
   (C) 0.48
   (D) 0.64
   (E) 0.6

GO ON TO THE NEXT PAGE

Questions 29–32 refer to the following diagram.

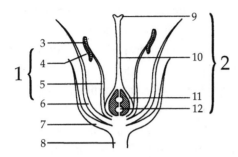

29. Which structure contains the female gameto-phyte?

(A) 1
(B) 6
(C) 7
(D) 9
(E) 12

30. Which is the part of the flower that catches the pollen?

(A) 3
(B) 4
(C) 9
(D) 5
(E) 6

31. Where do pollen grains develop?

(A) 4
(B) 5
(C) 6
(D) 7
(E) 8

32. A pollen grain fuses with the two polar bod-ies to form

(A) the flower
(B) the endosperm
(C) the pistil
(D) the sepal
(E) the stamen

Questions 33–35 refer to the figure below.

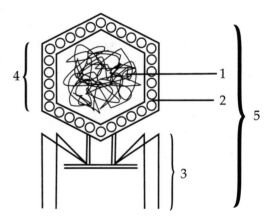

GO ON TO THE NEXT PAGE

33. Which structure attaches this organism to the cell surface?

    (A) 1
    (B) 2
    (C) 3
    (D) 4
    (E) 5

34. Structure 1 can be made up of

    I. DNA
    II. lipid
    III. protein

    (A) I only
    (B) II only
    (C) III only
    (D) I and III
    (E) I, II, and III

35. Structure 2 can be made up of

    I. DNA
    II. RNA
    III. protein

    (A) I only
    (B) II only
    (C) III only
    (D) I and II
    (E) I, II, and III

Part B

Each set of choices A–E below should be compared to the numbered statements that follow it. Choose the lettered choice that best matches each numbered statement. Fill in the correct oval on your answer sheet. Remember that a choice may be used once, more than once, or not at all in each set.

Questions 36–39:

    (A) simplex reflex
    (B) complex reflex
    (C) fixed action patterns
    (D) behavioral cycles
    (E) environmental rhythms

36. startle response

37. retrieval and maintenance of eggs of their species by female birds

38. simple response to simple stimuli

39. circadian rhythms

Questions 40–44:

    (A) simple diffusion
    (B) nephridia
    (C) malphigian tabules
    (D) flame cells
    (E) nephrons

GO ON TO THE NEXT PAGE

40.   grasshopper's method of excretion

41.   man's method of excretion

42.   planaria's method of excretion

43.   earthworm's method of excretion

44.   protozoa's method of excretion

Questions 45–49:

   (A)   binary fission
   (B)   budding
   (C)   spore formation
   (D)   vegetative propagation
   (E)   parthenogenesis

45.   reproduction of bacteria

46.   possible production of seedless fruit

47.   reproduction of yeast cells

48.   development of an egg without a sperm

49.   an example is an underground stem with buds

Part C

Each of the following sets of questions is based on a laboratory or experimental situation. Begin by studying the description of each situation. Next, choose the best answer to each of the questions that follow it. Fill in the corresponding oval on your answer form.

The following experiment describes the cyclical opening and closing of the stomata of a plant during a 24-hour period. Use this information to answer questions 50–53.

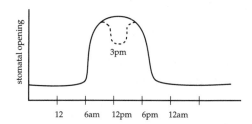

50.   The dotted line represents stomatal closing during the hottest part of the day. The stomata close because when the weather is hot

   (A)   the dark reactions begin
   (B)   chlorophyll begins to act
   (C)   increased CO_2 concentration occurs
   (D)   transpiration exceeds the uptake of water through the roots
   (E)   none of the above

GO ON TO THE NEXT PAGE

51. An open stomate leads to

    I.   increased photosynthesis
    II.   increased transpiration
    III. decreased chlorophyll

    (A) I only
    (B) II only
    (C) III only
    (D) I and II
    (E) I, II, and III

52. Guard cells open stomata in response to the light. What would happen if you kept the plant in a dark room until 2 hours after dawn?

    (A) The stomates would open anyway.
    (B) The stomates would stay closed until they received the light stimulus.
    (C) Stomates would stay closed all day.
    (D) Photosynthesis would increase.
    (E) Chlorophyll would degrade.

53. The rate of photosynthesis in some plants follows a 24-hour cycle even if light and other environmental factors are constant. This is an example of

    (A) a dark reaction
    (B) circadian rhythms
    (C) environmental rhythms
    (D) habituation
    (E) response to pheromones

A scientist was studying differential gene expression in a variety of tissues from the same organism. He investigated 8 different genes. The following chart describes which genes are turned on and producing their respective proteins in each of the different tissue types.

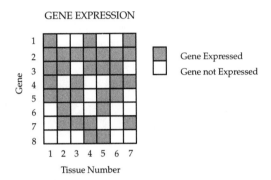

The chart will help you answer questions 54–57.

54. Which two cell types are the closest in the differentiation pathway?

    (A) 1 and 2
    (B) 2 and 5
    (C) 3 and 4
    (D) 4 and 5
    (E) 5 and 6

55. Which two are the farthest apart in the differentiation pathway?

    (A) 2 and 5
    (B) 2 and 3
    (C) 1 and 7
    (D) 5 and 7
    (E) 4 and 5

GO ON TO THE NEXT PAGE

56. Gene 8 is a protein that digests fats. The cell type that expresses this gene is probably found in the

    (A) gall bladder
    (B) stomach
    (C) liver
    (D) pancreas
    (E) salivary gland

57. Which is a housekeeping gene (a gene involved in protein synthesis in the ribosome that all cells must express to survive)?

    (A) 1
    (B) 2
    (C) 3
    (D) 4
    (E) 5

The graph below illustrates the data from an experiment in which antibody levels for two different antigens, A and B, were measured following exposure to these antigens. Antibodies play a crucial role in the human body's defense against disease.

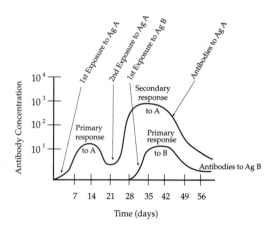

Questions 58–60 refer to this graph.

58. Antibodies are produced by

    (A) B cells
    (B) T cells
    (C) macrophages
    (D) neutrophils
    (E) natural killer cells

59. The large peak upon secondary stimulation is due to

    (A) primed memory cells secreting antibody
    (B) more antigen in the organism
    (C) T cell activation
    (D) passive immunity
    (E) none of the above

GO ON TO THE NEXT PAGE

60. The secondary response is what percent of the primary response?

    (A) .001%
    (B) .01%
    (C) 10%
    (D) 100%
    (E) 1000%

Ecological Section

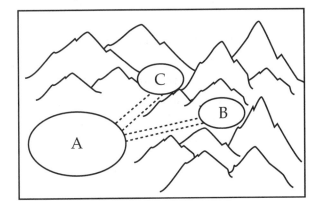

Map of Populations A, B, and C

A long-term research project studies a salamander species that inhabits several wet, cool valleys within a mountain range. Over a period of time the sizes of several populations are recorded, including Populations A, B, and C, shown on the map. The project also froze tissue samples from the populations and later analyzed for the samples for allele frequencies at Loci 1 and Loci 2. There are two alleles at Loci 1, dominant *T* and recessive *t*, and there are two alleles at Loci 2, dominant *W* and recessive *w*. The population sizes and allele frequencies over the 90-year study are presented in Table 1. The salamanders reach maturity in one year and reproduce once each year. Populations B and C were originally connected to Population A by a mountain pass through which the salamanders passed during the summer months. In 1915, a mine was developed in the mountain pass between populations A and B, preventing further migration of the salamanders between the two populations. The following year, the construction of a road prevented future migration of salamanders between populations A and C. There were no other changes in the environment, other than a significant warming and reduction of rain from 1939 to 1950 that affected the habitats of Populations A, B, and C equally.

Use this information to answer questions 61–65.

	Pop A Pop size	Pop A Loci 1	Pop A Loci 2	Pop B Pop size	Pop B Loci 1	Pop B Loci 2	Pop C Pop size	Pop C Loci 1	Pop C Loci 2
1900	46000	5% t	1% w	480	5% t	1% w	1000	5% t	1% w
1910	48000	5% t	1% w	450	5,5% t	1% w	900	5% t	1% w
1920	45000	5.5% t	1% w	460	7% t	1% w	850	5.5% t	1% w
1930	46000	6% t	1% w	400	18% t	3% w	700	7% t	.9% w
1940	48000	7% t	1% w	400	33% t	2% w	850	8% t	.8% w
1950	45000	8% t	1% w	200	45% t	5% w	800	9% t	.6% w
1960	44000	8% t	1% w	100	80% t	2% w	800	10% t	.6% w
1970	42000	7% t	1% w	50	100% t	0% w	700	12% t	.7% w
1980	42000	6% t	1% w	20	100% t	0% w	600	11% t	.8% w
1990	40000	5% t	1% w	0	ND	ND	550	10% t	1% w

Table 1: Population sizes and allele frequencies over time for Populations A, B, and C

GO ON TO THE NEXT PAGE

61. Which of the following occurs after the development of the road?

    A. By definition, populations A and C each become a separate species since they are geographically isolated.

    B. Population C undergoes adaptive radiation to form several new species.

    C. Population C evolves to have an increased number of heterozygotes at Loci 1.

    D. Population A will evolve a mechanism to migrate through the developed mountain pass.

    E. Population C will be eliminated due to competition from other salamander populations.

62. After many generations, the road is removed and the region is restored to its original habitat. The salamanders migrate through the pass once again, but Populations A and C do not interbreed. The populations cannot be distinguished based on physical examination alone. The most likely explanation for this observation is that:

    A. The males of Population A developed a new breeding behavior that females of C do not respond to.

    B. The animals of Population C have different allele frequencies than Population A.

    C. Population A has an additional chromosome.

    D. Selective pressure caused Population A to have less fitness.

    E. Population C has less fitness than Population A.

GO ON TO THE NEXT PAGE

63. Which of the following is the most likely description of changes in Population B over the course of the study?

A. Essential nutrients were missing from the diet of Population B that the animals had gained previously from the mountain pass.

B. Population B suffered from the lack of competition with other members of the species.

C. Population B evolved into a new species.

D. New mutations accumulated in the genome of Population B, causing increased negative selection pressure.

E. Inbreeding in Population B caused increased expression of harmful recessive alleles.

64. Population A contains multiple alleles at many of the loci studied. After the development of the mine, there is no further migration into or out of Population A, there is random mating, and there is no detectable mutation. Over 100 generations, the climate continues to grow warmer and drier. Which of the following can be predicted about allele frequencies in Population A?

A. Hardy-Weinberg equilibrium will be maintained and allele frequencies will not change.

B. The allele frequencies will change in the generation after migration ends, but then change no further.

C. The population will tend toward reduced diversity of alleles in the gene pool.

D. Allele frequencies will change as the environment changes.

E. There can be no changes in the allele frequencies of a closed population as long as random mating is maintained.

GO ON TO THE NEXT PAGE

65. A naturalist observes that the salamander habitat in the entire region has become fragmented by roads, land development, and flood control efforts. In 1998 the total population size is estimated to be 20 percent lower than twenty years earlier when the region was relatively undeveloped. The total number of square miles inhabited by salamanders is also decreased by 20 percent. The population of 200,000 individuals that was once fully interbreeding, however, is now divided into 40 distinct populations that no longer interbreed with each other. Which of the following would be the most likely recommendation by the naturalist to ensure the long-term survival of the species?

A. Collect all remaining salamanders and relocate them together in one of the remaining habitats.

B. Collect all remaining wild specimens and place them in zoos for breeding programs.

C. Remove a few individuals and relocate them to a new habitat that appears similar to the existing habitat, although the salamanders are not found there.

D. Trap a few salamanders and periodically relocate them between the existing populations.

E. Provide additional food sources for the species in the region.

A naturalist studies a species of primate that inhabits the rain forest. The females live together in large social groups and compete for position in the social hierarchy, with seniority and food-gathering ability as two of the determining factors for the leadership position. Males live within the group until they reach sexual maturity, at which time they leave the group and live on their own, mostly in isolation. Mature males only interact with the females during periods of mating which occur once a year. During mating season, males compete for mating opportunities with aggressive displays and battles between males that occasionally lead to significant and even fatal injury. Males are commonly observed during mating season interacting with social groups other than those of their birth. Females give birth an average of 5 times during their lives, with two children that reach maturity. In addition to studying the behavior of the species, the naturalist identifies the following pedigree:

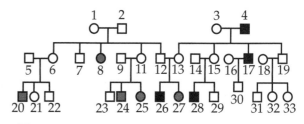

□ Male

O Female

Black shading indicates death occurred as a result of aggressive behavior initiated by the individual who died. Gray shading indicates death as a result of disease or as a result of aggressive behavior by another individual.

Use this information to answer questions 66–68.

GO ON TO THE NEXT PAGE

66. Females of this species are often observed feeding and caring for the offspring of other females. This behavior would be most likely the result of:
    A. fear of males
    B. the hierarchical structure of the primate social group
    C. kin selection
    D. instincts triggered by insufficient nutrition
    E. mutualism

67. The aggressive behavior of the males with each other probably has what affect on the gene pool of the population?

    A. This behavior will be selected against and will disappear from the species.
    B. This behavior will be selected for and maintained.
    C. Competition between females only is important for selection, and not selection between males.
    D. The behavior does not increase the size of the group, so it has no affect on natural selection.
    E. Since males live outside of the group for most of the year, they do not play a role in the fitness of the group.

68. Which of the following has the greatest fitness?

    A. Female 11, who cares for young with out regard for whether they are her own
    B. Male 7, who stays with the group to gather food and avoids aggressive behavior
    C. Male 4, who has killed a cousin and three other males and is himself killed in an aggressive battle initiated by himself
    D. Male 2, who lives twice as long as the average life span for males of this species
    E. Female 1, who is the dominant female of her social group

A large volcanic island in the South Atlantic 500 miles from the coast of Africa is inhabited by 12 species of bird that each have distinct diets, behavior, coloration, and beak shapes, and are not found on the mainland. Two species are flightless herbivores. All appear related in form to a species of bird on the mainland that does not generally fly over water but can occasionally survive at sea in a storm. There are no reptiles, mammals, or amphibians on the island, although there are many insect and plant species.

Use this information to answer questions 69–72.

GO ON TO THE NEXT PAGE

69. The number of different related bird species is probably due to which of the following?

    A. genetic instability
    B. the availability of many ecological niches not filled by competing species
    C. symbiosis
    D. recolonization of the island on several different occasions
    E. frequent storms in the region

70. A grass species on the island is homozygous at 40 different loci that are heterozygous on the mainland. These loci affect leaf shape, stem height, the opening and closing of stomata, and root length, among other traits. Which of the following is true?

    A. The plant reproduces only in an asexual manner.
    B. The plant is less able to evolve to fit a new environment in the event of a climate change.
    C. Migration of seeds from the mainland eliminates genetic diversity in the gene pool of the plant.
    D. The plant is polyploid.
    E. The grass species requires pollination by birds.

71. The evolution of two flightless bird species on the island indicates which of the following?

    A. Flight improved the ability of the species to escape predators, but impaired food gathering on the ground.
    B. Flight was originally evolved only to escape predators.
    C. Mimicry is a highly effective evolutionary strategy.
    D. The species had greater fitness without the ability to fly.
    E. Flight is not an effective means to escape predators.

72. Two of the bird species on the island are herbivores that both inhabit the same species of tree and eat the fruit the tree produces year-round. Which of the following might enable both species to inhabit the tree without competing for resources?

    A. One species lives in the top 30 percent of the tree and the other species lives in the lower 70 percent.
    B. One species is nocturnal and the other is active during the day.
    C. Reproductive isolation exists.
    D. The species have differing beak shapes.
    E. One species must become extinct.

GO ON TO THE NEXT PAGE

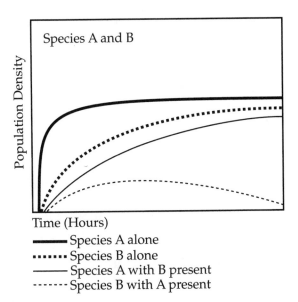

Figure 1

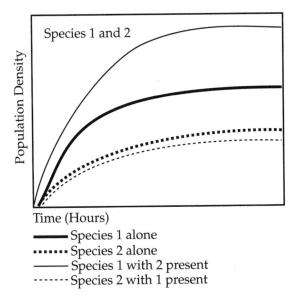

Figure 2

In the laboratory, a scientist examines the growth of two species of plankton, A and B, grown either together or separately (Figure 1). Both plankton feed on the same algae and are cultured in the presence of the algae, and the densities of the populations are examined over time. The culture is started in each case with the same number of plankton, and in the mixtures with a 1:1 ratio of each species. In another experiment, the growth of two algaes, species 1 and species 2, is examined, with the algae grown either separately or together (Figure 2).

Use this information to answer questions 73–75.

73. Which of the following is represented by the relationship between Species A and B in Figure 1?

    A.   commensalism

    B.   mutualism

    C.   symbiosis

    D.   competition

    E.   predation

74. Which of the following is best represented by the relationship between Species 1 and 2 in Figure 2?

    A.   commensalism

    B.   mutualism

    C.   parasitism

    D.   competition

    E.   predation

GO ON TO THE NEXT PAGE

75. According to the data in Figure 2, which of the following is most likely to occur between these two species?

   A. Species 1 blocks all of the sunlight from Species 2 and prevents Species 2 from growing as rapidly.

   B. Species 1 is unable to synthesize an essential amino acid that is missing from the nutrient solution but can be provided by Species 2.

   C. Species 1 secretes a toxic waste product that slows the growth of Species 2.

   D. Species 2 secretes a nutrient that is limiting in the culture medium for Species 1.

   E. Species 1 is better adapted to growth in culture.

76. In the pyramid of energy, the least amount of stored chemical energy is found in

   (A) primary producers
   (B) decomposers
   (C) primary consumers
   (D) secondary consumers
   (E) tertiary consumers

77. When a tick bird consumes insect pests from a rhino's back, which of the following processes is taking place?

   (A) mutualism
   (B) parasitism
   (C) saprophytism
   (D) commensalism
   (E) autotrophism

78. Which organism is correctly matched to its trophic level?

   (A) shark: primary consumer
   (B) cattle: primary consumer
   (C) fungi: producer
   (D) cyanobacteria: decomposer
   (E) butterflies: scavenger

79. A stable ecosystem

   I. is self-sustaining
   II. cycles materials between biotic and abiotic components
   III. requires an energy source
   IV. will have a high mutation rate

   (A) I only
   (B) I and II
   (C) I, II, and III
   (D) I, II, and IV
   (E) I, II, III, and IV

GO ON TO THE NEXT PAGE

80. Which of the following biomes are correctly paired?

    (A)   tundra: treeless frozen plain

    (B)   taiga: higher temperatures, torrential rains

    (C)   savanna: less than ten inches of rain per year

    (D)   desert: cold winters, warm summers, moderate rainfall

    (E)   none of the above

STOP

Biology E/M Test Two: E-Option Answer Key

1.	C	17.	A	33.	C	49.	D	65.	D
2.	E	18.	E	34.	A	50.	D	66.	C
3.	B	19.	C	35.	C	51.	D	67.	B
4.	A	20.	B	36.	B	52.	B	68.	C
5.	A	21.	E	37.	C	53.	B	69.	B
6.	C	22.	E	38.	A	54.	B	70.	B
7.	D	23.	B	39.	D	55.	D	71.	D
8.	A	24.	D	40.	C	56.	D	72.	A
9.	B	25.	C	41.	E	57.	B	73.	D
10.	D	26.	E	42.	D	58.	A	74.	A
11.	D	27.	E	43.	B	59.	A	75.	D
12.	A	28.	D	44.	A	60.	E	76.	E
13.	D	29.	E	45.	A	61.	C	77.	A
14.	C	30.	C	46.	D	62.	A	78.	B
15.	E	31.	A	47.	B	63.	E	79.	C
16.	D	32.	B	48.	E	64.	D	80.	A

Compute Your Practice Test Score

Step 1: Figure out your raw score. Refer to your answer sheet for the number right and the number wrong on the practice test you're scoring. (If you haven't checked your answers, do that now, using the answer key that follows the test.) You can use the chart below to figure out your raw score. Multiply the number wrong by 0.25 and subtract the result from the number right. Round the result to the nearest whole number. This is your raw score.

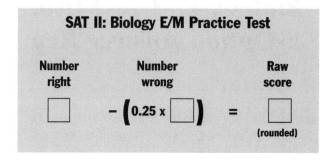

Step 2: Find your practice test score. Find your raw score in the left column of the table below. The score in the right column is an approximation of what your score would be on the SAT II: Biology E/M Test.

A note on your practice test scores: Don't take these scores too literally. Practice test conditions cannot precisely mirror real test conditions. Your actual SAT II: Biology E/M Subject Test score will almost certainly vary from your practice test scores. Your scores on the practice tests will give you a rough idea of your range on the actual exam.

Find Your Practice Test Score

Raw	Scaled	Raw	Scaled	Raw	Scaled	Raw	Scaled	Raw	Scaled
80	800	63	720	46	620	29	530	12	400
79	800	62	710	45	620	28	530	11	390
78	800	61	700	44	610	27	520	10	380
77	800	60	690	43	610	26	520	9	370
76	790	59	690	42	600	25	510	8	360
75	790	58	680	41	600	24	510	7	340
74	790	57	680	40	590	23	500	6	330
73	780	56	670	39	590	22	500	5	320
72	780	55	670	38	580	21	490	4	300
71	770	54	660	37	580	20	490	3	280
70	770	53	660	36	570	19	480	2	260
69	760	52	650	35	570	18	460	1	240
68	760	51	650	34	560	17	450	0	200
67	750	50	640	33	560	16	440		
66	750	49	640	32	550	15	430		
65	740	48	630	31	550	14	420		
64	730	47	630	30	540	13	410		

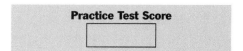

KAPLAN

Biology E/M Test Two: E-Option
Answers and Explanations

1. (C) This equation describes the formation of carbohydrates from carbon dioxide and water by plants. This process occurs in the chloroplasts and is part of photosythesis.

2. (E) The fetus has several features that help it to get oxygen from the mother, such as the ductus arteriosus, which shunts blood from the pulmonary artery to the aorta. The foramen ovale, meanwhile, is an opening that shunts blood from the right atrium to the left atrium. Both of these keep blood from traveling to the underdeveloped lungs. Fetal hemoglobin has a higher affinity for oxygen than adult hemoglobin, enabling it to become oxygenated in the placenta from the mother's blood.

3. (B) The ectoderm develops into the skin, eyes, and the nervous system. The endoderm develops into the lungs, the gastrointestinal tract, and the bladder lining. Finally, the mesoderm develops into everything else, including the musculoskeletal system, the circulatory system, and the reproductive organs.

4. (A) Flexors bend a joint to an acute angle, while extensors extend or straighten the bones at a joint.

5. (A) Aldosterone causes the active absorption of sodium ions in the nephron. Chloride ions and water passively follow.

6. (C) In the $AA \times aa$ cross, the F_1 generation will be 100 percent Aa, a heterozygous genotype with a dominant phenotype. In the F_2 generation, there will be a 1:2:1 ratio of $AA:Aa:aa$. Therefore, the F_1 generation is entirely dominant heterozygous. Answer choices I and II can be eliminated, leaving only III.

7. (D) In glycolysis and the Krebs cycle, energy is extracted from glucose and transferred to NADH and $FADH_2$. In electron transport, the energy of NADH and $FADH_2$ is transferred to a series of proteins in the inner mitochondrial membrane, with O_2 as the final electron acceptor at the end of the chain. This electron transport chain pumps protons out of the mitochondria, creating a proton gradient. The energy of the gradient drives the ATP synthesis. Each NADH leads to three ATP and each $FADH_2$ creates two. Without O_2 there is no electron transport or oxidative phosphorylation.

8. (A) The taiga have the most northern latitude of the choices available. Since growing season decreases as latitude increases, it must also have the shortest growing season.

9. (B) The Hardy-Weinberg Law states that gene ratios and allelic frequencies remain constant through the generations in a nonevolving population. Five criteria must be met in order for this to occur: random mating, a large population, no migration into or out of the population, no natural selection, and a lack of mutation. If all five of these criteria are met, gene frequencies will remain constant. Any time these five are not met, gene frequencies will change and evolution may occur.

10. (D) When a person takes too much antacid, HCl in the stomach is neutralized and pH rises from its normal very acidic levels. Pepsin is a stomach enzyme that works best in acidic conditions. When pH is raised, pepsin becomes nonfunctional. Therefore, pepsin activity would be most affected by an antacid overdose.

11. (D) Protistans may be unicellular or colonial, free living or symbiotic. They also can be heterotrophic or photosynthetic. Flagellates move via flagella, while sporozoans are responsible for malaria and spend part of their life cycle in mosquitoes. Bacilli, however, are rod-shaped bacteria, prokaryotes, not protists.

12. (A) Lactase breaks lactose into glucose and galactose. In (B), kinase is an enzyme that phosphorylates its substrate. As for (C), zymogen is an enzyme that is secreted in an inactive form. The zymogen is cleaved under certain physiological conditions to the active form of the enzyme. Important examples of zymogens include pepsinogen, trypsinogen, and chymotrypsinogen, which are cleaved in the digestive tract to yield the active enzymes pepsin, trypsin, and chymotrypsin. Finally, in (D), lipase breaks down lipids into free fatty acids, while in (E), phosphorylase removes a phosphate from its substrate.

13. (D) The least oxygenated blood would be found in the pulmonary artery. This artery transports blood that is deoxygenated after passage through all the tissues of the body back to the lungs to become oxygenated again.

14. (C) Snails are mollusks. Their characteristic mantle secretes a calcium-mineral shell. Snails also breathe through gills and have chambered hearts, blood sinuses, and a pair of ventral nerve cords. As for (D), insects have a tubular heart and an open circulatory system, like all arthropods. The cardinal and other birds have a chambered heart, like all vertebrates, but do not have gills. Sea urchins (echinoderms) and earthworms (annelids) do not have chambered hearts.

15. (E) Nondisjunction is a failure of a homologous pair of chromosomes to separate when gametes are formed by meiosis. The result is an extra chromosome or a missing chromosome for a given pair. For example, Down's syndrome can be attributed to an extra chromosome number 21. The number of chromosomes in a case of single nondisjunction is $2n + 1$ (47) or $2n - 1$ (45). In Down's syndrome, it is 47. Most of these embryos are aborted early in their development and only a few, like Down's syndrome (Trisomy 21), Trisomy 13, and Trisomy 18, make it to

term, albeit with developmental disorders. Hence breakage near the centromere might be induced by environmental factors such as mutagens, but would not be caused by nondisjunction.

16. (D) GH, FSH, LH, and ACTH are all secreted by the anterior pituitary, and would therefore be affected by its surgical removal. Insulin is produced by the pancreas.

17. (A) You can disregard *bb* and *DD*; all gametes will be *bD*. Your remaining options are now *ACbD*, *AcbD*, *aCbD*, and *acbD*, giving you a total of four different gametes. Or, 2 alleles at the A locus × 2 alleles at the the C locus = 4 possible gametes.

18. (E) Polysomes are defined as a group of ribosomes that attach to a strand of mRNA and simultaneously translate it. Histones are proteins that help to package DNA, allowing it to coil tightly in the nucleus, and the nucleolus is the region of the nucleus where rRNA is produced.

19. (C) Imprinting is a process in which environmental patterns or objects presented to a developing organism during a brief "critical period" in early life become accepted as permanent elements in the organism's life. In this case, the location at which the salmon was born becomes a permanent behavioral element, demonstrated when the salmon returns to this location to spawn.

20. (B) The definition of a species states that members of a particular species are able to mate and produce fertile, viable offspring.

21. (E) AB is known as the universal acceptor. It does not have antibodies to either the A or B antigens. Therefore, AB patients can receive blood from A, B, AB, or O people.

22. (E) A female has two *X* chromosomes, one inherited from her mother and one inherited from her father, while a given male has one *X* chromosome inherited from his mother and one *Y* chromosome inherited from his father. If a male expresses an *X*-linked trait, he must have inherited it from his mother. If normal parents have a color-blind son, he *must* have inherited the color blind gene, which is *X*-linked, from his mother. His mother *must* be a carrier of the color blind allele. The probability that a color blind son inherited the gene for color blindness from his mother is 100 percent.

23. (B) The Golgi apparatus consists of a stack of membrane-enclosed sacs. The Golgi receives vesicles and their contents from the smooth ER, modifies them (as in glycosylation), repackages them into vesicles, and distributes them. In (A), mitochondria are involved in cellular respiration, and in (C), the ER transports polypeptides around the cell and to the Golgi apparatus for packaging. The ribosome (D) is the site of protein synthesis, while lysosomes (E) are membrane-bound organelles that contain digestive enzymes and typically have a low pH.

24. (D) Spermatogenesis and oogenesis are both examples of gametogenesis in that both produce haploid gametes through reductional division (meiosis) of diploid cells. These processes occur in the gonads. They differ in that in spermatogenesis, the cytoplasm is equally divided during meiosis and four viable sperm are produced from one diploid cell. In oogenesis, on the other hand, the cytoplasm is divided unequally, and only one ovum, with the bulk of the cytoplasm, is produced in addition to two or three inert polar bodies. Spermatogenesis is also continuous, meaning that it occurs throughout life and not only during puberty. Meanwhile, oogenesis freezes at the end of meiosis I and does not complete meiosis II until fertilization.

25. (C) Fat contains approximately 9 calories/gram, while carbohydrates and proteins contain only 4 calories/gram. Sugar and starch are two forms of carbohydrates, and vitamins are coenzymes that are typically not metabolized.

26. (E) Oogenesis produces only one viable egg and two or three polar bodies. This is a result of unequal distribution of the cytoplasm during meiosis. Interstitial cells (A) are stimulated by LH to produce testosterone. FSH and testosterone then initiate the development of sperm in the seminiferous tubules. As for (B), eggs develop in follicles in the ovaries under the control of FSH. It is obvious in (C) that FSH plays a role in gamete production in both sexes. Finally, in (D), gametes become haploid through reductional division (meiosis) in which a diploid cell gives rise to either four haploid sperm or one haploid egg and two or three polar bodies.

27. (E) Progesterone readies the uterus for implantation by thickening and vascularizing the uterine lining. In (A), FSH secreted by the anterior pituitary, not progesterone, stimulates follicle growth. As for (B), FSH and LH are repressed by feedback inhibition from estrogen and progesterone. Testosterone and estrogen are responsible for the secondary sex characteristics of males and females respectively (C). As for choice (D), FSH and LH are produced by the anterior pituitary.

28. (D) To answer this question, we must use the Hardy Weinberg equation, $p^2 + 2pq + q^2 = 1$, in which p equals the gene frequency for the dominant allele, and q equals the gene frequency of the recessive allele. Hence, p^2 is the frequency of homozygous dominants in the population, $2pq$ is the frequency of heterozygotes, and q^2 is the frequency of homozygous recessives. For a trait with only two alleles, $p + q$ must equal 1, since the combined frequencies of the alleles must total 100 percent. In this problem, we are told that the frequency of the recessive allele for a particular trait is 0.6; hence $q = 0.6$. Since $p + q = 1$, $p = 0.4$. You're asked to determine the frequency of individuals expressing the dominant phenotype, not the dominant genotype. So, you're looking for $p^2 + 2pq$; $p^2 = 0.4$ squared $= 0.16$. $2pq = 2(0.6)(0.4) = 0.48$. Finally, $0.16 + 0.48 = 0.64$, which means that 64 percent of the individuals in the question express the dominant phenotype.

29. (E) The ovules are contained in the enlarged base of the pistil known as the ovary. Each ovule contains a monoploid egg nucleus, the female gametophyte.

30. (C) The stigma is the sticky top part of the flower that catches pollen.

31. (A) Pollen grains develop in the sac (called the anther) at the top of the thin, stalklike filament.

32. (B) One pollen grain (sperm) fuses with the egg to form the zygotes, while another fuses with two polar bodies to form the endosperm that nourishes the growing zygote.

33. (C) Tail fibers attach the bacteriophage to the cell surface, where it injects DNA into the cell.

34. (A) Bacteriophages contain only DNA, although some other viruses, such as HIV, use RNA as their genetic material. Proteins are found only in the coat and tail fibers, and lipids are not involved.

35. (C) The viral capsid and tail fibers are always made up of protein.

36. (B) Complex reflexes involve neural integration at higher levels such as the brain stem or the cerebrum. An example of this kind of reflex is the startle response that is provoked when you hear your name called.

37. (C) Fixed action patterns are complex, coordinated, innate behavioral responses to specific stimuli from the environment. Female birds will take care of eggs of their own species more vigilantly than they would an egg that doesn't resemble one of their own.

38. (A) Simple reflexes are simple, automatic responses to simple stimuli. One example is the knee-jerk reflex, with no direct involvement of the brain; this relex is a neural loop that only involves the spinal cord.

39. (D) Daily behavioral cycles are called circadian rhythms. These cycles are initiated intrinsically but modified by external factors such as day length.

40. (C) Grasshoppers, like all insects, remove nitrogenous wastes through Malphigian tubules which collect the waste from the body liquids and move them into the gut.

41. (E) Humans remove waste from their blood stream through filtration in the nephron, the functional unit of the kidney.

42. (D) Planaria have a primitive excretory system that consists of flame cells, which propel waste through tubules leading to the exterior.

43. (B) Two pairs of nephridia in each body segment of earthworms (annelids) excrete water, mineral salts, and urea wastes.

44. (A) Because protozoans are in contact with their external aqueous environment and they are very small, ammonia and carbon dioxide exit via simple diffusion through the cell membrane.

45. (A) Binary fission is a form of asexual reproduction found in bacteria and other one-celled organisms, in which the cell doubles its size, copies its DNA, and splits in two to make two cells, without exchanging genetic information.

46. (D) Vegetative propagation occurs when undifferentiated meristem cells provide for the development of a new plant from the parent. Plants that practice this form of propagation are able to produce seedless fruit, and daughter organisms are genetically identical to their parents. Examples of vegetative propagation include underground stems with buds (found in potatoes) and runners, which are stems running above or below the ground found in strawberries and some grasses.

47. (B) Yeast cells and some hydra reproduce by budding. This involves the unequal division of cytoplasm, making a smaller "daughter" cell which buds from the parent, although equal division of the genetic material is maintained, with each cell receiving one copy of the genome.

48. (E) Parthenogenesis occurs when an egg develops without fertilization by sperm. Some species can reproduce in the absence of sex this way. These organisms are typically haploid and contain only one copy of each locus.

49. (D) See the explanation for question 46 for a description of vegetative propagation.

50. (D) During the hottest part of the day, the loss of water (transpiration) through stomata exceeds the uptake of water through the roots. Therefore, to conserve water, the plant closes its stomata.

51. (D) An open stomate allows carbon dioxide to reach the chloroplasts so that photosynthesis can occur. This would therefore increase the rate of photosynthesis. The open stomata also leads to the loss of water through the opening, but does not affect chlorophyll at all.

52. (B) Stomata open when they receive light stimulus; therefore, they would stay closed in the dark room until they received light stimulus when put back into the sun.

53. (B) Twenty-four-hour daily cycles, such as sleep patterns, are maintained in the absence of external stimuli, although they will be affected by external stimuli. These cycles are examples of circadian rhythms.

54. (B) Tissue types 2 and 5 express the same genes, except that 2 expresses gene 7 and 5 expresses gene 8; therefore, 2 and 5 are probably very similar cell types.

55. (D) Tissue types 5 and 7 express completely opposite genes, except for gene 2, which all cells express. Therefore, based on their protein secretion, 5 and 7 are probably very different cells.

56. (D) The pancreas produces lipase, the enzyme that digests fats.

57. (B) Gene 2 is expressed in all cell types and is probably associated with something that all cells do, such as protein synthesis or DNA packing by histones.

58. (A) B cells produce antibodies; T cells, meanwhile, may be either cytotoxic (that is, they kill other cells) or serve as helper T cells that secrete proteins to stimulate other cell types.

59. (A) The large peak the second time you are exposed to an antigen or pathogen is due to the memory response of cells that have been primed for that antigen.

60. (E) The secondary response contained 10,000 antibodies per unit of blood, compared with 10 antibodies per unit of blood in the primary response. So, $10,000/10 = 1,000$ percent of the primary response.

Ecological Section

61. (C) Through nonrandom mating, random events in a small population, mutation, and natural selection, the allele frequencies in C will change over time and be different than A since the populations are isolated from each other. Examination of the table confirms this, with an increase in the *t* allele from 5 percent to 10 percent in 1990. Since the number of heterozygotes is $2pq$, according to Hardy-Weinberg, the percentage of heterozygotes goes from $2(.05)(.95) = 9.5\%$ to $2(.10)(.9) = 18\%$ ((C) is correct). Geographic isolation may lead to speciation, but it does not itself define a species (A). The salamander inhabits the same environment, so there is no reason to believe that it will evolve to several different species (B). There is no selective pressure driving the salamanders to develop the means to migrate (D). There are no other salamander populations that C interacts with, so there can be no competition (E).

62. (A) Geographic isolation often precedes the reproductive isolation that defines speciation. Many species have behaviors involved in mate selection that help to maintain reproductive isolation between closely related species. It is more likely that these behaviors will change more rapidly than significant physical differences or genetic differences that will lead to reproductive isolation (A). An additional chromosome, for example, is a dramatic evolutionary change and is *not* likely to occur and be tolerated. It is also likely to cause a change in the appearance of the animals (C). Two isolated populations will have different allele frequencies at many different foci, but this will not necessarily cause reproductive isolation (B). Natural selection will not select for less fit animals, generally, since reduced fertility equates with reduced fitness (D). The fitness of the populations does not relate to the failure to interbreed (E).

63. (E) Population B becomes extinct by the end of the study. Population B starts the study with only 500 individuals, 1 percent the size of Population A. One of the biggest problems in maintaining small populations is inbreeding. It is likely that a breeding pair in a small population will be related, increasing the number of homozygotes for harmful recessive alleles. If nutrients were missing from the pass, this would make itself known in the first generation, not many generations later (A). Geographic isolation plays a role in isolating the population, but is not itself directly responsible for the reduced viability of the population (B). There is no change in the mutation rate and no reason for mutations to accumulate (D).

64. (D) When the assumptions of Hardy-Weinberg are met, allele frequencies in a population reach a stable equilibrium in one generation. The assumptions are random mating, no net mutation, no natural selection, large population size, and no migration. The question states that some of the assumptions are met, but it does not say that there will be no natural selection. The population contains genetic variability, the climate is changing slowly, and salamanders are likely to be sensitive to a drier climate. Natural selection is inevitable, with some salamanders surviving better than others, altering allele frequencies in the gene pool (D). Hardy-Weinberg conditions are not met since there is natural selection (A). Allele frequencies change as the climate changes through a long process that takes longer than one generation (B). We have no information about the specific alleles involved, so we cannot make a statement about diversity (C). The fact that the population is isolated removes the possibility that allele frequencies are changed by migration, but does not eliminate natural selection (E).

65. (D) When a single large population is fragmented into many small ones, each small population is at much greater risk of extinction even if the total population size has not changed greatly. Allowing the salamanders to interbreed between the populations recreates the interbreeding from before their isolation and maintains the diversity of the gene pool as a whole. Gathering all of the salamanders and putting them into one habitat would probably lead to overcrowding and would make the entire population much more sensitive to any changes in that single region (A). Placing animals in zoos or captive breeding programs is a last resort when the population in the wild has dropped to such a small number that extinction is inevitable without this intervention (B). Putting the salamanders into related habitats is not ideal, since the salamanders are adapted to a specific habitat and it is not known how they will be adapted to the new environment. They exist as a part of the ecosystem (C). The danger to the salamanders is not lack of food, but restricted access to the gene pool (E).

66. (C) The social group is composed of many closely related individuals, particularly the females. Since the female is probably related to the children of other females in the group and shares the same genes, she is indirectly contributing to the passage of her own genes to the next generation when she cares for the young of a related female. This form of behavior is called kin selection, and is a behavior favored by natural selection in social animals.

67. (B) The aggressive behavior of males is stated by the passage to be part of the selection process for mates. Aggressive males have the best chances of finding a mate and passing their genes on to next generation, giving them good fitness (A). Males do not need to live within the group throughout the year for these traits to be selected for, since, strictly speaking, it is the genes the males carry and not the males themselves that are of concern in terms of fitness (E).

68. (C) By examining the pedigree, it is possible to determine how many offspring each of the individuals has parented. The fitness will be determined by this, while most of the description of each individual is irrelevant. Male 4 may be responsible for deaths, and may himself die in an aggressive conflict, but he is the parent of more individuals and grandchildren than the others who are described, giving him the highest fitness. Female 11 has only one surviving child and her behavior of caring for other children without preference for her own would cause her to have a lower fitness than if she had many children of her own. Male 7 may help others, but his lack of children, probably caused in part by his lack of aggressive behavior, gives him

poor fitness. Male 2 may be long-lived, but this is irrelevant for fitness, which depends only on offspring and is unaffected by the length of life. Male 2 has only three children who have offspring, compared to four for male 4, and has only three surviving grandchildren, compared to five for male 4. For female 1, being the dominant member of the group is irrelevant to her fitness, which is the same as for male 2.

69. (B) All life found on the island has evolved from a few organisms that were able to colonize the island originally. It is most likely that all of the bird species were derived from the single species of bird found on the mainland. On the mainland, the bird species had evolved to fit a specific ecological niche, competing with other species for food and resources, and contending with predators. On the island, the bird found no predators, no competition, and many potential ecological niches that were not occupied. A species in this situation will often undergo adaptive radiation, evolving into several species to fill the available ecological niches.

70. (B) The plant species has much less genetic variability in its gene pool than the same species on the mainland. Natural selection requires genetic variability in a population. If all of the individuals are the same, then they will all have the same fitness and there can be no selection. The grass species will not be able to evolve in the presence of climate change or other situations.

71. (D) There is no purpose or direction to evolution other than increasing fitness. We cannot state based on the information provided that flight serves a specific purpose on the island or otherwise. It is not possible to make any statement about the purpose of flight in other birds based on these flightless species in this specific environment ((A), (B) and (E) are wrong). Clearly however, for these two species they had higher fitness without flight since this is how they evolved ((D) is correct). This is not an example of mimicry (C).

72. (A) For the two species not to compete, they must somehow each occupy a different niche, thereby avoiding competing for the same food resource. If they live in different parts of the same tree, they will not be competing for the same pieces of fruit (A). If they eat the same fruit, being active during the day or at night will not prevent them from competing (B). Reproductive isolation is present by definition in different species, but not does not prevent competition (C). Differing beak shapes could allow them to eat different fruits, but this does not address the question (D). For a species to drive another to extinction, they must occupy the same niche and the surviving species must outcompete the other. We know from the question that they do not share the same niche (E).

73. (D) In Figure 1, there are two species competing for the same resource—a particular type of algae. Whenever two species are trying to occupy the same niche and compete for the same resource, one species will prove to have higher fitness and will over time dominate the niche in the ecosystem. In this experiment, Species B grows more slowly than species A, and cannot compete effectively in the culture as a result. Species A takes over, and Species B declines with A present. Commensalism, mutualism, and symbiosis all describe relationships in which one or both members benefit by the presence of the other, which is not the case here. Although both species are predators, they eat the algae and not each other (E).

74. (A) In this experiment, Species 2 grows at the same rate with species 1 present as it did on its own. Species 1, however, grows more rapidly in the presence of Species 2 than it did not its own, indicating that it benefited from the presence of the other species. Since the benefit is one-sided, the relationship is commensalism and not mutualism, which describes relationships in which both sides are benefited.

75. (D) Choices (A) and (C) describe relationships in which one member is harmed, which is not the case here, since Species 2 grows at the same rate either with or without Species 1. In (B), Species 1 would be absolutely dependent on Species 2 to survive, which is not the case. (D) is the best choice. Species 1 can grow well on its own, but with Species 2 it can grow even better, suggesting that Species 2 is altering the growth medium in some way to make it more favorable for Species 1.

76. (E) Each level in the pyramid of energy loses some energy from the level before it, due to the loss of heat and energy costs of maintenance of the organism at each level. Also, there are fewer organisms and less biomass in each level than in the one before it. Hence the least amount of stored energy would be found at the highest point of the chain.

77. (A) The tick bird helps the rhino by keeping it clean of parasites, eating the insects that infect the rhino, while the rhino in its turn provides food for the tick bird. This +/+ relationship is known as mutualism.

78. (B) Cattle, being herbivorous, are primary consumers; sharks are secondary consumers or tertiary consumers. Fungi are decomposers and cyanobacteria are producers. Butterflies, meanwhile, are primary consumers.

79. (C) A stable ecosystem is by nature self-sustaining (I). It needs no outside input of materials. It cycles its components, such as nitrogen and carbon, between its biotic and abiotic components as both elemental and "fixed" components (II). It also requires an energy source, such as the sun (III). Mutation rates, however, are not affected by an ecosystem's stability or instability (IV).

80. (A) The tundra is a treeless, frozen plain between the taiga and the northern ice sheets. The taiga (B) is a spruce-filled forest inhabited by moose and bear, while the savanna (C) is a grassland with low rainfall and populated with hoofed herbivores and carnivorous predators. Finally, the desert (D) receives less than ten inches per year of rainfall and is populated by small plants and animals.

ANSWER SHEET FOR BIOLOGY E/M
TEST THREE: M-OPTION

1 Ⓐ Ⓑ Ⓒ Ⓓ Ⓔ   21 Ⓐ Ⓑ Ⓒ Ⓓ Ⓔ   41 Ⓐ Ⓑ Ⓒ Ⓓ Ⓔ   61 Ⓐ Ⓑ Ⓒ Ⓓ Ⓔ
2 Ⓐ Ⓑ Ⓒ Ⓓ Ⓔ   22 Ⓐ Ⓑ Ⓒ Ⓓ Ⓔ   42 Ⓐ Ⓑ Ⓒ Ⓓ Ⓔ   62 Ⓐ Ⓑ Ⓒ Ⓓ Ⓔ
3 Ⓐ Ⓑ Ⓒ Ⓓ Ⓔ   23 Ⓐ Ⓑ Ⓒ Ⓓ Ⓔ   43 Ⓐ Ⓑ Ⓒ Ⓓ Ⓔ   63 Ⓐ Ⓑ Ⓒ Ⓓ Ⓔ
4 Ⓐ Ⓑ Ⓒ Ⓓ Ⓔ   24 Ⓐ Ⓑ Ⓒ Ⓓ Ⓔ   44 Ⓐ Ⓑ Ⓒ Ⓓ Ⓔ   64 Ⓐ Ⓑ Ⓒ Ⓓ Ⓔ
5 Ⓐ Ⓑ Ⓒ Ⓓ Ⓔ   25 Ⓐ Ⓑ Ⓒ Ⓓ Ⓔ   45 Ⓐ Ⓑ Ⓒ Ⓓ Ⓔ   65 Ⓐ Ⓑ Ⓒ Ⓓ Ⓔ
6 Ⓐ Ⓑ Ⓒ Ⓓ Ⓔ   26 Ⓐ Ⓑ Ⓒ Ⓓ Ⓔ   46 Ⓐ Ⓑ Ⓒ Ⓓ Ⓔ   66 Ⓐ Ⓑ Ⓒ Ⓓ Ⓔ
7 Ⓐ Ⓑ Ⓒ Ⓓ Ⓔ   27 Ⓐ Ⓑ Ⓒ Ⓓ Ⓔ   47 Ⓐ Ⓑ Ⓒ Ⓓ Ⓔ   67 Ⓐ Ⓑ Ⓒ Ⓓ Ⓔ
8 Ⓐ Ⓑ Ⓒ Ⓓ Ⓔ   28 Ⓐ Ⓑ Ⓒ Ⓓ Ⓔ   48 Ⓐ Ⓑ Ⓒ Ⓓ Ⓔ   68 Ⓐ Ⓑ Ⓒ Ⓓ Ⓔ
9 Ⓐ Ⓑ Ⓒ Ⓓ Ⓔ   29 Ⓐ Ⓑ Ⓒ Ⓓ Ⓔ   49 Ⓐ Ⓑ Ⓒ Ⓓ Ⓔ   69 Ⓐ Ⓑ Ⓒ Ⓓ Ⓔ
10 Ⓐ Ⓑ Ⓒ Ⓓ Ⓔ   30 Ⓐ Ⓑ Ⓒ Ⓓ Ⓔ   50 Ⓐ Ⓑ Ⓒ Ⓓ Ⓔ   70 Ⓐ Ⓑ Ⓒ Ⓓ Ⓔ
11 Ⓐ Ⓑ Ⓒ Ⓓ Ⓔ   31 Ⓐ Ⓑ Ⓒ Ⓓ Ⓔ   51 Ⓐ Ⓑ Ⓒ Ⓓ Ⓔ   71 Ⓐ Ⓑ Ⓒ Ⓓ Ⓔ
12 Ⓐ Ⓑ Ⓒ Ⓓ Ⓔ   32 Ⓐ Ⓑ Ⓒ Ⓓ Ⓔ   52 Ⓐ Ⓑ Ⓒ Ⓓ Ⓔ   72 Ⓐ Ⓑ Ⓒ Ⓓ Ⓔ
13 Ⓐ Ⓑ Ⓒ Ⓓ Ⓔ   33 Ⓐ Ⓑ Ⓒ Ⓓ Ⓔ   53 Ⓐ Ⓑ Ⓒ Ⓓ Ⓔ   73 Ⓐ Ⓑ Ⓒ Ⓓ Ⓔ
14 Ⓐ Ⓑ Ⓒ Ⓓ Ⓔ   34 Ⓐ Ⓑ Ⓒ Ⓓ Ⓔ   54 Ⓐ Ⓑ Ⓒ Ⓓ Ⓔ   74 Ⓐ Ⓑ Ⓒ Ⓓ Ⓔ
15 Ⓐ Ⓑ Ⓒ Ⓓ Ⓔ   35 Ⓐ Ⓑ Ⓒ Ⓓ Ⓔ   55 Ⓐ Ⓑ Ⓒ Ⓓ Ⓔ   75 Ⓐ Ⓑ Ⓒ Ⓓ Ⓔ
16 Ⓐ Ⓑ Ⓒ Ⓓ Ⓔ   36 Ⓐ Ⓑ Ⓒ Ⓓ Ⓔ   56 Ⓐ Ⓑ Ⓒ Ⓓ Ⓔ   76 Ⓐ Ⓑ Ⓒ Ⓓ Ⓔ
17 Ⓐ Ⓑ Ⓒ Ⓓ Ⓔ   37 Ⓐ Ⓑ Ⓒ Ⓓ Ⓔ   57 Ⓐ Ⓑ Ⓒ Ⓓ Ⓔ   77 Ⓐ Ⓑ Ⓒ Ⓓ Ⓔ
18 Ⓐ Ⓑ Ⓒ Ⓓ Ⓔ   38 Ⓐ Ⓑ Ⓒ Ⓓ Ⓔ   58 Ⓐ Ⓑ Ⓒ Ⓓ Ⓔ   78 Ⓐ Ⓑ Ⓒ Ⓓ Ⓔ
19 Ⓐ Ⓑ Ⓒ Ⓓ Ⓔ   39 Ⓐ Ⓑ Ⓒ Ⓓ Ⓔ   59 Ⓐ Ⓑ Ⓒ Ⓓ Ⓔ   79 Ⓐ Ⓑ Ⓒ Ⓓ Ⓔ
20 Ⓐ Ⓑ Ⓒ Ⓓ Ⓔ   40 Ⓐ Ⓑ Ⓒ Ⓓ Ⓔ   60 Ⓐ Ⓑ Ⓒ Ⓓ Ⓔ   80 Ⓐ Ⓑ Ⓒ Ⓓ Ⓔ

Use the answer key following the test to count up the number of questions you got right and the number you got wrong. (Remember not to count omitted questions as wrong.) The "Compute Your Score" section that follows the Answer Key will show you how to find your score.

BIOLOGY E/M
TEST THREE: M-OPTION

Part A

Directions: Each question or incomplete statement below is followed by five possible answers or completions, lettered A–E. Choose the answer that is the best in each case. Fill in the corresponding oval on your answer sheet.

1. Which of the following is found in all forms of life?

    I.   genetic material
    II. protein
    III. water

    (A) I only
    (B) II only
    (C) III only
    (D) I and III
    (E) I, II, and III

2. Which of the following is NOT essential for blood clotting?

    (A) sodium ions
    (B) calcium ions
    (C) prothrombin
    (D) vitamin K
    (E) platelets

3. Which substance does NOT act as a digestive enzyme for proteins?

    (A) pepsin
    (B) trypsin
    (C) carboxypeptidase
    (D) chymotrypsin
    (E) gastrin

4. Which of the following is a function of bone?

    I.   formation of blood cells
    II. protection of vital organs
    III. framework for movement

    (A) I only
    (B) II only
    (C) III only
    (D) I and II
    (E) I, II, and III

GO ON TO THE NEXT PAGE

5. The hypothesis that chloroplasts and mitochondria were originally prokaryotic organisms living within eukaryotic hosts is supported by the fact that mitochondria and chloroplasts

   I.   possess protein synthetic capability
   II. possess genetic material
   III. possess a lipid bilayer membrane
   IV. possess characteristic ribosomes

   (A) II only
   (B) IV only
   (C) II and IV
   (D) III and IV
   (E) I, II, and IV

6. A culture of white blood cells is grown on a nutrient media containing a poison that blocks the electron transport chain. Under these conditions

   (A) ATP production will remain the same
   (B) ATP production will decrease
   (C) oxygen consumption will increase
   (D) ethanol production will increase
   (E) none of the above

7. Many animals use panting as a means of cooling themselves down. The mechanism behind panting is to

   (A) rapidly increase carbon dioxide expiration
   (B) moisten the mucosa of the respiratory passages
   (C) minimize the movement of respiratory muscles
   (D) decrease body heat via evaporation
   (E) none of the above

8. Glucose in the glomerular filtrate is partially resorbed in the

   (A) Bowman's capsule
   (B) glomerulus
   (C) proximal convoluted tubule
   (D) villi
   (E) ureter

9. Which are correctly related?

   (A) white blood cell: no nucleus
   (B) smooth muscle cell: multinuclear
   (C) smooth muscle: voluntary action
   (D) cardiac muscle: involuntary action
   (E) smooth muscle: striations

GO ON TO THE NEXT PAGE

10. A process that cannot take place in a haploid cell is

(A) mitosis
(B) meiosis
(C) ATP production
(D) DNA replication
(E) transcription

11. To test whether a tall plant (in which the tallness trait is dominant) is homozygous or heterozygous, you could

    I.   cross it with a tall plant that had a short parent
    II.   cross it with a tall plant that had two tall parents
    III.   cross it with a short plant

(A) I only
(B) II only
(C) III only
(D) I and II
(E) I and III

12. Strontium is preferentially incorporated into growing long bone. Therefore, if a child were exposed to strontium, where would the highest concentration of strontium most likely be found?

(A) in the cartilage lining the joints
(B) in the center of long bones
(C) in the skull bones
(D) near the epiphyseal plates of long bones
(E) evenly distributed throughout the body

13. Venous blood en route from the kidneys to the heart must pass through the

(A) iliac vein
(B) inferior vena cava
(C) liver
(D) hepatic vein
(E) pulmonary vein

14. Humans cannot produce Vitamin K. However, even when their diet is lacking in Vitamin K, they have plenty of it in their bloodstream. How can this be possible?

(A) Vitamin K is synthesized in the liver using hormones absorbed from ingested plant cells.
(B) Vitamin K is a byproduct of protein degradation during digestion.
(C) Vitamin K is synthesized by bacteria that inhabit the colon.
(D) Vitamin K can be produced by exposure to sunlight.
(E) none of the above

15. The bicarbonate ion in the digestive tract

(A) neutralizes stomach acid
(B) promotes phagocytosis by white blood cells
(C) carries oxygen to the lungs
(D) functions in the blood clotting mechanism
(E) none of the above

GO ON TO THE NEXT PAGE

16. Which of the following is NOT characteristic of fermentation?

    (A) It is anaerobic.
    (B) It requires glucose.
    (C) It produces energy.
    (D) It requires oxygen.
    (E) It produces ethanol.

17. If a tracer substance is injected into a patient's superior vena cava, which of the following structures would the tracer reach last?

    (A) the right ventricle
    (B) the left ventricle
    (C) the pulmonary veins
    (D) the left atrium
    (E) the right atrium

18. Which of the following is the pacemaker of the heart?

    (A) the foramen ovale
    (B) the sinoatrial node
    (C) the ductus arteriosus
    (D) the bundle of His
    (E) the vagus nerve

19. If a diabetic accidentally overdosed on insulin, which of the following would be likely to occur?

    (A) increased levels of glucose in the blood
    (B) increased glucose concentration in urine
    (C) dehydration due to increased urine excretion
    (D) increased conversion of glycogen to glucose
    (E) increased conversion of glucose to glycogen

20. The best description of identical twins is that they are

    (A) twins of the same sex
    (B) twins from a single egg
    (C) twins from two eggs that have been fertilized by the same sperm
    (D) twins from two eggs fertilized by two separate sperm
    (E) twins from a single egg fertilized by two separate sperm

21. Which of the following lacks vertebrae?

    (A) the duckbill platypus
    (B) the turtle
    (C) the amphioxus
    (D) the trout
    (E) the rabbit

GO ON TO THE NEXT PAGE

KAPLAN

22. Ingestion of the insecticide Parathion, which blocks acetylcholinesterase function, would cause

(A) a decrease in postsynaptic signals
(B) a halt to all synaptic nervous transmissions
(C) an increase in acetylcholine concentration in synapses
(D) a decrease in acetylcholine concentration in synapses
(E) levels of acetylcholine to remain the same

23. Straight tail (*T*) is dominant over bent tail (*t*) in mice, and long-tailed mice (*L*) are dominant over short-tailed mice (*l*). Which cross must produce all straight, long-tailed mice?

(A) *TtLl* × *TtLl*
(B) *Ttll* × *TTLl*
(C) *TtLL* × *ttLL*
(D) *TtLl* × *TTLL*
(E) none of the above

24. In adult humans, red blood cells

(A) have no nucleus
(B) are replaced in the liver
(C) are outnumbered by white blood cells in the circulatory system
(D) are made in the spleen
(E) are the sites of rapid protein synthesis

25. The notochord is

(A) present in all adult chordates
(B) present in all echinoderms
(C) present in chordates during embryonic development
(D) always a vestigial organ in chordates
(E) part of the nervous system of all vertebrates

26. A typical human gamete

I. contains a haploid number of genes
II. will always contain an *X* or a *Y* chromosome
III. is a result of mitosis
IV. has undergone genetic recombination

(A) I and II
(B) I and III
(C) II and III
(D) II, III, and IV
(E) I, II, and IV

27. The absorption of oxygen from the atmosphere into the blood takes place in the

(A) pulmonary artery
(B) pulmonary vein
(C) alveoli
(D) trachea
(E) bronchi

GO ON TO THE NEXT PAGE

28. Which of the following is a substance secreted by a member of a species that affects other members of the same species?

(A) gender-specific proteins
(B) lacrimal fluid
(C) enzymes
(D) hormones
(E) pheromones

29. The rate of breathing is controlled by involuntary centers in the

(A) cerebrum
(B) cerebellum
(C) medulla oblongata
(D) spinal cord
(E) hypothalamus

30. Enzymes

I. are proteins
II. typically work best at pH 7.2
III. are changed during a reaction
IV. are found in the nucleus only

(A) I only
(B) II only
(C) I and II
(D) I, II, and III
(E) I, II, III, and IV

31. Which of the following occurs in the cell nucleus?

I. RNA synthesis
II. protein synthesis
III. DNA synthesis

(A) I only
(B) II only
(C) III only
(D) I and III
(E) I, II, and III

32. Which statement about the plasma membrane is false?

(A) It serves as a selectively permeable barrier to the external environment.
(B) It serves as a mediator between the internal and external environments.
(C) In eukaryotes, it contains the cytochrome chain of oxidative phosphorylation.
(D) It contains phospholipids as a structural component.
(E) It contains proteins that in some cases span the membrane.

GO ON TO THE NEXT PAGE

KAPLAN

33. Which of the following is a type of genetic mutation?

   (A)   point
   (B)   silent
   (C)   insertion
   (D)   frameshift
   (E)   all of the above

34. Pancreatic exocrine secretions contain all of the following EXCEPT

   (A)   proteases
   (B)   lipases
   (C)   amylases
   (D)   glucagon
   (E)   bicarbonate ions

Questions 35–38 refer to the figure of the vascular system of a plant below.

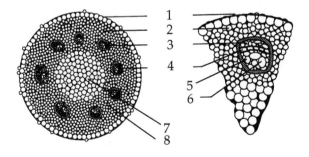

35. Which structure is responsible for the transport of nutrients?

   (A)   2
   (B)   3
   (C)   4
   (D)   5
   (E)   6

36. Which structure is responsible for the transport of water and minerals?

   (A)   2
   (B)   3
   (C)   4
   (D)   5
   (E)   1

37. Which structure is made up of rapidly dividing, undifferentiated cells?

   (A)   2
   (B)   3
   (C)   4
   (D)   5
   (E)   6

38. Which structure is used to support the plant?

   (A)   4
   (B)   5
   (C)   6
   (D)   7
   (E)   8

GO ON TO THE NEXT PAGE

Part B

Each set of choices A–E below should be compared to the numbered statements that follow it. Choose the lettered choice that best matches each numbered statement. Fill in the correct oval on your answer sheet. Remember that a choice may be used once, more than once, or not at all in each set.

Questions 39–42:

(A) virus
(B) bacteria
(C) amoeba
(D) planaria
(E) sponge

39. may contain plasmid DNA

40. has a cell membrane but lacks a true nucleus

41. genetic material can be either DNA or RNA

42. bilaterally symmetrical

Questions 43–47:

(A) glycolysis
(B) fermentation
(C) Krebs cycle
(D) electron transport chain
(E) photosynthesis

43. utilizes a proton pump

44. occurs in the mitochondrial matrix

45. is an anaerobic process that forms NAD^+

46. occurs in the inner membrane of the mitchondria

47. creates byproducts of ethanol or lactic acid

Questions 48–51:

(A) musculoskeletal system
(B) endocrine system
(C) circulatory system
(D) nervous system
(E) reproductive system

48. transports respiratory gases, nutrients, and wastes

49. enables organisms to receive and respond to stimuli

50. is a basic internal framework

51. performs internal communication through the circulatory system and coordinates the activities of the organ systems

Questions 52–55:

(A) adenine
(B) guanine
(C) thymine
(D) cytosine
(E) uracil

GO ON TO THE NEXT PAGE

KAPLAN

52. purine that binds with three hydrogen bonds

53. found only in RNA

54. pyrimidine that binds with three hydrogen bonds

55. purine found in both DNA and RNA that binds with two hydrogen bonds

Part C

Each of the following sets of questions is based on a laboratory or experimental situation. Begin by studying the description of each situation. Next, choose the best answer to each of the questions that follow it. Fill in the corresponding oval on your answer form.

A stable population exists in Hardy-Weinberg equilibrium with two alleles, T and t. TT and Tt individuals have the ability to curl their tongues, while tt individuals cannot curl their tongues. The allele frequency of T is 0.8.

The data above will help you answer questions 56–58.

56. What is the allele frequency of t?

(A) 0.04
(B) 0.2
(C) 0.8
(D) 0.16
(E) 0.64

57. What is the percentage of heterozygotes?

(A) 4%
(B) 32%
(C) 64%
(D) 50%
(E) 75%

58. What is the percentage of individuals that can curl their tongues?

(A) 32%
(B) 4%
(C) 64%
(D) 96%
(E) 50%

A scientist grew her bacterial colonies in a variety of different concentrations of nutrients to determine at which concentration they grew best. The results of her experiments are summarized in the following bar graph.

GROWTH BAR CHART

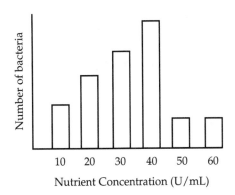

Now answer question 59.

GO ON TO THE NEXT PAGE

59. What is the optimum concentration of nutrient supplement?

(A) 10 U/mL
(B) 20 U/mL
(C) 30 U/mL
(D) 40 U/mL
(E) 50 U/mL

A mouse was put in a maze and timed to determine how long it took him to find the food pellet at the end. A total of eight trials were done until the researchers were confident that the mouse had learned the maze. The data follows in both tabular form in the table and graphical form.

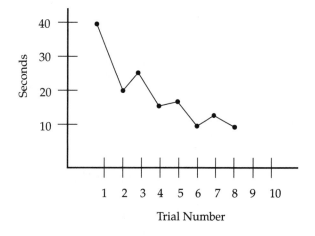

Trial Number	Number of Seconds
1	40
2	20
3	25
4	16
5	17
6	10
7	12
8	10

Use this information to answer question 60.

60. The greatest improvement is shown in the

(A) first trial
(B) second trial
(C) third trial
(D) fourth trial
(E) fifth trial

Molecular Section

The first step in glycolysis is catalyzed by the enzyme hexokinase. The reaction catalyzed by this enzyme is the phosphorylation of glucose to produce glucose 6-phosphate, with the consumption of one ATP and the production of one ADP. A scientist isolates hexokinase from bovine skeletal muscle and studies its activity in a variety of conditions. In Figure 1, the rate of glucose 6-phosphate production was studied as more enzyme was added, with a constant amount of glucose and ATP added. In Figure 2, the rate of glucose 6-phosphate was studied as more glucose was added to the mixture, with the amount of hexokinase held at a constant amount.

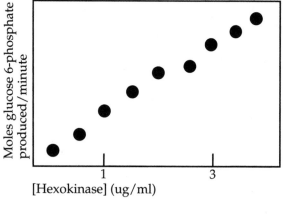

Figure 1

GO ON TO THE NEXT PAGE

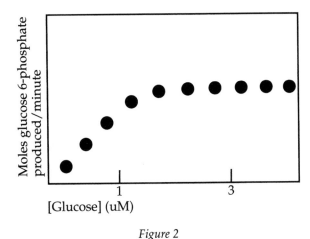

Figure 2

Use this data to answer questions 61–63.

61. Which of the following is true of hexokinase?

    A. Hexokinase increases the equilibrium ratio of [glucose 6-phosphate]/ [glucose].

    B. Hexokinase increases the rate at which equilibrium is reached, but does not change the equilibrium.

    C. Hexokinase increases the rate and changes the equilibrium.

    D. Hexokinase increases the rate until it is consumed, and the reaction ceases.

    E. Hexokinase can only catalyze the forward reaction.

62. If the amount of hexokinase in the mixture is doubled from 1 ug/ml to 2 ug/ml, which of the following will occur?

    A. The activation energy of the reaction will be 50 percent lower with 2 ug/ml than with 1 ug/ml.

    B. The rate of ADP production will double.

    C. Hexokinase will be consumed at half of the original speed.

    D. The rate of glucose 6-phosphate production will increase three-fold.

    E. The rate of glucose 6-phosphate production will be saturated and increase only slightly.

63. If the concentration of glucose in Figure 2 is increased to 6 uM from 3 uM, which of the following will be observed?

    A. The rate of glucose-6 phosphate production will double, and the rate of ATP consumption will go down.

    B. The rate of glucose-6 phosphate production and ATP consumption will both double.

    C. The rate of glucose-6 phosphate production will go down, and the rate of ATP consumption will increase.

    D. The rate of glucose-6 phosphate production will remain constant, and ATP consumption will increase.

    E. The rate of glucose-6 phosphate production will remain constant, and the rate of ATP consumption will remain constant.

GO ON TO THE NEXT PAGE

64. Which of the following is true regarding DNA replication?

I. DNA replication is semi-conservative.
II. DNA replication occurs during Prophase.
III. DNA synthesis occurs in a 3′ to 5′ direction.
IV. Purines hydrogen bond to pyrimidines.

(A) I and II
(B) I, II, and III
(C) I, III, and IV
(D) I, II, III, and IV
(E) I and IV

65. During the process of oxidative phosphorylation, oxygen serves as

(A) the initial acceptor of electrons
(B) the final acceptor of electrons
(C) a high-energy intermediate
(D) a phosphorylating agent
(E) a reducing agent

A researcher grew *E. coli* bacteria that either had a plasmid (an extranuclear piece of DNA) or did not have a plasmid. These two types of *E. coli* were grown in two different growth mediums—a normal nutrient growth medium, and a normal nutrient growth medium with tetracycline, an antibiotic. The results of these experiments are summarized in the following graphs of growth of bacteria versus time.

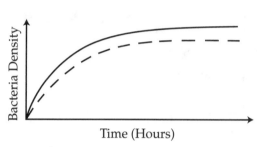

Graph 1: Bacterial Growth in Nutrient Media Alone

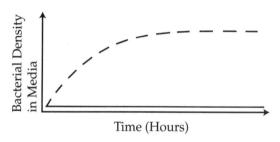

Graph 2: Growth of Bacteria in Media with Tetracycline

Use this data to answer questions 66–70.

GO ON TO THE NEXT PAGE

KAPLAN

66. When the two different *E. coli* strains, with and without the plasmid, were grown in nutrient media without antibiotics, which of the following occurred?

(A) The strain without the plasmid did not reach maximum growth potential.

(B) Both strains reached maximum growth potential.

(C) The strain with the plasmid did not reach maximum growth potential.

(D) Neither strain reached maximum growth potential.

(E) The strain with the plasmid had a higher population density than the strain without the plasmid.

67. When the two strains were cultured in nutrient media with the antibiotic tetracycline

(A) the strain without the plasmid was able to reach maximum population potential

(B) both strains reached maximum population potential

(C) the strain with the plasmid was able to reach maximum population potential

(D) neither strain was able to reach maximum population potential

(E) tetracycline did not affect either bacterial colony

68. The plasmid found in this *E. coli* strain codes for an enzyme that

(A) allows *E. coli* to grow in the presence of tetracycline

(B) digests the nutrient media

(C) causes the cells to autolyse

(D) makes this *E. coli* strain virulent

(E) none of the above

69. Which of the following would not be found in the plasmid?

(A) thymine

(B) adenine

(C) uracil

(D) cytosine

(E) guanine

70. According to Darwin's theory of natural selection, the *E. coli* strain with the plasmid

(A) is more fit in regular media than the strain without the plasmid

(B) is a result of genetic mutation

(C) is less fit than the strain without the plasmid in the antibiotic media

(D) has a competitive advantage when grown in regular media

(E) has a competitive advantage when grown in the media that contains tetracycline

GO ON TO THE NEXT PAGE

71. Which statement about glycolysis is NOT true?

   (A) Glycolysis converts a single molecule into two molecules of pyruvate.
   (B) Glycolysis can produce a net total of 2 ATPs for each glucose.
   (C) The end-product of glycolysis can form ethanol, lactate, or acetyl CoA.
   (D) During glycolysis, $FADH_2$ is produced.
   (E) During glycolysis, NADH is produced.

72. A cell is placed in a medium containing radioactively labeled thymine. If the cell undergoes two rounds of replication while in this medium, some of the radioactivity will appear

   (A) in each DNA double-helix
   (B) in half of the strands of DNA
   (C) in the proteins produced
   (D) in the mRNA
   (E) in the rRNA

73. Which statement regarding protein synthesis is false?

   (A) tRNA molecules shuttle amino acids that are incorporated into the protein.
   (B) Proteins are formed on the ribosomes.
   (C) mRNA is not necessary for proper protein synthesis.
   (D) Ribosomal RNA is needed for proper binding of the mRNA message.
   (E) Ribosomes are found either in the cytoplasm or attached to the endoplasmic reticulum.

74. Which of the following is an acceptable nitrogen base composition for double-stranded DNA?

   (A) 31% A; 19% T; 31% C; 19% G
   (B) 36% A; 36% U; 24% C, 24% G
   (C) 48% A; 48% T; 52% C; 52% G
   (D) 31% A; 31% T; 19% C; 19% G
   (E) 24% A; 24% U; 36% C; 36% G

75. The genetic code is considered degenerate because

   (A) more than one codon can code for a single amino acid
   (B) one codon can code for multiple amino acids
   (C) more than one anticodon can bind to a given codon
   (D) only one anticodon can bind to a given codon
   (E) none of the above

GO ON TO THE NEXT PAGE

In 1952, Alfred Hershey and Martha Chase conducted a set of experiments to determine whether DNA or proteins were the genetic material of living organisms. The original experiments took advantage of the fact that sulfur is not a component of DNA but is found in most proteins, while phosphorous is a component of DNA and not of proteins. Variations of the Hershey-Chase experiments have been performed by other researchers, as described below. These experiments used lambda bacteriophage, which is a DNA virus. Lambda bacteriophage has both a lytic cycle and a lysogenic cycle.

Experiment 1:

The protein coat of lambda bacteriophage was labeled with radioactive sulfur, ^{35}S. This phage culture was then allowed to infect a culture of *E. coli*. The phage carcasses, called ghosts, were separated from the bacterial cells before any virions could be produced. The bacterial cells were separated by centrifugation and the radioactivity was measured.

Experiment 2:

The phage DNA was labeled with radioactive phosphorus, ^{32}P. This phage culture was then allowed to infect *E. coli*. The phage carcasses, called ghosts, were separated from the bacterial cells before any virions could be produced. The bacterial cells were separated by centrifugation and the radioactivity was measured.

Use this information to answer questions 76–79 below.

76. In Experiment 1, the radio-labeled sulfur subsequently appeared in

(A) *E. coli* proteins
(B) *E. coli* chromosomes
(C) lambda ghost proteins
(D) lambda ghost DNA
(E) none of the above

77. In Experiment 2, most of the radioactivity was found in

(A) lambda ghosts
(B) *E. coli* cells
(C) *E. coli* progeny
(D) lambda progeny
(E) none of the above

78. ^{32}P-labeled phage infected a culture of *E. coli* and entered a lysogenic cycle. The radioactivity of a 1 mL sample of the culture with a concentration of 1 x 105 cells/mL was measured. The *E. coli* were then allowed to undergo 3 rounds of replication. What fraction of the initial radioactivity would be present in a 1 mL sample, which contains cells diluted to a final concentration of 1 x 105 cells/mL, from the culture containing the final generation of *E. coli*?

(A) $\dfrac{1}{8}$

(B) $\dfrac{1}{4}$

(C) $\dfrac{1}{10}$

(D) $\dfrac{1}{3}$

(E) $\dfrac{1}{2}$

GO ON TO THE NEXT PAGE

79.   After centrifugation of the *E. coli* using standard laboratory protocols, all of the following structures would be found EXCEPT

      (A)   ribosomes
      (B)   DNA
      (C)   mitochondria
      (D)   cell walls
      (E)   RNA

In your town, people who drank water from a particular well developed a serious bacterial infection. In order to isolate the causative agent, you perform the following experiment.

You take a water sample from the well and inoculate a nutrient agar plate (Plate I) and a nutrient agar plate containing tetracycline, an antibiotic (Plate II). You also inoculate a nutrient agar plate with a sample you isolated from one of your patients (Plate III). As a control, you inoculate a plate of nutrient agar with distilled water (Plate IV). The results are depicted as the figures below.

Use this information to answer question 80 on the following page.

bacteria 4 in speckled area

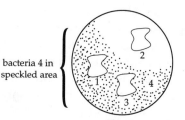

Plate I:
Well water and Nutrient agar

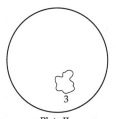

Plate II:
Well water and Nutrient agar + antibiotic (tetracycline)

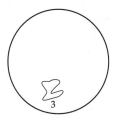

Plate III:
Patient sample on nutrient agar

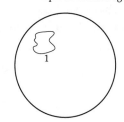

Plate IV:
Distilled water on nutrient agar

GO ON TO THE NEXT PAGE

80. Which of the following explains the growth pattern of Plate I?

   (A) Bacteria 2 produces an antibiotic to which Bacterias 1, 3, and 4 are susceptible.

   (B) Bacteria 2 produces an antibiotic to which Bacteria 4 is susceptible.

   (C) Bacteria 2 produces an antibiotic to which Bacterias 1 and 4 are susceptible.

   (D) Bacteria 4 requires a protein produced by Bacteria 1.

   (E) Bacteria 4 requires a protein produced by Bacteria 3.

STOP

Biology E/M Test Three: M-Option Answer Key

1.	E	17.	B	33.	E	49.	D	65.	B
2.	A	18.	B	34.	D	50.	A	66.	B
3.	E	19.	E	35.	A	51.	B	67.	C
4.	E	20.	B	36.	D	52.	B	68.	A
5.	E	21.	C	37.	C	53.	E	69.	C
6.	B	22.	C	38.	D	54.	D	70.	E
7.	D	23.	D	39.	B	55.	A	71.	D
8.	C	24.	A	40.	B	56.	B	72.	A
9.	D	25.	C	41.	A	57.	B	73.	C
10.	B	26.	E	42.	D	58.	D	74.	D
11.	E	27.	C	43.	D	59.	D	75.	A
12.	D	28.	E	44.	C	60.	B	76.	C
13.	B	29.	C	45.	B	61.	B	77.	B
14.	C	30.	C	46.	D	62.	B	78.	A
15.	A	31.	D	47.	B	63.	E	79.	C
16.	D	32.	C	48.	C	64.	E	80.	B

Compute Your Practice Test Score

Step 1: Figure out your raw score. Refer to your answer sheet for the number right and the number wrong on the practice test you're scoring. (If you haven't checked your answers, do that now, using the answer key that follows the test.) You can use the chart below to figure out your raw score. Multiply the number wrong by 0.25 and subtract the result from the number right. Round the result to the nearest whole number. This is your raw score.

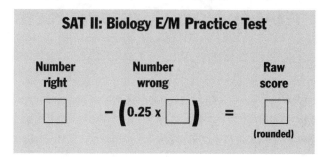

Step 2: Find your practice test score. Find your raw score in the left column of the table below. The score in the right column is an approximation of what your score would be on the SAT II: Biology E/M Test.

A note on your practice test scores: Don't take these scores too literally. Practice test conditions cannot precisely mirror real test conditions. Your actual SAT II: Biology E/M Subject Test score will almost certainly vary from your practice test scores. Your scores on the practice tests will give you a rough idea of your range on the actual exam.

Find Your Practice Test Score

Raw	Scaled	Raw	Scaled	Raw	Scaled	Raw	Scaled	Raw	Scaled
80	800	63	720	46	620	29	530	12	400
79	800	62	710	45	620	28	530	11	390
78	800	61	700	44	610	27	520	10	380
77	800	60	690	43	610	26	520	9	370
76	790	59	690	42	600	25	510	8	360
75	790	58	680	41	600	24	510	7	340
74	790	57	680	40	590	23	500	6	330
73	780	56	670	39	590	22	500	5	320
72	780	55	670	38	580	21	490	4	300
71	770	54	660	37	580	20	490	3	280
70	770	53	660	36	570	19	480	2	260
69	760	52	650	35	570	18	460	1	240
68	760	51	650	34	560	17	450	0	200
67	750	50	640	33	560	16	440		
66	750	49	640	32	550	15	430		
65	740	48	630	31	550	14	420		
64	730	47	630	30	540	13	410		

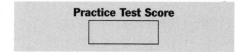

KAPLAN

Biology E/M Test Three: M-Option Answers and Explanations

1. **(E)** All forms of life, from bacteria to man, have certain things in common. They all have some form of nucleic acid as genetic material as well as proteins, and they are composed largely of water.

2. **(A)** Clotting occurs when platelets in an open wound release thromboplastin, which initiates a series of reactions ultimately leading to the formation of a fibrin clot. Thromboplastin, with the aid of calcium and vitamin K as cofactors, converts inactive plasma prothrombin to the active form thrombin in a series of steps. Thrombin in its turn converts fibrinogen (dissolved in plasma) into the fibrinous protein fibrin. Finally, threads of fibrin trap red blood cells to form clots.

3. **(E)** Pepsin, trypsin, carboxypeptidase, and chymotrypsin are all enzymes that digest proteins. Gastrin is a hormone released by the pyloric mucosa when food enters the stomach. It stimulates the secretion of gastric juices.

4. **(E)** The bony skeleton serves as a support system within all vertebrate organisms. Muscles are attached to the bones, permitting movement. The skeleton also provides protection for vital organs. For example, the rib cage protects the heart and the lungs, while the skull and vertebral column protect the brain and the spinal cord. The hollow cavity formed within many bones is filled with bone marrow, the site of formation of blood cells.

5. **(E)** The endosymbiotic hypothesis proposes that primative blue-green algae entered into a symbiotic arrangement with early eukaryotic cells to develop into chloroplasts, while a prokaryote ancestor entered into a similar arrangement with the precursors of eukaryotic cells to become mitochondria. Chloroplasts and mitochondria have circular DNA and genomes like modern prokaryotes. They do not have their own translation systems as well. The fact that they have lipid bilayer membranes does not necessarily support the hypothesis, since other organelles also are membrane bound with no evidence of endosymbiotic origin.

6. **(B)** If the electron transport chain is blocked in a cell that utilizes aerobic respiration to produce ATP, then the amount of ATP that can be produced is decreased. This is because the cell is only able to produce ATP through glycolysis.

7. **(D)** The large amounts of air that come into the upper respiratory passages during panting permit water evaporation from the mucosal surfaces, thereby allowing for heat loss. Panting is turned on by the thermoregulatory centers of the brain that monitor blood temperatures.

8. **(C)** As blood passes through the capillary tuft of the glomerulus, it is filtered as it passes through the capillary walls. Blood cells and protein remain in the blood, while water, salts, glucose, and amino acids are filtered into the Bowman's capsule. Resorption of amino acids, glucose, and salts occurs predominantly in the proximal convoluted tubules. The cells lining the tubules actively transport these materials out of the tubular lumen and into the peritubular capillary network. Movement of these materials produces an osmotic gradient that drives some water out of the tubules through simultaneous passive diffusion. In (D), villi are small projections in the walls of the small intestine that increase the surface area to facilitate absorption of nutrients. As for (E), the ureter is a duct that carries urine from the kidneys to the bladder.

9. **(D)** Cardiac cells have intercalated disk connections joining the cytoplasm between adjoining cardiac muscle cells. Although they have some striations, they are not voluntary. Smooth muscle, on the other hand, has no striations, is mononuclear, and is involuntary. White blood cells have nuclei. In adult humans, red blood cells lack nuclei, in order to make room for as much hemoglobin as possible.

10. (B) A cell that is n (haploid) cannot undergo meiosis to become $\frac{1}{2}n$. (A) and (D) are incorrect, however, because haploid organisms undergo mitosis to grow. An example of such an organism is a braconid wasp. These animals in the haploid form are males (n); females ($2n$) are only formed when a female mates. As for (E), an organism, whether diploid or haploid, must be able to transcribe genes and make ATP to maintain life.

11. (E) This question illustrates two types of test cross. A test cross is performed to determine if a particular dominant individual's phenotype is a homozygous or heterozygous genotype. In this case, there are two possible tall genotypes for the tall plant—TT, the pure homozygous tall, and Tt, the hybrid heterozygous tall. These two individuals would have the same phenotype. To determine the genotype, the unknown tall plant would be mated with a recessive, short plant. If the tall plant and the short plant produce only tall offspring, then it can be assumed that the original tall plant is homozygous TT. If the mating of the unknown tall organism and the short organism produces any short offspring at all, then we know that the original unknown tall was heterozygous. This is because short offspring can be produced only when one short gamete is produced from each parent. In I, crossing an unknown with a Tt will also give you the information you seek, because if any short progeny are produced at all, your unknown must be heterozygous. If all of the offspring are tall, your unknown must be homozygous dominant.

12. (D) The question is asking you where growth occurs in long bone. The answer is in the epiphyseal plates. The epiphyseal plates are regions of cartilaginous cells separating the shaft of the long bone, called the diaphysis, from its two dilated ends, called the epiphyses. The epiphyseal plates are located at either end of long bone, which are referred to as the proximal and distal ends. Since growth occurs only at the epiphyseal plates, the strontium would be incorporated near the plates.

13. (B) Blood leaving the kidneys travels through the inferior vena cava before entering the right atrium. All of the blood from the lower half of the body is collected in the inferior vena cava. This vessel merges with the superior vena cava (which collects blood from the upper half of the body) immediately before it enters the right atrium. In (A), the iliac veins circulate blood in each leg and return it to the more central portions of the body before joining to become the common iliac vein, which then enters the inferior vena cava. As for (C) and (D), blood enters the liver through the hepatic portal artery after absorbing monosaccharides and amino acids from the small intestine. It leaves the liver through the hepatic portal vein after filtration occurs. Finally, in (E), the pulmonary vein brings blood back from the lungs to the left atrium. This blood has already reached the heart via the right atrium.

14. (C) The vitamin K we need for blood coagulation is synthesized by the intestinal bacteria residing in our colons. Vitamin K is required by the liver to synthesize some important blood factors. Deficiency of Vitamin K leads to defective blood clotting.

15. (A) Bicarbonate ions are secreted by the pancreas into the small intestine and act to neutralize the acid of the stomach as it passes into the intestine. Bicarbonate ion is involved in transporting CO_2 to the lungs, but not O_2 (C). Also, this does not occur in the digestive tract.

16. (D) Fermentation is a process that occurs during anaerobic respiration in organisms such as yeast. Glucose is converted to pyruvic acid, producing ATP. Then pyruvic acid is changed into ethyl alcohol, a waste product of the fermentation process. Fermentation produces energy but does not require oxygen. The oxygen-requiring reactions of respiration occur in aerobic respiration, not in fermentation, and they occur as NADH and $FADH_2$ molecules produced in the Krebs cycle are sent to the electron-transport chain for the production of ATPs. The final electron acceptor in these reactions is oxygen. In fermentation, there is no aerobic stage, and ATP is never produced through a Krebs cycle and electron transport chain mechanism.

17. (B) This question is a matter of knowing the pathway that blood travels in the heart. Deoxygenated blood drains into the right atrium from both the inferior vena cava and the superior vena cava. From the right atrium, the blood flows into the right ventricle, which then pumps it to the lungs via the pulmonary arteries. Carbon dioxide is exchanged for oxygen in the alveoli of the lungs. Oxygenated blood is returned to the left atrium via the pulmonary veins. From the left atrium, the blood drains into the left ventricle, which pumps it into the aorta for circulation throughout the body. Hence if a tracer substance is injected into the superior vena cava, it would take the longest amount of time to reach the left ventricle.

18. (B) The heart internally generates the rhythmic stimulation that triggers each heartbeat. The impulse for each heartbeat begins in the sinoatrial node, located in the wall of the right atrium at the approximate point at which the vena cava enters and travels through the atria. It is then picked up by the AV nodes and carried to the AV bundle (also known as the bundle of His) and transported through the ventricles through the Purkinje fibers. (A) and (C) are structures in the fetal heart that ensure that blood is shunted away from the fetus' developing lungs. The vagus nerve (E) is a cranial nerve that regulates the heartbeat in accordance with signals from the parasympathetic nervous system, although it does not determine the heartbeat. The SA node is able to maintain the heartbeat without any stimulation from the nervous system.

19. (E) Insulin is the hormone secreted by the beta cells of the pancreas in response to high blood glucose levels. Insulin decreases blood glucose by stimulating cells to uptake glucose, and by stimulating the conversion of glucose into its storage form, glycogen, in the liver and muscle cells. An overdose of insulin can, and often does, lead to a sharp decrease in blood glucose concentration.

20. (B) Identical twins are produced when a zygote formed by one egg and one sperm splits during the four- or eight-cell stage to develop into two genetically identical organisms. Identical twins (A) will always be of the same sex, but fraternal twins can also be of the same sex. The twins in (D) are termed fraternal twins and are no more genetically alike than siblings. (C) and (E), meanwhile, are impossible events.

21. (C) An amphioxus is a chordate but not a vertebrate. Chordates have a stiff dorsal rod called the notochord during a certain period of their embryological development, and amphioxus and tunicate worms do not lose their notochords. Chordates also have paired gill slits, a dorsal hollow nerve cord, and a tail extending beyond the anus at some point during development. Vertebrates, meanwhile, have bones called vertebrae that form the backbone. Bony vertebrae replace the notochord of the embryo and protect the nerve cord. Mammals, amphibians, reptiles, birds, and fish all possess these vertebrae. Let's touch on the other choices here: the duckbilled platypus (A) is an example of an egg-laying mammal; the turtle (B) is an example of a reptile; the trout (D) is a fish, and rabbits (E) are mammals (all mammals are vertebrates).

22. (C) Acetylcholine is a neurotransmitter, which causes depolarization of the postsynaptic membrane of one neuron when released by the presynaptic terminal of another neuron. Acetylcholine is removed from the synapse by the enzyme acetyl cholinesterase. Anticholinesterase inhibitors like parathion disrupt this activity, meaning that the acetylcholine can't be degraded and the concentration of acetylcholine in the synapse will increase.

23. (D) If a homozygous dominant, *TTLL*, is crossed with any genotype, the offspring will be heterozygous and have the dominant phenotype. All the other options have potential for some homozygous recessives.

24. (A) Red blood cells are produced in the bone marrow. They lose their nuclei to make room for more hemoglobin, which means that they cannot reproduce, repair themselves, or make proteins. Red blood cells actually greatly outnumber leukocytes (white blood cells) (C); as for (D), the spleen stores a reservoir of red blood cells and acts as a biological and physical filter for the blood, but it does not make red blood cells.

25. (C) The notochord appears as a semirigid chord in the dorsal part of all chordates sometime during embryonic development. In lower chordates, this chord remains as a semirigid chord, although in higher chordates it is seen only in the embryo and not as a vestigial organ. Echinoderms (B) do not possess a notochord. As for (D), the notochord remains in the lower chordates, such as the amphioxus and the tunicate worm, while the notochord (E) is not part of the nervous system.

26. (E) During meiosis, the gamete reduces its genetic component from $2n$ to n, resulting in a haploid cell with half the normal chromosome number. When a haploid egg and sperm unite, they form a diploid organism known as a zygote. All ova will contain an X chromosome and all sperm will contain either an X or a Y chromosome. These gametes are formed during the two reductional divisions called meiosis. During metaphase I of meiosis I, tetrads form, and sister chromatids undergo the homologous recombination known as crossing over.

27. (C) Alveoli are thin air sacs that act as the sites of air exchange between the environment and the blood via passive diffusion. The bronchi (E) are the two main branches of the air intake pathway. One bronchus goes to each lung, and each bronchus is divided into smaller sections termed bronchioles. The trachea (D) is the region of the air intake pathway located between the glottis and the bronchi. It is also known as the windpipe.

28. (E) Pheromones like primer or releaser pheromones are substances secreted by organisms that result in long-term change in the behavior of other members of that organism's species.

29. (C) The breathing center in the medulla oblongata monitors the increase in CO_2 through its sensory cells. It will also detect a decrease in pH in the blood, which is indicative of an increase of CO_2 levels in the blood. A decrease in O_2 is monitored peripherally by chemoreceptors, located in the carotid bodies in the carotid arteries and in the aortic bodies in the aorta. In (A), the cerebrum is involved in sensory interpretation, memory, and thought, while the cerebellum (B) is involved in fine motor coordination, balance, and equilibrium. Finally, the spinal cord (D) relays sensory and motor information to and from the brain, and the hypothalamus (E) regulates hunger, thirst, body temperature, sex drive, and emotion.

30. (C) Enzymes are catalysts of biological reactions, increasing the rate of reactions without themselves changing and without changing the final equilibrium. Enzymes are typically proteins (I), although they may rarely be RNA, and they typically act most effectively at a physiological pH of 7–7.4 (except for the enzymes which break down protein in the stomach, which act most effectively in an acidic environment) (II). Enzymes are never changed during a reaction (III). They increase the rate of a reaction, but are not themselves affected by that reaction. Enzymes are found throughout the cell, not just in the nucleus.

31. (D) In the nucleus, DNA is produced during cell division, while RNA is produced by transcription of DNA. mRNA travels from the nucleus into the cytoplasm, where it is translated into polypeptides on the ribosomes.

32. (C) The plasma membrane separates the cellular contents from the environment. It is responsible for the permeability of the membrane—in other words, for what is allowed in and out of the cell.

The fluid mosaic model of the plasma membrane states that this membrane is a bilayer of phospholipid interspersed with proteins acting as receptors, pores, and channels. The pores and channels cross the entire membrane. The cytochrome chain referred to in the correct answer is actually located in the cristae (the inner membrane) of the mitochondria.

33. (E) A point mutation changes the sequence of one nucleotide; a silent mutation (B), on the other hand, affects the DNA sequence, but does not affect the protein produced from that DNA sequence. An insertion (C) is an addition of nucleotides which often leads to a frameshift mutation (D) in which the reading frame of the DNA is altered and the protein produced is rendered nonfunctional.

34. (D) As an exocrine gland, the pancreas secretes proteases, lipases, and amylases that aid in the digestion of food, as well as bicarbonate ion that buffers the pH of the chyme coming from the stomach. Glucagon, meanwhile, is an endocrine gland secreted by the pancreas in response to a low blood glucose level. It causes an increase in the levels of glucose by degrading glycogen and decreasing the uptake of glucose by muscles.

35. (A) The phloem, the thin-walled cells on the outside of the vascular bundle, transports nutrients down the stem of the plant.

36. (D) The xylem, the thick-walled, usually hollow cells on the inside of the vascular bundles, transports water and minerals up the plant stem.

37. (C) The cambium is two cells thick and located in between the xylem and phloem. These actively dividing, undifferentiated cells give rise to the xylem and the phloem.

38. (D) The pith is the innermost tissue layer that is used for the storage of nutrients and plant support.

39. (B) Bacteria contain extrachromosomal circular DNA called plasmids. They often code for genes that code for antibiotic resistance or other things that increase the fitness of the bacteria.

40. (B) Bacteria are prokaryotes. These organisms have a plasma membrane, and lack membrane-bound organelles and a true nucleus.

41. (A) Viruses are composed of genetic material, either DNA or RNA, enclosed in a protein coat called a capsid. They are obligate parasites in that they cannot replicate unless they have infected a host cell.

42. (D) Planaria are bilaterally symmetrical, which implies that they have two sides that are mirror images of each other.

43. (D) The electron transport chain uses a proton pump to drive the ATP synthase that produces ATP. It is found in the inner mitochondrial membrane of the mitochondria. The energy of the proton gradient is harvested by the cell through the production of large quantities of ATP.

44. (C) The Krebs cycle occurs in the mitochondrial matrix.

45. (B) Fermentation is anaerobic and regenerates NAD^+ for glycolysis.

46. (D) See the explanation to question 43.

47. (B) The byproducts of fermentation are ethanol (in prokaryotes and some yeast) and lactic acid (in most of the other eukaryotes, including humans). The buildup of lactic acid accounts for the soreness of your muscles after a tough workout.

48. (C) The function of a circulatory system is to transport gases, nutrients, and wastes throughout the body.

49. (D) The nervous system receives input from the five senses. This input is analyzed by the central nervous system and acted upon by the motor neurons. This allows organisms to receive and respond to stimuli.

50. (A) The musculoskeletal system forms the basic framework for the human body. Everything else is either built upon this framework or protected within it.

51. (B) The hormones secreted by the various endocrine organs allow tissues and organs to communicate. This is crucial for the maintenance of homeostasis, reproduction, and other life functions.

52. (B) Guanine (along with adenine) is a purine that binds cytosine (a pyrimidine) via three hydrogen bonds.

53. (E) Uracil is a pyrimidine found only in RNA.

54. (D) See the explanation for question 52.

55. (A) Adenine (along with guanine) is a purine that binds thymine with two hydrogen bonds in DNA and uracil with two hydrogen bonds in RNA.

56. (B) In order for Hardy-Weinberg equilibrium to be maintained, a large population, no net mutations, no migration, and random mating must all occur. If these criteria are met, then the allele frequencies for two alleles must equal 1, $p + q = 1$. The phenotypic frequencies must also equal one in the following equation: $p^2 + 2pq + q^2 = 1$. In this case, $T = 0.8$, so $p = 0.8$. $0.8 + q = 1$, so $q = 0.2$. Therefore, the allele frequency of t is 20 percent, or 0.2.

57. (B) The percentage of heterozygous individuals is defined by the $2pq$ term. So, $2(0.8)(0.2) =$ the percentage of heterozygous individuals, which comes to 0.32 or 32 percent.

58. (D) Individuals that can curl their tongues have at least one dominant allele. This means that you are looking for the number of p^2 individuals, who have two dominant alleles, and the number of $2pq$ individuals, who have only one dominant allele. These two numbers added together will give you the number of individuals that can curl their tongues. Therefore, $(0.8)2 + 2(0.8)(0.2) = 0.96$ or 96 percent.

59. (D) The bacteria that have the largest population are growing most optimally. In this experiment, the optimum concentration is at 40 U/mL.

60. (B) The second time the mouse tried the maze, he cut his time in half, from 40 to 20 seconds or a 50 percent increase. The next largest drop in time is after the third trial, when the mouse shaved 9 seconds off his previous time.

Molecular Section

61. (B) Hexokinase, like all enzymes, reduces the activation energy of a reaction to make it go faster, but it does not change the end point, the equilibrium, of the reaction. Enzymes are not consumed in a reaction (D) and catalyze both the forward and the reverse reaction (E).

62. (B) In the graph, the reaction rate increases linearly with increasing enzyme. Doubling the amount of enzyme will double the rate of the reaction. ADP is one of the products of the reaction, along with glucose 6-phosphate, so its production will double with twice the enzyme present.

63. (E) Enzymes catalyze reactions at a specific active site in the protein structure. As more substrate is added (glucose), the active sites become full, or saturated. In the graph, the reaction rate increases as more glucose is added at first, then levels off. The graph indicates that there is no increase in the reaction rate above 3 uM glucose, because the active sites of the enzyme are full. If the reaction is saturated, then adding more glucose will not change either glucose 6-phosphate production or ATP consumption.

64. (E) DNA replication is semiconservative in that the two daughter double helices each receive one strand from the parent DNA. DNA replication occurs at Interphase, but only in a 5′ to 3′ direction, not 3′ to 5′. This forms Okasaki fragments on the leading strand. Purines (such as adenine and guanine) always bind with pyrimidines (like thymine and cytosine).

65. (B) In glycolysis and the Krebs cycle, energy is extracted from glucose and transferred to NADH and $FADH_2$. In electron transport, the energy of NADH and $FADH_2$ is transferred to a series of proteins in the inner mitochondrial membrane, with O_2 as the final electron acceptor at the end of the chain. This electron transport chain pumps protons out of the mitochondria, creating a proton gradient. The energy of the gradient drives the ATP synthesis. Each NADH leads to three ATP and each $FADH_2$ creates two. Without O_2 there is no electron transport or oxidative phosphorylation.

66. (B) Without antibiotics, both strains reached their maximum growth potential.

67. (C) With the antibiotics in Graph 2, only the *E. coli* strain that has the plasmid reached its maximum growth potential. The strain that did not have the plasmid could not grow.

68. (A) The plasmid must allow the bacteria to grow in the presence of an antibacterial compound such as tetracycline, probably by activating the antibiotic or giving the bacteria a way to get around the inhibition of the antibiotic.

69. (C) Plasmids are circular pieces of extra chromosomal DNA, and uracil is a component of RNA not found in DNA.

70. (E) Natural selection states that fitter organisms have a competitive advantage over unfit organisms where fitness is defined by the ability to reproduce. Therefore, a bacteria that can grow in the presence of an antibiotic such as tetracycline is more fit and has a competitive advantage over a strain that does not contain this plasmid when both are grown in the presence of tetracycline.

71. (D) The first set of reactions are anaerobic and involve the breakdown of glucose into pyruvate (A). Two ATP molecules are required and four are produced, leaving a net total of 2 ATP/glucose (B). Pyruvic acid will either become acetyl CoA and enter the Krebs cycle during aerobic respiration, or will become lactic acid or ethanol in anaerobic respiration (C). Two NADH molecules are produced by glycolysis; these will donate their electrons to compound Q in the electron transport chain (E). $FADH_2$, however, is not produced until the Krebs cycle.

72. (A) DNA replication is semiconservative. In consequence, after the DNA has been replicated in the medium containing a radioactive tracer, every helix of DNA will have at least one strand that will have incorporated some of the radioactivity. After two rounds of DNA replication, there will be four copies of DNA, two that have the radioactive tracer in both strands and two that have the tracer in only one of the strands.

73. (C) Protein synthesis does require mRNA. tRNA (A) brings the amino acid to the ribosome, where it interacts with the mRNA that has the appropriate sequence. As for (B), tRNA molecules do in fact have an amino acid bound to their 3' end. The mRNA is read from 5' to 3' as the ribosome moves along the message. (D) and (E) are also both true.

74. (D) Because DNA is double stranded and because of the rules of complementary base pairing, the quantity of adenine must equal the quantity of thymine, and the quantity of cytosine must equal the quantity of guanine. The reason for this is that in DNA, adenine will bind only with thymine, while cytosine binds only with guanine. After eliminating the answer choices that do not follow this rule, you can eliminate any answer choices that mention uracil, since it is found only in RNA. Finally, to reach the correct answer, you must add up your percentages and make sure that they equal 100 percent of the bases in the DNA strand.

75. (A) Genetic codes are referred to as "degenerate" because more than one codon can code for each of an amino acid. For example, both UAU and UAC code for tyrosine. Since there are three bases in a codon, there are 64 possible codons and only 20 amino acids. (B) is the opposite of answer choice (A) and therefore untrue. It is essential that the anticodon be the exact base pair complement of the codon to maintain protein integrity. (D) is true, but it does not account for the degeneracy of the genetic code.

76. (C) When the ^{32}P labeled phage were used in Experiment 2, most of the radioactivity ended up inside the bacterial cells, indicating that the phage DNA entered the cells. If the phage had been allowed to enter a lytic cycle, the ^{32}P would also have been recovered in the phage progeny. In Experiment 1, when the ^{35}S-labeled phage were used, most of the radioactive material ended up in the phage ghosts, indicating that the phage protein never entered the bacterial cell. The conclusion to be drawn from these experiments is that DNA is the hereditary material, and that phage proteins are mere structural packaging that is discarded after delivering the viral DNA to the bacterial cell. The entire "life purpose" of a virus is devoted to finding a host cell and getting its nucleic acid inside it. This is crucial for a virus, since it must use the genetic machinery of a host cell to replicate. Since the protein was labeled with ^{35}S, you would expect to find this isotope of sulfur only in the viral ghost, which you're told is the phage carcass. And since the DNA was labeled with ^{32}P, you would expect to find this radioactive isotope only in the bacterial cells.

77. (B) See the explanation to question 76 above.

78. (A) We know that the radio labeled phage DNA has been inserted into the bacterial chromosome and will undergo replication just as if it were native bacterial DNA. During DNA replication, the double helix unwinds, and each strand acts as a template for complementary base-pairing in the synthesis of two new daughter helices. Each new daughter helix contains an intact strand from the parent helix, which in this problem contains radioactivity, and a newly synthesized strand, which contains no radioactivity, since all of its nucleotides were synthesized by the bacterial cell. Thus DNA replication is semiconservative—half of the original DNA is conserved from one generation to the next. Since every bacterial cell division doubles the number of cells and the total amount of DNA, three rounds of replication will increase the total number of cells and the total amount of DNA by a factor of eight, while the amount of radioactivity remains constant. In essence, the ^{32}P is diluted with each round of replication. This means that the fraction of radioactivity in a 1 mL sample taken from the culture of the final generation will be one-eighth the amount of radioactivity taken from a culture of the parental generation, after the cultures are diluted to have the same number of cell/ml.

79. (C) This questions tests your ability to differentiate between prokaryotic and eukaryotic cells. A prokaryote is a unicellular organism that lacks membrane-bound organelles. Mitochondria, which are membrane bound, would not be found in a prokaryotic cell.

80. (B) Bacteria 4 grows all over the nutrient plate, except in the regions surrounding Bacteria 2. The logical conclusion is that Bacteria 2 produces an antibiotic that kills Bacteria 4 (B). We don't know whether or not this antibiotic can also kill Bacterias 1 and 3, since they are not in close enough proximity to Bacteria 2 to be affected by it.

APPENDIX

GLOSSARY

abiotic   nonliving, as in the physical environment

absorption   the process by which water and dissolved substances pass through a membrane

acetylcholine   a transmitter substance released from the axons of nerve cells at the synapse

active immunity   protective immunity to a disease in which the individual produces antibodies as a result of previous exposure to the antigen

adaptation   a behavioral or biological change that enables an organism to adjust to its environment

adaptive radiation   the production of a number of different species from a single ancestral species

adenosine phosphate   adenosine diphosphate (ADP) and adenosine triphoshate (ATP), which are energy storage molecules

ADH (vasopressin)   a hormone that regulates water reabsorption

adipose   fatty tissue, fat-storing tissue, or fat within cells

adrenal cortex   the outer part of the adrenal gland that secretes many hormones, including cortisone and aldosterone

adrenal medulla   the inner part of the adrenal gland that secretes adrenalin

adrenaline (epinephrine)   an "emergency" hormone stimulated by anger or fear; increases blood pressure and heart rate in order to supply the emergency needs of the muscles

adrenocorticotrophic hormone   usually referred to as ACTH and secreted by the anterior lobe of the pituitary gland; stimulates the adrenal cortex to produce its characteristic hormones

aerobe   an organism that requires oxygen for respiration and can live only in the presence of oxygen

aerobic   requiring free oxygen from the atmosphere for normal activity and respiration

aldosterone   hormone active in osmoregulation; a mineral corticoid produced by the adrenal cortex; stimulates reabsorption of Na^+ and secretion of K^+

alimentary canal   an organ centrally involved in the human digestive system

allantois   the extraembryonic membrane of birds, reptiles, and mammals that serves as an area of gaseous exchange and as a site for the storage of noxious excretion products

allele   one of two or more types of genes, each representing a particular trait; many alleles exist for a specific gene locus

alternation of generations   the description of a plant life cycle that consists of a diploid, asexual, sporophyte generation and a haploid, sexual, gametophyte generation

alveolus   an air sac in the lung; the site of respiratory exchange, involving diffusion of oxygen and carbon dioxide between the air in the alveolus and the blood in the capillaries (plural = *alveoli*)

amnion   the extraembryonic membrane in birds, reptiles, and mammals that surrounds the embryo, forming an amniotic sac

amoeboid movement   movement involving the flowing of cytoplasm into pseudopods, as in amoeba

anaerobe   an organism that does not require free oxygen in order to respire

anaerobic   living or active in the absence of free oxygen; pertaining to respiration that is independent of oxygen

analogous   describes structures that have similar function but different evolutionary origins; e.g., a bird's wing and a moth's wing

anaphase   the stage in mitosis that is characterized by the migration of chromatids to opposite ends of the cell; the stage in meiosis during which homologous pairs migrate (Anaphase I), and the stage in meiosis during which chromatids migrate to different ends of the cell (Anaphase II)

androgen   a male sex hormone (e.g., testosterone)

angiosperm   a flowering plant; a plant of the class Angiospermae that produces seeds enclosed in an ovary and is characterized by the possession of fruits and flowers

Annelida   the phylum to which segmented worms belong

anther   the part of the male reproductive organ (the stamen) that produces and stores pollen

antibiotic   an antipathogenic substance (e.g., penicillin)

antibody   globular proteins produced by tissues that destroy or inactivate antigens

antigen   a foreign protein that stimulates the production of antibodies when introduced into the body of an organism

aorta   the largest artery; carries blood from the left ventricle

aortic arch   blood vessels located between ascending and descending aortas that deliver blood to most of upper body

appendage   a structure that extends from the trunk of an organism and is capable of active movements

aqueous humor   fluid in the eye, found between the cornea and the lens

Arachnida   a class of arthropods that includes scorpions, spiders, mites, and ticks

artery   a blood vessel that carries blood away from the heart

Arthropoda   the phylum to which jointed-legged invertebrates belong, including insects, arachnids, and crustaceans

asexual reproduction   the production of daughter cells by means other than the sexual union of gametes (as in budding and binary fission)

assimilation   the conversion of digested foods and other materials into forms usable by the body (i.e., the conversion of amino acids into proteins)

assortative mating   the type of mating that occurs when an organism selects a mating partner that resembles itself

atrium   the thin-walled anterior chamber of the heart (also called the auricle)

autolysis   self-digestion occurring in plant and animal tissues, particularly after they have ceased to function properly

autonomic nervous system   the part of the nervous system that regulates the involuntary muscles, such as the walls of the alimentary canal; includes the parasympathetic and sympathetic nervous systems

autosome   any chromosome that is not a sex chromosome

autotroph   an organism that utilizes the energy of inorganic materials such as water and carbon dioxide or the sun to manufacture organic materials; examples of autotrophs include plants

auxin   a plant growth hormone

axon   a nerve fiber

bacillus   bacteria that are rod shaped

bacteriophage   a type of virus that can destroy bacteria by infecting, parasitizing, and eventually killing them

bile   an emulsifying agent secreted by the liver

bile salts   compounds in bile that aid in emulsification

binary fission   asexual reproduction; in this process, the parent organism splits into two equal daughter cells

binomial nomenclature   the system of naming an organism by its genus and species names *biome*   a habitat zone, such as desert, grassland, or tundra

biotic   living, as in living organisms in the environment

blastula   a stage of embryonic development in which the embryo consists of a hollow ball of cells

Bowman's capsule   part of the nephron in the kidney; involved in excretion

bud   in plants, an area of undifferentiated tissue covered by embryonic leaves

budding   a process of asexual reproduction in which the offspring develop from an outgrowth of the plant or animal

buffer   a substance that prevents appreciable changes in pH in solutions to which small quantities of acids or bases are added

calorie   a unit of heat; the amount of heat required to raise the temperature of one gram of water by one degree centigrade (Note: a large Calorie (food calorie) = 1000 calories)

Calvin cycle   cycle in photosynthesis that reduces fixed carbon to carbohydrates through the addition of electrons (also known as the "dark cycle")

cambium   undifferentiated tissue in the stem of a plant that aids growth in width

capillary   a tube one cell thick that carries blood from artery to vein; the site of material exchange between the blood and tissues of the body

carapace   a bony or chitinous case or shield covering the back or part of the back of an animal (e.g., the shell of a crab)

carbohydrate   an organic compound to which hydrogen and oxygen are attached; the hydrogen and oxygen are in a 2:1 ratio; examples include sugars, starches, and cellulose

carbon cycle   the recycling of carbon from decaying organisms for use in future generations

carnivore   a flesh-eating animal; a holotrophic animal that subsists on other animals or parts of animals

carotene   an orange plant pigment that is the precursor of Vitamin A

cation   an ion with a positive charge, or an ion that migrates towards the cathode (negative electrode) in an electric field

cell wall   a wall composed of cellulose that is external to the cell membrane in plants; it is primarily involved in support and in the maintenance of proper internal pressure

cell wall plate   in mitosis of higher plants, the structure that forms between the divided nuclei of the two daughter cells and eventually becomes the cell wall

central nervous system (CNS)   encompasses the brain and the spinal cord

centriole   the small granular body within the centrosome to which the spindle fibers attach

centromere   the place of attachment of the mitotic fiber to the chromosome

centrosome   a structure in animal cells containing centrioles from which the spindle fibers develop

cephalic   pertaining to the head

cerebellum   the hindbrain region that controls equilibrium and muscular coordination

cerebral cortex   the outer layer of cerebral hemispheres in the forebrain, consisting of gray matter

cerebral hemisphere   one of the paired lateral divisions of the forebrain

cerebrum   the largest portion of the human brain; it is believed to be the center of intelligence, conscious thought, and sensation

chemosynthesis   the process by which carbohydrates are formed through chemical energy; found in bacteria *chemotropism*   the orientation of cells or organisms in relation to chemical stimuli; the growth or movement response of organisms to chemical stimuli

chitin   a white or colorless, amorphous, horny substance that forms part of the outer integument of insects, crustaceans, and some other invertebrates; it also occurs in certain fungi

chlorophyll   a green pigment that performs essential functions as an electron donor and light "entrapper" in photosynthesis

chloroplast   a plastid containing chlorophyll

Chordata   an animal phylum in which all members have a notochord, dorsal nerve cord, and pharyngeal gill slits at some embryonic stage; includes the Cephalochordata and the Vertebrates

chorion   the outermost, extra-embryonic membrane of reptiles and birds

chromatid   one of the two strands that constitute a chromosome; chromatids are held together by the centromere

chromatin   a nuclear protein of chromosomes that stains readily

chromosome   a short, stubby rod consisting of chromatin that is found in the nucleus of cells; contains the genetic or hereditary component of cells (in the form of genes)

chyme   partially digested food in the stomach

circadian rhythms   daily cycles of behavior

cleavage   the division in animal cell cytoplasm caused by the pinching in of the cell membrane

climax community   the stable, biotic part of the ecosystem in which populations exist in balance with each other and with the environment

clotting   the coagulation of blood caused by the rupture of platelets and the interaction of fibrin, fibrinogen, thrombin, prothrombin, and calcium ions

cloaca   the chamber in the alimentary canal of certain vertebrates located below the large intestine, into which the ureter and reproductive organs empty (as in frogs)

cochlea   the sensory organ of the inner ear of mammals; it is coiled and contains the organ of Corti

codominant   the state in which two genetic traits are fully expressed and neither dominates

Coelenterata (Cnidarians)   an invertebrate animal phylum in which animals possess a single alimentary opening and tentacles with stinging cells; examples are jellyfish, corals, sea anemones, and hydra

coelom   the space between the mesodermal layers that forms the body cavity of some animal phyla

coenzyme   an organic cofactor required for enzyme activity

colon   the large intestine

commensal   describes an organism that lives symbiotically with a host; this host neither benefits nor suffers from the association

conditioning   the association of a physical, visceral response with an environmental stimulus with which it is not naturally associated; a learned response

cone   a cell in the retina that is sensitive to colors and is responsible for color vision

consumer   organism that consumes food from outside itself instead of producing it (primary, secondary, and tertiary)

contractile vacuole   a specialized structure that controls osmotic pressure by removing water from the cell; exists in protozoans

cornea   the outer, transparent layer of the eye

corpus callosum   a tract of nerve fibers connecting the two cerebral hemispheres

corpus luteum   a remnant of follicle after ovulation that secretes the hormone progesterone

cortex   in plants, the tissue between the epidermis and the vascular cylinder in the roots and stems of plants; in animals, the outer tissue of some organs *cortisone*   a hormonal secretion of the adrenal cortex

cotyledon   a "seed leaf"; responsible for food digestion and storage in a plant embryo

cretinism   a thyroid deficiency that results in stunted growth and feeblemindedness

crossing over   the exchange of parts of homologous chromosomes during meiosis

cross-pollination   the pollination of the pistil of one flower with pollen from the stamen of a different flower of the same species

Crustacea   crustaceans; a large class of arthropods, including crabs and lobsters

cuticle   a waxy protective layer secreted by the outer surface of plants, insects, etcetera

cytochrome   a hydrogen carrier containing iron that functions in many cellular processes, including respiration

cytokinesis   a process by which the cytoplasm and the organelles of the cell divide; the final stage of mitosis

cyton   the cell body of a neuron

cytoplasm   the living matter of a cell, located between the cell membrane and the nucleus

cytoskeleton   the organelle that provides mechanical support and carries out motility functions for the cell

cytosine   a nitrogen base that is present in nucleotides and nucleic acids; it is paired with guanine

deamination   the removal of an amino group from an organism, particularly from an amino acid

deletion   the loss of all or part of a chromosome

deme   a small, local population

dendrite   the part of the neuron that transmits impulses to the cell body

deoxyribose   a five carbon sugar that has one oxygen atom less than ribose; a component of DNA (deoxyribose nucleic acid)

diastole   the passive, rhythmical expansion or dilation of the cavities of the heart (atria or ventricles) that allows these organs to fill with blood; preceded and followed by systole (contraction)

dicotyledon   a plant that has two seed leaves or cotyledons

diencephalon   the hind portion of the forebrain of vertebrates

differentiation   a progressive change from which a permanently more mature or advanced state results; for example, a relatively unspecialized cell's development into a more specialized one

diffusion   the movement of particles from one place to another as a result of their random motion

digestion   the process of breaking down large organic molecules into smaller ones

dihybrid   an organism that is heterozygous for two different traits

dimorphism   the instance of polymorphism in which there is a difference of form between two members of a species, as between males and females

diploid   describes cells that have a double set of chromosomes in homologous pairs ($2n$)

disaccharide   a sugar composed of two combined monosaccharides (e.g., sucrose, lactose)

disjunction   the separation of homologous pairs of chromosomes following meiotic synapsis

DNA   deoxyribonucleic acid; found in the cell nucleus, its basic unit is the nucleotide; contains coded genetic information; can replicate on the basis of heredity

dominance   a dominant allele suppresses the expression of the other member of an allele pair when both members are present; a dominant gene exerts its full effect regardless of the effect of its allelic partner

dorsal root   the sensory branch of each spinal nerve

duodenum   the most anterior portion of the small intestine of vertebrates, adjacent to the stomach; the continuation of the stomach into which the bile duct and pancreatic duct empty

ecological succession   the orderly process by which one biotic community replaces another until a climax community is established

ecology   the study of organisms in relation to their environment

ectoderm   the outermost embryonic germ layer that gives rise to the epidermis and the nervous system

egg (ovum)   the female gamete; it is nonmotile, large in comparison to male gametes, and stores nutrients

electron transport chain   a complex carrier mechanism located on the inside of the inner mitochondrial membrane of the cell; releases energy, and is used to form ATP

embolus   a blood clot that is formed within a blood vessel

emulsion   a colloidal system involving the dispersion of a liquid within a liquid

endemic   pertaining to a restricted locality; ecologically, occurring only in one particular region

endocrine gland   a ductless gland that secretes hormones directly into the bloodstream

endocytosis   a process by which the cell membrane is invaginated to form a vesicle which contains extracellular medium

endoderm   the innermost embryonic germ layer that gives rise to the lining of the alimentary canal and to the digestive and respiratory organs

endoplasmic reticulum   a network of membrane-enclosed spaces connected with the nuclear membrane; transports materials through the cell; can be soft or rough

enzyme   an organic catalyst and protein

endoplasm   the inner portion of the cytoplasm of a cell or the portion that surrounds the nucleus

endosperm   the triploid tissue in some seeds that contains stored food and is formed by the union of one sperm nucleus with two nuclei of the female's gametophyte

epidermis   the outermost surface of an organism

epididymis   the coiled part of the sperm duct, adjacent to the testes in mammals

epiglottis   in mammals, a flap of tissue above the glottis; it folds back over the glottis in swallowing to close the air passages of the lungs; contains elastic cartilage

epicotyl   the portion of seed plant embryo above the cotyledon

epinephrine   see *adrenalin*

epithelium   the cellular layer that covers external and internal surfaces

epiphyte   a plant that lives on another plant commensalistically

erythrocyte   an anucleate red blood cell that contains hemoglobin

esophagus   the portion of alimentary canal connecting the pharynx and the stomach

estrogen   a female sex hormone secreted by the follicle

ethanol fermentation   a form of anaerobic respiration found in yeast and bacteria

ethylene   a hormone that ripens fruit and induces aging

eukaryote   multicellular organism

Eustachian tube   an air duct from the middle ear to the throat that equalizes external and internal air pressure

excretion   the elimination of metabolic waste matter

exocrine   pertaining to a type of gland that releases its secretion though a duct; e.g., the salivary gland, the liver

exocytosis   a process by which the vesicle in the cell fuses with the cell membrane and releases its contents to the outside

exoskeleton   describes arthropods and other animals whose skeletal or supporting structures are outside the skin

eye   a sensory organ capable of detecting light

F_1 the first filial generation (first offspring)

F_2 the second filial generation; offspring resulting from the crossing of individuals of the F_1 generation

fallopian tube   the mammalian oviduct that leads from the ovaries to the uterus

feedback mechanism   the process by which a certain function is regulated by the amount of the substance it produces

femur   the thigh bone of vertebrates

fermentation   anaerobic respiration that yields 2 molecules of ATP, lactic acid, ethyl alcohol and carbon dioxide, or some similar compound via the glycolytic pathway

fertilization   the fusion of sperm and the egg to produce a zygote

fibrin   protein threads that form in the blood during clotting

fibrinogen   blood protein that is transformed to fibrin upon clotting

fitness   the ability of an organism to contribute its alleles and therefore its phenotypic traits to future generations

flagellate   an organism that possesses one or more whiplike appendages called flagella

flagellum   a microscopic, whiplike filament that serves as a locomotor structure in flagellate cells

follicle   the sac in the ovary in which the egg develops

food vacuole   a vacuole in the cytoplasm in which digestion takes place (in protozoans)

frame shift mutation   a mutation involving the addition or loss of nucleotides

fruit   a mature ovary

FSH   an anterior pituitary hormone that stimulates the follicles in females and the function of the seminiferous tubules in males

functional groups   chemical groups attached to carbon skeletons that give compounds their functionality

gall bladder   an organ that stores bile

gamete   a sex or reproductive cell that must fuse with another of the opposite type to form a zygote, which subsequently develops into a new organism

gametophyte   the haploid, sexual stage in the life cycle of plants (alternation of generations)

ganglion   a grouping of neuron cell bodies that acts as a coordinating center

gastrula   a stage of embryonic development characterized by the differentiation of the cells into the ectoderm and endoderm germ layers and by the formation of the archenteron; can be two-layer or three-layer

gene   the portion of a DNA molecule that serves as a unit of heredity; found on the chromosome

gene frequency   a decimal fraction that represents the presence of an allele for all members of a population that have a particular gene locus

genetic code   a four-letter code made up of the DNA nitrogen bases A, T, G, and C; each chromosome is made up of thousands of these bases

genetic drift   random evolutionary changes in the genetic makeup of a (usually small) population

genotype the genetic makeup of an organism without regard to its physical appearance; a homozygous dominant and a heterozygous organism may have the same appearance but different genotypes

genus in taxonomy, a classification between species and family; a group of very closely related species, e.g., homo, felis

geographical barrier any physical feature that prevents the ecological niches of different organisms (not necessarily different species) from overlapping

geotropism any movement or growth of a living organism in response to the force of gravity

germ cell a reproductive cell

germ layer one of the primary tissues of the embryo

gibberellin a hormone that stimulates plant stem elongation

gill slit a perforation leading from the pharynx to the outside environment that is a characteristic of chordates at one stage of their development

glomerulus a network of capillaries in the Bowman's capsules of the kidney

glottis in mammals, the slitlike opening formed by the vocal folds in the larynx

glycogen a starch form in animals; glucose is converted to glycogen in the liver

glycolysis the anaerobic respiration of carbohydrates

goiter (simple) an enlargement of the thyroid gland due to lack of iodine

Golgi apparatus membranous organelles involved in the storage and modification of secretory products

gonads the reproductive organ that produces sex cells (e.g., ovary, testes)

Graffian follicle the cavity in the mammalian ovary in which the egg ripens

granum the smallest particle that is capable of carrying out photosynthesis; the functional unit of a chloroplast

gray matter a portion of the CNS consisting of cytons (cell bodies), their dendrites, and synaptic connections

guanine a purine (nitrogenous base) component of nucleotides and nucleic acids; it links up with cytosine in DNA

guard cell one of a pair of kidney-shaped cells that surround a stomate and regulate the size of the stomate in a leaf

gymnosperm a plant that belongs to the class of seed plants in which the seeds are not enclosed in an ovary; includes the conifers

haploid describes cells (gametes) that have half the chromosome number typical of the species (n chromosome number)

hemoglobin a protein compound containing iron that is found in red blood cells; hemoglobin pigment combines with oxygen and gives the red blood cells their respiratory function

hepatic portal system the veins that carry blood from the digestive organs to the liver

herbivore a plant-eating animal

hermaphrodite an organism that possesses both the male and the female reproductive organs

heterotroph an organism that must get its inorganic and organic raw materials from the environment; a consumer

heterozygous describes an individual that possesses two contrasting alleles for a given trait (*Tt*)

homeotherm an animal with a constant body temperature

homologous describes two or more structures that have similar forms, positions, and origins despite the differences between their current functions; examples are the arm of a human, the flipper of a dolphin, and the foreleg of a horse

homozygous describes an individual that has the same gene for the same trait on each homologous chromosome (*TT* or *tt*)

hormone a chemical messenger that is secreted by one part of the body and carried by the blood to affect another part of the body, usually a muscle or gland

host any organism that is the victim of a parasite

humerus a bone of the upper arm

hybrid an offspring that is heterozygous for one or more gene pairs

hydrostatic skeleton fluid skeleton of annelids

hyperthyroidism an oversecretion of thyroid that leads to high metabolism and exophthalmia goiter

hypertonic describes a fluid that has a higher osmotic pressure than another fluid it is compared to; it exerts greater osmotic pull than the fluid on the other side of a semipermeable membrane; hence, it possesses a greater concentration of particles, and acquires water during osmosis

hypocotyl the portion of the embryonic seed plant below the point of attachment of the cotyledon; forms the root

hypothalamus a section of the posterior forebrain associated with the pituitary gland

hypotonic describes a fluid that has a lower osmotic pressure than a fluid it is compared to; it exerts lesser osmotic pull than the fluid on the other side of a semipermeable membrane; hence, it possesses a lesser concentration of particles, and loses water during osmosis

ilium the dorsal part of the hip girdle

immunity a resistance to disease developed through immune system

imprinting the process by which environmental patterns or objects presented to a developing organism during a "critical period" of its growth is accepted as a permanent element of its behavior.

incomplete dominance genetic blending; each allele exerts some influence on the phenotype (for example, red and white parents may yield pink offspring)

independent assortment the law by which genes on different chromosomes are inherited independently of each other

ingestion the intake of food from the environment into the alimentary canal

inner ear a fluid-filled sensory apparatus that aids in balance and hearing

insulin a hormone produced by the Islets of Langerhans in the pancreas; regulates blood sugar concentration by converting glucose to glycogen (in the process lowering glucose level)

integument refers to protective covering, such as the covering of an ovule, that develops into the seed coat, or an animal's skin

interphase a metabolic stage between mitoses in which genetic material is reproduced

interstitial cells cells which in the female are located between the ovarian follicles, and in the male are located between the seminiferous tubules of the testes; in both cases, these cells produce male sex hormones

inversion occurs when a segment of genetic material on a chromosome becomes reversed

iris the colored part of the eye that is capable of contracting and regulating the size of the pupils

irritability the ability to respond to a stimulus

isolation the separation of some members of a population from the rest of their species; prevents interbreeding and may lead to the development of a new species

isomer one of a group of compounds that is identical in atomic composition, but different in structure or arrangement

isotonic   describes a fluid that has the same osmotic pressure as a fluid it is compared to; it exerts the same osmotic pull as the fluid on the other side of a semipermeable membrane; hence it neither gains nor loses net water during osmosis, and possesses the same concentration of particles before and after osmosis occurs

Krebs cycle   process of aerobic respiration that fully harvests the energy of glucose; also known as the citric acid cycle

lactase   the enzyme that acts upon lactose

lacteal   a lymph tubule located in the villus that absorbs fatty acids

lactid acid fermentation   a type of anaerobic respiration found in fungi, bacteria, and human muscle cells

larva   a period in the development of animals between the embryo and adult stages; starts at hatching and ends at metamorphosis

legume   a flowering plant with simple dry fruit, characterized by nodes on their roots that contain nitrogen-fixing bacteria (e.g., beans, clover)

lens   a structure of the eye that focuses images on the retina by changing its convexity

levels of structure   different relationships that are formed in proteins between the original sequence of amino acids and more complex three-dimensional compounds

lichen   an association between an algae and a fungus that is symbiotic and mutualistic in nature

linkage   occurs when different traits are inherited together more often than they would have been by chance alone; it is assumed that these traits are linked on the same chromosome

lipase   a fat-digesting hormone

lipid   a fat or oil

littoral zone   a marine biome; a region on the continental shelf that contains an ocean area with depths of up to 600 ft

Loop of Henle   the thin, bent part of the renal tubule that is the site of the counter-current flow and the sodium gradient

luteinizing hormone (LH)   secreted by the anterior pituitary gland, this hormone stimulates the conversion of a follicle into the corpus luteum and the secretion of progesterone by the corpus luteum; it also stimulates the secretion of sex hormones by the testes

lymph   a body fluid that flows in its own circulatory fluid in lymphatic vessels separate from blood circulation

lymph capillary   one of many tubules that absorb tissue fluid and return it to the bloodstream via the lymphatic system

lymphocyte   a kind of white blood cell in vertebrates that is characterized by a rounded nucleus; involved in the immune response

lysosome   an organelle that contains enzymes that aid in intracellular digestion

macula   a sensory hair structure in the utriculus and the sacculus of the inner ear; orients the head with respect to gravity

malleus   the outermost bone of the middle ear (hammer)

malpighian tubules   tubules that excrete metabolic wastes into the hindgut in arthropods

maltase   an enzyme that acts upon maltose and converts it into glucose

maltose   a 12-carbon sugar that is formed by the union of two glucose units (a disaccharide)

marsupial   a pouched mammal, such as the kangaroo or opossum

medulla   the inner layer of an organ surrounded by the cortex

medulla oblongata   the posterior part of the brain that controls the rate of breathing and other autonomic functions

medusa   a jellyfish; the bell-shaped, free-swimming stage in the life cycle of coelenterates

meiosis   a process of cell division whereby each daughter cell receives only one set of chromosomes; the formation of gametes

Mendelian laws   laws of classical genetics established through Mendel's experiments with peas

meninges   three membranes that envelop the brain and spinal cord (pia mater, dura mater, and arachnoid)

meristem   an undifferentiated, growing region of a plant that is constantly undergoing cell division and differentiation

mesoderm   the primary germ layer, developed from the lip of the blastopore, that gives rise to the skeleton, the circulatory system, and many organs and tissues between the epidermis and the epithelium

metabolism   a group of life-maintaining processes that includes nutrition, respiration (the production of usable energy), and the synthesis and degradation of biochemical substances

metamorphosis   the transformation of an immature animal into an adult; a change in the form of an organ or structure

metaphase   a stage of mitosis; chromosomes line up at the equator of the cell

microbodies   organelles that serve as specialized containers for metabolic reactions

micron (micrometer)   one-thousandth of a millimeter; a unit of microscopic length

mitochondria   cytoplasmic organelles that serve as sites of respiration; a rod-shaped body in the cytoplasm known to be the center of cellular respiration

mitosis   a type of nuclear division that is characterized by complex chromosomal movement and the exact duplication of chromosomes; occurs in somatic cells

monocotyledon   a plant that has a single cotyledon or seed-leaf

monohybrid   an individual that is heterozygous for only one trait

monosaccharide   a simple sugar; a 5- or 6-carbon sugar (e.g., ribose or glucose)

morphology   the study of form and structure

morula   the solid ball of cells that results from cleavage of an egg; a solid blastula that precedes the blastula stage

mucosa   a mucus-secreting membrane, such as the inner intestinal lining

mutagenic agent   agent that induces mutations; typically carcinogenic

mutation   changes in genes that are inherited

mutualism   a symbiotic relationship from which both organisms involved derive some benefit

myelin sheath   a fatty sheath surrounding the axon of a neuron that aids in stimulus transmission; it is secreted by the Schwann cells

NAD   an abbreviation of nicotinamide-adenine-dinucleotide, also called DPN; a respiratory oxidation-reduction molecule

NADP   an abbreviation of nicotinamide-adenine-dinucleotide-phosphate, also called TPN; an organic compound that serves as an oxidation-reduction molecule

nephron   functional urinary tubules responsible for excretion in the kidney of vertebrates

nerve   a bundle of nerve axons

nerve cord   a compact linear organization of nerve tissues with ganglia in the CNS

nerve net   a multidirectional sensory system of lower animals such as the hydra, consisting of nerve fibers spread throughout the ectoderm

neural tube   an embryonic structure that gives rise to the central nervous system *neuron*   a nerve cell

niche   the functional role and position of an organism in an ecosystem; embodies every aspect of the organism's existence

nictitating membrane   a thin, transparent, eyelid-like membrane that opens and closes laterally across the cornea of many vertebrates (the third eyelid)

nitrogen cycle   the recycling of nitrogen from decaying organisms for use in future generations

nondisjunction   the failure of some homologous pairs of chromosomes to separate following meiotic synapsis

notochord   a flexible, supportive rod running longitudinally through the dorsum ventral to the nerve cord; found in lower chordates and in the embryos of vertebrates

nuclear membrane   a membrane that envelopes the nucleus and separates it from the cytoplasm; present in eukaryotes

nucleolus   a dark-staining small body within the nucleus; composed of RNA

nucleotide   an organic molecule consisting of joined phosphate, 5-carbon sugar (deoxyribose or ribose), and a purine or a pyrimidine (adenine, guanine, uracil, thymine, or cytosine)

nucleus   an organelle that regulates cell functions and contains the genetic material of the cell

olfactory   related to the sense of smell

oogenesis   a process of formation of ova

organelle   a specialized structure that carries out particular functions for eukaryotic cells; examples include the plasma membrane, the nucleus, and ribosomes

osmoregulation   the ways in which organisms regulate their supply of water

osmosis   the diffusion of water through a semipermeable membrane, from an area of greater concentration to an area of lesser concentration

ovary   the female gonad in animals; the base of the pistil in plants

oviduct   a tube connecting the ovaries and the uterus

oxidation   the removal of hydrogen or electrons from a compound or addition of oxygen; half of a redox (oxidation or reduction) process

pairing (synapsis)   an association of homologous chromosomes during the first meiotic division

parasitism   a relationship in which one organism benefits at the expense of another

parasympathetic   pertaining to a subdivision of the autonomic nervous system of vertebrates

parathyroid   an endocrine gland of vertebrates, usually paired, and located near or within the thyroid; it secretes parathormone, which controls the metabolism of calcium

parenchyma   plant tissue consisting of large thin-walled cells for storage

passive immunity   a resistance to disease produced through the injection of antibodies

parthenogenesis   a form of asexual reproduction in which the egg develops in the absence of sperm

pathogen   a disease-causing organism (*pathogenic* = disease inducing)

pedigree   a family tree depicting the inheritance of a particular genetic trait over several generations

pelagic zone   a marine biome typical of the open seas

pepsin   a stomach enzyme that partially digests proteins

peptide   the kind of bond formed when two amino acid units are jointed end to end; a double unit is called a dipeptide; the joining of many amino acid units into a chain results in a polypeptide that is the structural unit of a protein molecule

peripheral nervous system comprises somatic and autonomic nervous systems; consists of cranial nerves and spinal nerves

peristalsis waves of contraction and relaxation passing along a tubular structure, such as the digestive tube

permeability degree of penetrability, as in membranes that allow given substances to pass through; the ability to penetrate

pH a symbol that denotes the relative concentration of hydrogen ions in a solution: the lower the pH, the more acidic a solution; the higher the pH, the more basic is a solution; pH is equal to $-\log (H^+)$

phagocyte any cell capable of ingesting another cell

pharynx the part of the alimentary canal between the mouth and the esophagus

phenotype the physical appearance or makeup of an individual, as opposed to its genetic makeup

pheromone substance secreted by organisms that influences the behavior of other members of the same species

phloem the vascular tissue of a plant that transports organic materials (photosynthetic products) from the leaves to other parts of the plant

photolysis a process of photosynthesis in which water is split into H^+ and OH^-; the hydrogen ion is then joined to NADP

photoperiodism a response by an organism to the duration and timing of light and dark conditions

photosynthesis the process by which light energy and chlorophyll are used to manufacture carbohydrates out of carbon dioxide and water; an autotrophic process using light energy

phototropism plant growth stimulated by light (stem: +, towards light; root: −, away from light)

phylogeny the study of the evolutionary descent and interrelations of groups of organisms

phylum a category of taxonomic classification that is ranked above class; kingdoms are divided into phyla

physiology the study of all living processes, activities, and functions

pineal body a structure found between the cerebral hemispheres of vertebrates; secretes melatonin, which may help regulate the pituitary by regulating hypothalamic releasing factors

pinocytosis the intake of fluid droplets into a cell

pistil the part of the flower that bears the female gametophyte

pith the central tissue of a stem, used for food storage

pituitary a gland composed of two parts, anterior and posterior, each with its own secretions; called the "master gland" because its hormones stimulate secretion by other glands

placenta a structure formed by the wall of uterus and the chorion of embryo; serves as the area in which the embryo obtains nutrition from the parent

planaria the class of free-living flatworms

plankton passively floating or drifting flora and fauna of a body of water; consists mainly of microscopic organisms

plasma the liquid part of blood

plasma membrane the cell membrane

plasmodium a motile, multinucleate mass of protoplasm resulting from fusion of uninuclear amoeboid cells; an organism consisting of such a structure, e.g., a slime mold

plastid cytoplasmic bodies within a plant cell that are often pigmented (e.g., chloroplasts)

platelet small, disc-shaped bodies in the blood that play a chief role in coagulation

pleural cavity the cavity between the lungs and the wall of the chest *plexus*   a network, particularly of nerve or blood vessels

point mutation   a mutation in which a single nucleotide base is substituted for another nucleotide base

polar body   nonfunctional haploid cells created during meiosis in females; they have very little cytoplasm—most has gone into the functional egg cell

pollen   the microspore of a seed plant

pollination   the transfer of pollen to the micropyle or to a receptive surface that is associated with an ovule (such as a stigma)

polymer   a large molecule that is composed of many similar molecular units (e.g., starch)

polymorphism   the individual differences of form among the members of a species

polyp   a typical coelenterate individual with a hollow tubular body whose outer ectoderm is separated from its inner endoderm by mesoglea

polyploidy   a condition in which an organism may have a multiple of the normal number of chromosomes (4*n*, 6*n*, etcetera)

polysaccharide   a carbohydrate that is composed of many monosaccharide units joined together, such as glycogen, starch, and cellulose

pons   the part of the hindbrain located in the brain stem

population   all the members of a given species inhabiting a certain locale

Porifera   the phylum of sponges

primary oocyte   a cell that divides to form the polar body and the secondary oocyte

primary spermatocyte   a cell that divides to form two secondary spermatocytes

producer   organism that produces its own food; first stage in the food chain

progesterone   the hormone secreted by the corpus luteum of vertebrates and the placenta of mammals; its function is to maintain the endometrium

prokaryote   unicellular organism with simple cell structure

prophase   a mitotic or meiotic stage in which the chromosomes become visible and during which the spindle fibers form; synapsis takes place during the first meiotic prophase

protein   one of a class of organic compounds that is composed of many amino acids; contains C, H, O, and N

prothrombin   a constituent of the plasma of the blood of vertebrates; it is converted to thrombin by thrombokinase in the presence of calcium ions, thus contributing to the clotting of blood

Protista   a kingdom of unicellular living organisms that are neither animals nor plants; includes some groups of algae, slime molds, and protozoa

ptyalin   a digestive enzyme of the saliva that turns starch into maltose (salivary amylase)

pulmonary   relating to the lung

pupil   an opening in the eye whose size is regulated by the iris

purine   a nitrogenous base such as adenine or guanine; when joined with sugar and phosphate, a component of nucleotides and nucleic acids

pyrimidine   a nitrogen base such as cytosine, thymine, and uracil; when joined with sugar and phosphate, a component of nucleotides and nucleic acids

pyloric valve   a muscular valve regulating the flow of food from the stomach to the small intestine

recessive   pertains to a gene or characteristic that is masked when a dominant allele is present

recombinant DNA technology   technology that allows for manipulation of genetic material

reduction   a change from a diploid nucleus to a haploid nucleus, as in meiosis

regeneration   the ability of certain animals to regrow missing body parts

respiration   a chemical action that releases energy from glucose to form ATP

respiratory center   the area of medulla that regulates the rate of breathing

reticulum   a network or mesh of fibrils, fibers, or filaments, as in the endoplasmic reticulum

retina   the innermost tissue layer of the eyeball that contains light-sensitive receptor cells

Rh factor   an antigen in blood; can cause erythroblastosis fetalis when the mother is Rh⁻ and the fetus is Rh⁺

rhizome   an underground stem

ribosome   an organelle in the cytoplasm that contains RNA; serves as the site of protein synthesis

rhodopsin   the pigment in rod cells that causes light sensitivity

rickettsia   a kind of microorganism that is between a virus and a bacterium; parasitic within the cells of insects and ticks

RNA   an abbreviation of ribonucleic acid, a nucleic acid in which the sugar is ribose; a product of DNA transcription that serves to control certain cell activities; acts as a template for protein translation; types include mRNA (messenger), tRNA (transfer), and rRNA (ribosomal)

rod   a cell in the retina that is sensitive to weak light

root hair   outgrowths of a root's epidermal cells that allow for greater surface area for absorption of nutrients and water

saprophyte   an organism that obtains its nutrients from dead organisms

secondary tissue   tissue formed by the differentiation of cambium that causes a growth in width of a plant stem

selective breeding   the creation of certain strains of specific traits through control of breeding

self-pollination   the transfer of pollen from the stamen to the pistil of the same flower

semicircular canals   fluid-filled structures in the inner ear that are associated with the sense of balance

seminal fluid   semen

seminiferous tubules   structures in the testes that produce sperm and semen

sensory neuron   a neuron that picks up impulses from receptors and transmits them to the spinal cord

serum   the fluid that remains after fibrinogen is removed from the blood plasma of vertebrates

sex chromosome   there are two kinds of sex chromosomes, X and Y; XX signifies a female and XY signifies a male; there are fewer genes on the Y chromosome than on the X chromosome

sex linkage   occurs when certain traits are determined by genes on the sex chromosomes

sinus   a space in the body (e.g., blood sinus or maxillary sinus)

small intestine   the site of most digestion of nutrients and absorption of digested nutrients (e.g. the wall of the alimentary canal)

smooth muscle   involuntary muscle (e.g., the wall of the alimentary canal)

somatic cell   any cell that is not a reproductive cell

species   a group of populations that can interbreed

spermatogenesis   the process of forming the sperm cells from primary spermatocytes

spindle   a structure that arises during mitosis and helps separate the chromosomes; composed of tubulin

spiracle   the external opening of the trachea in insects, opening into respiratory system

sphincter   a ring-shaped muscle that is capable of closing a tubular opening by constriction; one example is the orbicularis oris muscle around the mouth

spore   a reproductive cell that is capable of developing directly into an adult

sporophyte   an organism that produces spores; a phase in the diploid-haploid life cycle that alternates with a gametophyte phase

stamen   the part of the flower that produces pollen

steroid   one of a class of organic compounds that contains a molecular skeleton of four fused rings of carbon; includes cholesterol, sex hormones, adrenocortical hormones, and Vitamin D

stigma   the uppermost portion of pistil upon which pollen grains alight

stoma (stomate)   a microscopic opening located in the epidermis of a leaf and formed by a pair of guard cells; the guard cells interact physically and regulate the passage of gas between the internal cells and the external environment

stomach   the portion of alimentary canal in which some protein digestion occurs; its muscular walls of stomach churn food so that it is more easily digested; its low pH environment activates certain protein-digesting enzymes

stroma   a dense fluid within the chloroplast; the site at which CO_2 is converted into sugars in photosynthesis

style   a stalklike or elongated body part, usually pointed at one end; part of the pistil of the flower

substrate   a substance that is acted upon by an enzyme

sucrase   an enzyme that acts upon sucrose

symbiosis   the living together of two organisms in an intimate relationship; includes commensalism, mutualism, and parasitism

sympathetic   pertaining to a subdivision of the autonomic nervous system

synapse   the junction or gap between the axon terminal of one neuron and the dendrites of another neuron

synergistic   describes organisms that are cooperative in action, such as hormones or other growth factors that reinforce each other's activity

synaptic terminal   the swelling at the end of an axon

synapsis   the pairing of homologous chromosomes during meiosis

systole   the contraction of the atria or ventricles of the heart

taiga   a terrestrial habitat zone that is characterized by large tracts of coniferous forests, long and cold winters, and short summers; bounded by tundra in the north and found particularly in Canada, northern Europe, and Siberia

taxonomy   the science of classification of living things

telophase   a mitotic stage in which nuclei reform and nuclear membrane reappears

test cross   the breeding of an organism with a homozygous recessive in order to determine whether an organism is homozygous dominant or heterozygous dominant for a given trait

testes   the male gonads that produce sperm and male hormones

tetrad   a pair of chromosome pairs present during the first metaphase of meiosis

thalamus   a lateral region of the forebrain

thermoregulation   the ways in which organisms regulate their internal heat

thoracic duct   a major lymphatic that empties lymph into a vein in the neck

thorax   the part of the body of an animal that is between the neck or head and the abdomen

thrombin   a substance that participates in the clotting of blood in vertebrates; formed from pro-thrombin, it converts fibrinogen into fibrin

thrombokinase   the enzyme released from the blood platelets in vertebrates during clotting; transforms prothrombin into thrombin in the presence of calcium ions; also known as thromboplastin

thymine   a pyrimidine component of nucleic acids and nucleotides; pairs with adenine in DNA

thymus   a ductless gland in upper chest region concerned with immunity and the maturation of lymphocytes

thyroid   an endocrine gland located in the neck that produces thyroxin

thryoxin   a hormone of the thyroid that regulates basal metabolism

tissue   a mass of cells that have similar structures and perform similar functions

trachea   an air-conducting tube, e.g., the windpipe of mammals or the respiratory tubes of insects

transcription   the first stage of protein synthesis, in which the information coded in the DNA base is transcribed onto a strand of mRNA

translation   the final stages of protein synthesis in which the genetic code of nucleotide sequences is translated into a sequence of amino acids

translocation   the transfer of a piece of chromosome to another chromosome

transpiration   the evaporation of water from leaves or other exposed surfaces of plants

trilobite   a marine arthropod, now extinct, that lived during the Paleozoic era

trypsin   an enzyme from the pancreas that digests proteins in the small intestine

tundra   the biome located between the polar region and the taiga; characterized by a short growing season, no trees, and frozen ground

turgor pressure   the pressure exerted by the contents of a cell against the cell membrane or cell wall

umbilicus   the navel; the former site of connection between the embryo and the umbilical cord

ungulate   a hoofed animal

uracil   a pyrimidine found in RNA (but not in DNA); pairs with DNA adenine

urea   an excretory product of protein metabolism

ureter   a duct that carries urine from the kidneys to the bladder

urethra   a duct through which the urine passes from the bladder to the outside

urinary bladder   an organ that stores urine temporarily before it is excreted

urine   fluid excreted by the kidney containing urea, water, salts, etcetera

uterus   the womb in which the fetus develops

vacuole   a space in the cytoplasm of a cell that contains fluid

vagus nerve   the tenth cranial nerve that innervates digestive organs, heart, and other areas

vein   a blood vessel that carries blood back to the heart from the capillaries

ventral root   the basal branch of each spinal nerve; carries motor neurons

ventricle   the more muscular chamber(s) of the heart that pump blood to the lungs and to the rest of the body

vestigial organ   an organ that is not functional in an organism, but was functional at some period in its evolution

villus   a small projection in the walls of the small intestine that increases the surface area available for absorption (plural: *villi*) *vitamin*　an organic nutrient required by organisms in small amounts to aid in proper metabolic processes; may be used as an enzymatic cofactor; since it is not synthesized, it must be obtained prefabricated in the diet

white matter　an accumulation of axons within the CNS that is white because of its fatty, myelin sheath

wood　xylem that is no longer being used; gives structural support to the plant

xylem　vascular tissue of the plant that aids in support and carries water

yolk sac　a specialized structure that leads to the digestive tract of a developing organism and provides it with food during early development

zygote　a cell resulting from the fusion of gametes

INDEX

F

KAPLAN

Want more information about our services, products or the nearest Kaplan center?

1 **Call our nationwide toll-free numbers:**

1-800-KAP-TEST for information on our test prep courses, private tutoring and admissions consulting

1-800-KAP-ITEM for information on our books and software

1-888-KAP-LOAN* for information on student loans

2 **Connect with us online:**

On the web, go to:
www.kaptest.com

Via email:
info@kaplan.com

3 **Write to:**

Kaplan
888 Seventh Avenue
New York, NY 10106

KAPLAN®

*Kaplan is not a lender and does not participate in determination of loan eligibility.